Magnetorheology
Advances and Applications

RSC Smart Materials

Series Editor:
Hans-Jörg Schneider, *Saarland University, Germany*
Mohsen Shahinpoor, *University of Maine, USA*

Titles in the Series:
1: Janus Particle Synthesis, Self-Assembly and Applications
2: Smart Materials for Drug Delivery: Volume 1
3: Smart Materials for Drug Delivery: Volume 2
4: Materials Design Inspired by Nature
5: Responsive Photonic Nanostructures: Smart Nanoscale Optical Materials
6: Magnetorheology: Advances and Applications

How to obtain future titles on publication:
A standing order plan is available for this series. A standing order will bring delivery of each new volume immediately on publication.

For further information please contact:
Book Sales Department, Royal Society of Chemistry, Thomas Graham House, Science Park, Milton Road, Cambridge, CB4 0WF, UK
Telephone: +44 (0)1223 420066, Fax: +44 (0)1223 420247
Email: booksales@rsc.org
Visit our website at www.rsc.org/books

Magnetorheology
Advances and Applications

Edited by

Norman Wereley
University of Maryland, Maryland, U.S.A.
Email: wereley@umd.edu

RSC Publishing

RSC Smart Materials No. 6

ISBN: 978-1-84973-667-1
ISSN: 2046-0066

A catalogue record for this book is available from the British Library

Published by The Royal Society of Chemistry,
Thomas Graham House, Science Park, Milton Road,
Cambridge CB4 0WF, UK

Registered Charity Number 207890

For further information see our web site at www.rsc.org

Preface

The field of magnetorheology has expanded and advanced to such a great extent over the past two decades that a book offering key perspectives on the field was in order. Magnetorheological (MR) fluids were invented over 60 years ago and there remain many challenges in the fundamental understanding of MR fluids, yet the development of applications using this technology continues to be robust. One only has to look to the millions of MR shock absorbers installed in primary suspensions of automobiles. The first installation of these devices was in the Cadillac STS sedan, and was pioneered by General Motors, Lord Corporation and Adelphi in the late 1990's and early 2000's. The 2014 Corvette continues to offer Magnetic Selective Ride Control (MSRC), and MSRC is offered in the premium primary suspension packages. Magnetorheological shock absorbers are available for the primary suspensions of many vehicles offered by Ferrari, Audi, Acura, and GM.

The genesis of this book project was the 12th International Conference on Electrorheological Fluids and Magnetorheological Suspensions that was held in Philadelphia, Pennsylvania from August 16-20, 2010. This conference was organized by Prof. Rongjia Tao, Professor and Chair of the Dept. of Physics at Temple University, who served as the conference chair. As the 12th in a series of biennial conferences with a focus on magnetorheology, I was particularly intrigued by many of the 15 plenary and invited lectures that were presented at this conference. Dr. Leanne Marle, Commissioning Editor at the Royal Society of Chemistry, approached me, at the suggestion of Professor Mohsen Shahinpoor, to put together an edited book on magnetorheology. I then corresponded with several of the plenary and other speakers from this conference to write chapters based on their own research in the field of magnetorheology and applications. The annual SPIE Smart Structures and Materials also has recurring sessions each year on magnetorheological fluids and applications, and I also asked some contributors to this conference to write a chapter. Because of the enthusiastic support of the authors of the book chapters, I agreed to undertake this editing project.

RSC Smart Materials No. 6
Magnetorheology: Advances and Applications
Edited by Norman Wereley
© The Royal Society of Chemistry 2014
Published by the Royal Society of Chemistry, www.rsc.org

The scope of the book is broad in the sense that the first part of the book is devoted to advances in the physics and chemistry of magnetorheological fluids, while the second part is devoted to applications of magnetorheological fluids across domains ranging from aerospace, to medical devices, to optical polishing.

The first eight chapters of the book present significant advances in our understanding of magnetorheology. Topics range from the use of nanofibers to enhance MR fluid properties, to magnetoelasticity, magnetorheological tribology, the use of coatings to improve MR fluid performance, ensuring performance of MR fluids at temperature extremes, super-strong MR fluids. These first eight chapters demonstrate that the physics and chemistry of magnetorheological fluids continues to be an interesting and challenging field for innovative and ground breaking research.

One hallmark of a successful technology is the transition of that technology from the laboratory to prototypical and/or commercial applications. The second group of eight chapters focuses on applications of MR fluids, such as hybrid magnetic circuits that offer improved fail-safe and multi-functional MR damper capability, novel magnetorheological damper valves exploiting flow through a porous media, aerospace applications such as helicopter lag dampers and mitigation of shock loads, high precision optical polishing, and haptic devices for medical applications.

This volume of the Smart Materials book series, published by the Royal Society of Chemistry, could only have completed with the help of many enthusiastic supporters. I would like to thank Professor Mohsen Shahinpoor, who serves as one of the editors of this book series, and Dr. Leanne Marle, who served as the commissioning editor, for encouraging me to edit this volume in the RSC Smart Materials book series. As this book progressed, I received patient encouragement from Mrs. Alice Toby-Byant, who served as the Commissioning Administrator. During the final stage of editing the book, I stayed for a week at a house in Fenwick Island, Delaware, and interspersed editing with walks on its Atlantic beaches. Mrs. Toby-Brant seemed to instantaneously answer every query that I had during that week and made the editing process an efficient one. The production staff at Royal Society of Chemistry did a marvelous job at every stage, and I thank all of them for their efforts. I also want to thank all of the authors who contributed to this book and for their strong support of this project. As a result of the many opportunities for technical and social interactions at the International Conference on ER Fluids and MR Suspensions, the SPIE Smart Structures and Materials, and other conferences, I am pleased to count these contributing authors as among not only my colleagues, but also my friends.

The intent is that this volume will spur new research and innovation in the field of magnetorheology, and provide insights to current practitioners as well inspiring new researchers and practitioners to contribute to this exciting field. I am sure that I speak for all of the contributing authors when I say that we sincerely hope the readers of this volume will enjoy reading these pages as much as we have had putting this volume together.

Norman M. Wereley

Contents

RSC Smart Materials No. 6
Magnetorheology: Advances and Applications
Edited by Norman Wereley
© The Royal Society of Chemistry 2014
Published by the Royal Society of Chemistry, www.rsc.org

Chapter 3   Magnetoelasticity   56
M. Zrinyi

**Chapter 4   MR Fluids at the Extremes: High-Energy and
              Low-Temperature Performance of LORD® MR
              Fluids and Devices   74**
Daniel E. Barber

CHAPTER 1

Importance of Interparticle Friction and Rotational Diffusion to Explain Recent Experimental Results in the Rheology of Magnetic Suspensions

G. BOSSIS,[a] P. KUZHIR,*[a] M. T. LÓPEZ-LÓPEZ,[b]
A. MEUNIER[a] AND C. MAGNET[a]

[a] CNRS UMR7336, Laboratory of Condensed Matter Physics, University of Nice – Sophia Antipolis, 28 avenue Joseph Vallot, 06108 Nice Cedex 2, France; [b] Department of Applied Physics, University of Granada, Avda. Fuentenueva s/n, 18017, Granada, Spain
*Email: kuzhir@unice.fr

1.1 Introduction

Magnetorheological (MR) fluids are suspensions of magnetized micron-sized particles in a dispersing liquid. When an external magnetic field is applied, the particles acquire magnetic moments, attract to each other due to dipolar forces and form anisotropic aggregates aligned preferably with the magnetic field direction. Thus, upon a field application MR fluids undergo a reversible jamming responsible for a several order of magnitude increase in effective viscosity and appearance of a yield stress – threshold mechanical stress required for onset of flow.[1,2] This phenomenon, referred to as magnetorheological effect, is

RSC Smart Materials No. 6
Magnetorheology: Advances and Applications
Edited by Norman Wereley

being effectively used in numerous smart engineering applications.[3,4] Enhancement of the MR effect and/or reduction of the size of the MR devices are important problems for these applications. One of the possible solutions of such problems consists of using rod-like magnetic particles, which produce a higher MR response as compared to spherical particles.[5–7] Another solution consists of changing the orientation of an external magnetic field relative to the direction of the MR fluid flow. In this chapter we aim to describe physical mechanisms of the MR effect in the suspensions of rod-like magnetic particles (called hereinafter magnetic fiber suspensions) as well as in conventional MR suspensions (composed of spherical particles) subjected to a magnetic field longitudinal to the flow direction.

New MR fluids based on magnetic micro- and nano-fibers have been developed during last few years using different techniques, such as iron electrodeposition in alumina membranes,[5,8] chemical precipitation of an iron salt followed by aging in the presence of a magnetic field,[9,10] reduction of cobalt and nickel ions in polyols.[6,11] The magnetic fiber suspensions have shown better sedimentation stability[12] and developed a yield stress much larger than the one of the suspensions of spherical particles at the same magnetic field intensities and the same particle volume fraction.[7,8,10,11,13–15] Such enhanced magnetorheological effect in fiber suspensions can be explained in terms of the interfiber solid friction[16,17] and by enhanced magnetic permeability of these suspensions as compared to the permeability of conventional MR fluids.[7,15] Both these effects are reviewed in detail in the present publication. Note that the similar particle shape effect has been observed in electrorheological (ER) fluids[18–21] and was attributed to both the physical overlapping of the elongated particles (unavoidably leading to the interparticle friction) and to their strong dielectric properties.[22–24]

Concerning the effect of the magnetic field orientation on the MR response of conventional MR fluids, it should be mentioned that most of the studies were focused on their flows in the presence of the magnetic field perpendicular to the flow – presumably, the case of the largest practical interest. In such geometry, the particle structures are formed perpendicularly to the flow direction, they oppose a large hydraulic resistance to the flow and generate a relatively high dynamic yield stress.[2,25] In magnetic fields parallel to channel walls, the particle aggregates are expected to be oriented along the stream-lines and be (in theory) infinitely long because they are not subjected to tensile hydrodynamic forces. In such conditions, the suspension should undergo a Newtonian behavior and a certain decrease of its viscosity could be expected. This expectation is only confirmed for the suspensions composed of weakly paramagnetic particles,[26] such as human red blood cells, which do not belong to the class of MR fluids. However, for conventional MR fluids, composed of strongly magnetizable particles, the stress level in parallel fields is relatively high and the MR fluid develops a strong Bingham behavior,[25,27] which does not corroborate with the assumption of alignment of aggregates in flow direction. Such a strong "longitudinal" MR effect has recently been explained by stochastic rotary oscillations of the aggregates caused by many-body magnetic interactions with neighboring aggregates.[28] The inter-aggregate interactions are accounted for by an effective rotational diffusion process with a diffusion constant proportional to the mean square interaction torque – a net magnetic torque exerted to a given aggregate by all the neighboring aggregates. Such a mechanism is reviewed in details in the present chapter.

The present chapter is organized as follows. In Section 1.2, we consider the microstructure (Section 1.2.1) and the rheology of magnetic fiber suspensions. Both effects of interparticle solid friction (Section 1.2.2) and the hydrodynamic interactions in the fiber suspension (Section 1.2.3) are thoroughly reviewed. The non-linear viscoelastic response of these suspensions developed in a large amplitude oscillatory shear (LAOS) flow is described in Section 1.2.4. Section 1.3 is devoted to the flow of a conventional MR fluid (composed of spherical particles) in the longitudinal magnetic field. A rotational diffusion concept is employed to explain an unexpectedly strong MR response in such geometry. Finally the conclusions and perspectives are outlined in Section 1.4.

1.2 Magnetic Fiber Suspensions

In this section, we consider shear deformation and shear flow of suspensions composed of cobalt micron-sized fibers synthesized *via* the polyol method described in detail by López-López *et al.*[6] Anisotropic growth in the synthesis of cobalt fibers was induced by means of the application of a magnetic field during the whole synthesis time. Cobalt fibers were polydisperse with average length and width of 60 ± 24 μm and 4.8 ± 1.9 μm, respectively, as shown by SEM microscopy [Figure 1.1]. Cobalt spheres with an average diameter of 1.34 ± 0.40 μm were also synthesized in order to compare their MR response to the one of the cobalt fibers. The important feature of both types of particles is that their bulk magnetic properties are essentially the same, independently of their morphology. So, an enhanced magnetic permeability of the fiber suspensions, mentioned in

Figure 1.1 SEM image of the cobalt fibers (with kind permission from the *Journal of Rheology*).

Section 1.1, is explained by a weaker demagnetizing field inherent to fibers (as compared to spherical particles) due to their elongated shape. It is clear that the rheological response of the magnetic fiber suspension depends on its micro-structure developed under magnetic fields. So, the starting point of the present section will be visualization and analysis of the suspension microstructure in the absence of flows.

1.2.1 Microstructure

Some photos of planar structures of diluted suspensions of cobalt fibers (solid concentration 0.1 vol%) confined between two parallel glass slides (the gap was fixed to 0.15 mm) are shown in Figure 1.2. As is seen in Figure 1.2(a), in the absence of magnetic field the fibers form an entangled network with approxi-mately isotropic orientation of fibers, and even at low fiber concentration (0.1 vol%), each fiber seems to have at least a few contact points with the neighboring ones. It can also be observed that individual fibers are gathered together in aggregates. Such aggregation in the absence of magnetic field could be due to the combination of different effects: (1) magnetic attraction between fibers because of their remnant magnetization [$M_r = 53$ kA m^{-1}]; (2) short range van-der-Waals interaction; and (3) mechanical cohesion between rough fiber surfaces. Such cohesion is likely due to the solid friction between fibers and could

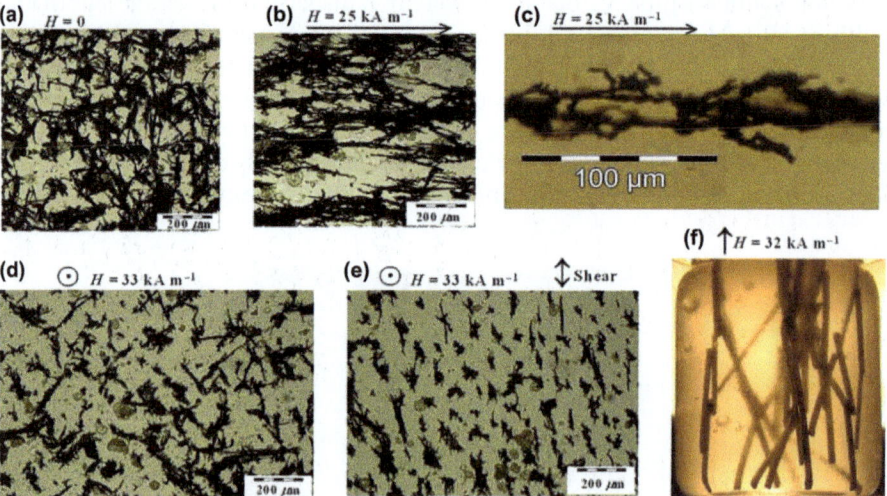

Figure 1.2   Photos of planar structures of diluted suspensions of cobalt fibers (solid concentration 0.1 vol%) confined between two parallel glass slides (the gap was fixed at 0.15 mm). (a) In the absence of applied magnetic field; (b,c) in the presence of an applied magnetic field parallel to the glass slides; (d,e) in the presence of an applied magnetic field normal to the glass slides: unstrained suspension (d) and strained suspension (e). (f) Photo of a 3D structure of a model fiber suspension under the presence of applied magnetic field (with kind permission from the *Journal of Rheology*).

involve an important contribution to the flocculation of the fiber suspension, as reported by Mason,[29] Schmid *et al.*[30] and Switzer and Klingenberg.[31]

When a magnetic field parallel to the glass slides is applied, the fiber network becomes deformed and approximately aligned with the field direction [Figure 1.2(b)]. Notice that the fiber network remains entangled, the fibers are linked to the neighboring ones and, therefore, there is no complete alignment with the field. This can be explained by appearance of the solid friction between fibers, which hinders their motion and does not allow them to get completely aligned with the field. Hence, the structure observed is not at equilibrium. Otherwise, without friction, the free energy of the fiber suspension would have been minimized, and a structure with all the fibers aligned completely with the magnetic field, joined end by end with the neighboring ones, would have been observed. A zoomed view of the fiber network upon magnetic field application is presented in Figure 1.2(c). As observed, the fibers are rather polydisperse and have an irregular rough surface. They are linked to each other either by their extremities or by their lateral sides. In the latter situation, two contacting fibers either are attached by their lateral sides or cross each other at some angle. It seems that any type of interfiber contact is equiprobable.

Alternatively, when a magnetic field normal to the glass slides is applied, the fibers tend to become aligned in the vertical plane, *i.e.* transversely to the glass slides [Figure 1.2(d)]. However, as can be observed, some fiber aggregates are so big that they cannot be aligned in the vertical plane because their movement is restricted by the gap between the glass slides. And even smaller fiber aggregates do not get strictly perpendicular to the glass slides – fibers are always attached to the neighboring ones by magnetic and friction forces. Note that this structure is rather different from the column-like structure observed in suspensions of spherical magnetic particles.[2] Notice also that when this fiber suspension is sheared (the upper glass slide is displaced horizontally), under the presence of a vertical magnetic field, the fiber aggregates get more oriented in the direction of shear [Figure 1.2(e)]. Thus, we believe that upon magnetic field application, the fibers gather into aggregates, which span the gap between the glass slides, and they are tilted, when sheared, in the direction of the shear.

Finally, a photo of a 3D structure of a model fiber suspension consisting of steel rods (15 mm in length and 1 mm in diameter) in silicone oil, in the presence of an applied magnetic field, is shown in Figure 1.2(f). Similarly to the planar structures discussed above, the fibers form a dendrite-like structure oriented preferably along the magnetic field lines. As seen in Figure 1.2(f), most of the contacts between fibers are either side-by-side or side-by-end, while end-by-end contacts are infrequent. In fact, this model structure shown in Figure 1.2(f) is quite similar to the structure shown in Figure 1.2(c). In both cases the fibers can either attach to neighboring ones by their lateral side (line contact) or cross each other at a certain angle (point contact).

The existence of different types of interfiber contacts is an essential point that must be taken into account to theoretically model the magnetorheology of suspensions of magnetic fibers. This is done in Section 1.2.2, where we introduce a microstructural model for magnetic fiber suspensions and explain the enhanced MR response of these suspensions in terms of interfiber solid friction. Theoretically determined static yield stress of the fiber suspension is compared to the measured one obtained from experiments on quasi-static shear deformation of the suspension.

1.2.2 Rheology: Interparticle Friction and Static Yield Stress

Let us consider a suspension of identical magnetic fibers confined between two infinite plates. The distance between these plates is supposed to be much larger than the fiber length. When the magnetic field is applied normally to the plates, the fibers attract each other and form some kind of anisotropic network. Precise details of such a network may only be predicted by particle level numerical simulations. To gain the first insight into the rheology of the magnetic fiber suspension, we impose artificially a stochastic near-planar suspension structure, which seems to be rather close to the one observed in experiments [Figure 1.2]. In more detail, we suppose that all the fibers lie more or less in planes parallel to the shear plane. Thus, the fiber suspension can be represented as a series of sheets, each one parallel to the shear plane, and containing stochastically oriented fibers, as depicted in Figure 1.3(a). The suspension is sheared by a displacement of the upper plate, and the strain angle is Θ. We shall calculate the stress *vs.* strain dependency and the suspension yield stress under the following considerations:

1. The fibers are supposed to not to slip over the plates.
2. The magnetic dipolar forces acting between fibers are negligible [according to Kuzhir *et al.*[17]] and the only forces exerted on the fibers are the contact forces.
3. Most of the contact points are located on the lateral fiber surface rather than at the fiber extremities.

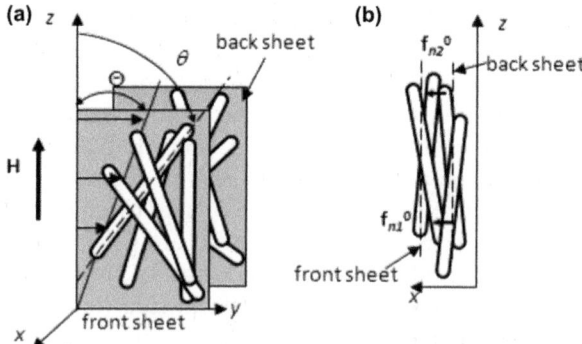

Figure 1.3   Geometry of the near-planar structure. (a) The fiber network can be "sliced" into sheets parallel to the shear *yz*-plane. (b) Projection of the fiber network onto the *xz*-plane (with kind permission from the *Journal of Rheology*).

4. The surface of the fibers is rough [*cf.* Figure 1.1]. When the suspension is sheared, all the fibers slide over each other and exert friction forces on the neighboring fibers. In general, the value of these forces should depend on the shear rate. However, at low shear rates, considered in this section, a boundary lubrication regime between rough fiber surfaces is expected. In this regime, the friction forces appear to be independent of speed[32] and are supposed to follow the Coulomb's friction law, $f_\tau = \xi f_n$, with ξ being the friction coefficient and f_n the normal force exerted by a neighboring fiber to a given fiber. At higher shear rates, considered in Section 1.2.3, the surface roughness generates a lifting force leading to hydrodynamic lubrication between fibers with the friction force proportional to the shear rate.

5. The contact forces between fibers belonging to different sheets are entirely defined by interparticle magnetic forces. Since the latter are neglected, the former should also be negligible. Therefore both the normal force f_n and the friction force f_τ are supposed to belong to the shear yz-plane and the friction force is assumed to be longitudinal with the fiber major axis.

The mechanical stresses arising in strained fiber suspensions are due to contact forces acting on fibers and the latter are, to a large extent, determined by the balance of torques. The projection of torques (exerted to a given fiber) onto the shear yz-plane reads:

$$-T_m + \sum_{\text{contacts}} s f_n = 0, \qquad (1.1)$$

where s is the distance between the center of the given fiber and the contact point; the summation in second term of the left-hand side of eqn (1.1) is performed over all contact points of a given fiber; T_m is the magnetic torque exerted by the external magnetic field to a given fiber; the expression for this torque reads:[17]

$$T_m = \frac{1}{2} \mu_0 \frac{\chi_f^2}{2 + \chi_f} V_f\, H^2 \sin 2\theta \qquad (1.2)$$

where H is the internal magnetic field, θ is the angle between a given fiber and the magnetic field vector [Figure 1.3(a)], $V_f = 2\pi a^2 l$ is the volume of the fiber, a and l are the fiber radius and semi-length, respectively, χ_f is the fiber magnetic susceptibility, and $\mu_0 = 4\pi\, 10^{-7}$ Henry m^{-1} is the magnetic permeability of vacuum.

Since there is no any significant flow in the quasi-static deformation regime, the only contribution to the suspension stress tensor is the particle stress. This is a volume average of the stresses contributed by each fiber. In our particular case, the forces acting on the fibers are concentrated in single points (point-wise interactions), and the expression for the suspension shear stress (yz-component of the stress tensor) is given by Larson[33] and Toll and Manson:[34]

$$\sigma = \frac{1}{V} \sum_{\text{fibers}} \sum_{\text{contacts}} r_z f_y \tag{1.3}$$

Here V is the total volume of the suspension, $\mathbf{r}$ is the vector connecting the fiber center with the contact point, $r_z = s \cos \theta$ is the projection of the vector $\mathbf{r}$ onto the z-axis, $f_y = f_n \cos \theta + f_\tau \sin \theta$ is the projection of the contact force $\mathbf{f}$ onto the y-axis (flow direction); the sum is taken over all the contact points on every particle of the suspension. Taking into account eqn (1.1) and the relation $f_\tau = \xi f_n$, we arrive to the following expression for the shear stress:

$$\sigma = \frac{1}{V} \sum_{\text{fibers}} \left(T_m \cos^2 \theta + \frac{1}{2} \xi \, T_m \sin 2\theta \right) \tag{1.4}$$

Replacing the magnetic torque T_m by the expression (1.2) and averaging the stress over all possible fiber orientations, we get the final expression for the suspension stress:

$$\sigma = \underbrace{\frac{1}{2} \Phi \mu_0 H^2 \left\langle \frac{\chi_f^2}{2 + \chi_f} \sin 2\theta \, \cos^2 \theta \right\rangle}_{\text{contribution from magnetic torque}} + \underbrace{\frac{1}{4} \xi \Phi \mu_0 H^2 \left\langle \frac{\chi_f^2}{2 + \chi_f} \sin^2 2\theta \right\rangle}_{\substack{\text{contribution from friction forces} \\ \text{induced by the magnetic torque}}} \tag{1.5}$$

The angular mean in eqn (1.5) is calculated *via* the angular distribution function $F(\theta)$ of the near-planar structure: $\langle ... \rangle = \int_{-\pi/2}^{\pi/2} F(\theta) \, (...) \, d\theta$. The fiber orientation is supposed to be strongly influenced by the shear deformation, and the angular distribution function is assumed to be Gaussian and centered at the strain angle, Θ [Figure 1.3(a)]:

$$F(\theta) = \alpha_1 \exp[-\alpha_2 (\theta - \Theta)^2] \tag{1.6}$$

Here α_1 and α_2 are the parameters of the distribution function. In the absence of shear, the fiber distribution is considered to be isotropic in the yz-plane. When the strain is progressively increased, the fibers incline with the strain and get more aligned. At a threshold strain angle, Θ_a, the structure is supposed to be completely stretched, the straight fiber chains making the angle Θ_a with the magnetic field. Under these conditions, the coefficient α_2 of the distribution function must be zero at zero shear, and infinite at the strain angle Θ_a. A simple function, $\alpha_2(\Theta)$ respecting the above conditions and adopted in our model is $\alpha_2 = \Theta/(\Theta_a - \Theta)$. The first coefficient, α_1, is found from the normalization condition: $\int_{-\pi/2}^{\pi/2} F(\theta) \, d\theta = 1$, *i.e.* $\alpha_1 = \left(\int_{-\pi/2}^{\pi/2} \exp\left[-\alpha_2 (\theta - \Theta)^2\right] d\theta \right)^{-1}$. Finally, the threshold strain angle is set at $\Theta_a = 60°$ [*cf.* Kuzhir *et al.*[17]].

The stress–strain curve obtained by this model is plotted in Figure 1.4 and compared with the corresponding curves obtained for other two micro-structural models – the models of the column and zigzag structures described in

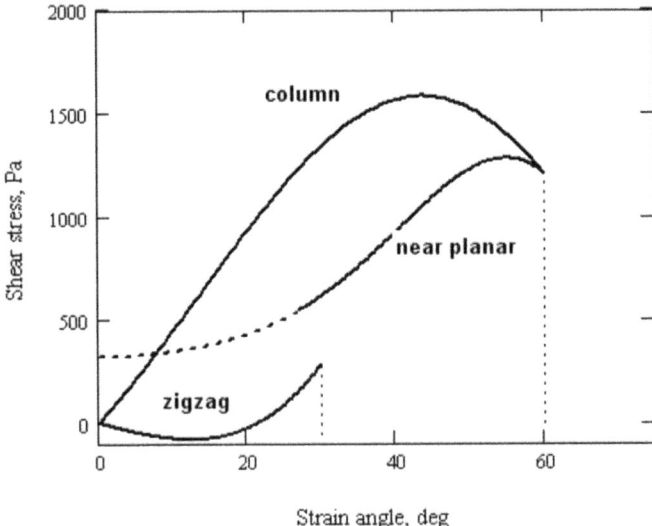

Figure 1.4 Stress–strain curve for column, zigzag and near-planar stochastic structure of a fiber suspension at magnetic field intensity $H_0 = 100$ kA m^{-1}, fiber volume fraction $\Phi = 0.05$, and friction coefficient $\xi = 1$ (with kind permission from the *Journal of Rheology*).

detail in Kuzhir *et al.*[17] As is seen in this figure, the stress–strain relation for the near-planar structure presents a local maximum, which corresponds to the yield stress. This maximum is observed at a strain angle close to the angle Θ_a of complete alignment of the structure. Note that the stress–strain curve departs from non-zero shear stress at zero strain. This is not surprising because we have assumed that, at any strain, all fibers slide over each other and experience the friction force, $f_\tau = \xi f_n$. At zero strain, the normal forces between randomly oriented fibers are not zero, leading to non-zero friction forces. In reality, when the fibers do not slide, the friction forces between them can take any value within the range: $-\xi f_n \le f_\tau \le \xi f_n$. Consequently, at small strain angles our model cannot predict with confidence the shear stress of the near-planar structure.

Let us now compare the predictions of the theoretical models with the experimental values of the static yield stress of the fiber suspensions. Figure 1.5 shows the magnetic field dependency of the yield stress for four solid volume fractions, $\Phi = 0.01$, 0.03, 0.05 and 0.07. The five curves in each graph correspond to the theoretical results using both the model of the near-planar stochastic structure described in the present chapter and the models of the column or zigzag structures described in Kuzhir *et al.*;[17] the solid circles correspond to the experimental results. As observed in Figure 1.5, the highest estimation of the yield stress is given by the model of the column structure with friction (upper solid curve), and the lowest estimation by the model of the zigzag structure (lower solid curve). At magnetic fields $H_0 \ge 100$ kA m^{-1}, the experimental points lie between these two curves. At lower magnetic fields the

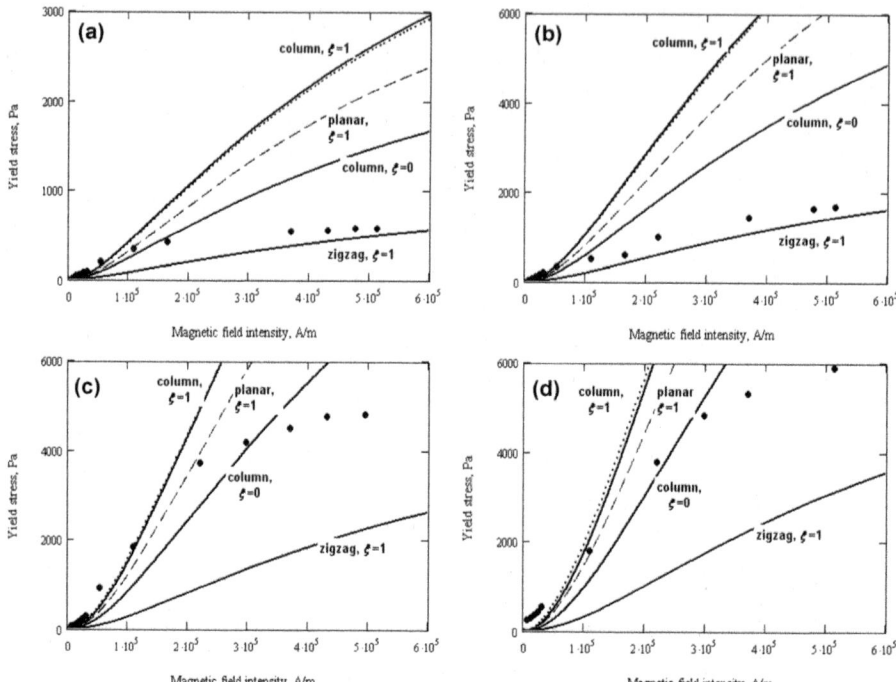

Figure 1.5   Yield stress of fiber suspensions versus external magnetic field intensity, H_0, for different fiber volume fractions, Φ: (a) $\Phi = 0.01$; (b) $\Phi = 0.03$; (c) $\Phi = 0.05$; and (d) $\Phi = 0.07$. The upper and the middle solid lines correspond to the model of the column structure with $\xi = 1$ and $\xi = 0$, respectively; lower solid line – model of the zigzag structure with $\xi = 1$; dashed line – model of the near-planar stochastic structure [eqn (1.5)] with $\xi = 1$; solid circles: experimental data (with kind permission from the *Journal of Rheology*).

experimental yield stress is higher than the one given by the highest theoretical estimation. This is possibly due to the underestimated value of the initial magnetic susceptibility of the fibers used in our calculations, $\chi_i = 17.3$.

Comparing different theoretical predictions, we note that the yield stress for the column structure with the friction coefficient $\xi = 1$, is roughly two times higher than the yield stress for the same structure without friction. By analyzing the zigzag model we can identify the two reasons why this model predicts the lowest yield stress. Firstly, the strained zigzag chains act as compressed springs that push upward against the rheometer plate. Secondly, this structure has a relatively low anisotropy compared to the column structure. The most realistic model – the model of the near-planar stochastic structure – gives a reasonable correspondence with the experiments at fiber volume fractions $\Phi = 0.05$ and 0.07 [Figure 1.5(c)–(d)]. This model takes into account the friction between fibers as well as the progressive alignment of the fiber network with increasing strain.

Let us now analyze the effect of solid concentration on the yield stress of fiber suspensions. The three inspected models – column structure, zigzag structure and near-planar structure – give almost linear concentration dependence of the yield stress, which comes from the assumption (5) that the friction force between fibers is longitudinal and always equal to ξf_n. In this case, in eqn (1.3), the sum $\sum_{contacts} r_z f_y$ over all the contact points on a given fiber is simply proportional to the magnetic torque acting on a considered fiber, whatever the number of contact points is. Consequently, the theoretical yield stress is linear in the number of fibers per unit volume (*i.e.* in the concentration) rather than in the total number of contact points. Such a linear trend is inconsistent with a power-law concentration dependence of the yield stress observed experimentally:[14] $\tau_Y \propto \Phi^{1.5}$. In a real situation of a 3D stochastic structure, the friction term of the stress is not necessarily proportional to the magnetic torque and can hide a stronger concentration dependence. This is the case of isotropic suspensions of non-magnetic elastic fibers, for which the yield stress is proportional to the number of contact points per unit volume, which varies as the square of the solid volume fraction.[34,35]

Note finally, that experiments show that the static yield stress of the cobalt fiber suspensions is approximately three times larger than that of the suspension of cobalt spheres at the same volume fraction and the same magnetic field.[14] The microstructural model presented in this section reveals an importance of the interparticle friction in fiber suspensions. According to eqn (1.5), the solid friction gives a contribution to the total stress, which is, at least, comparable with the magnetic torque contribution. So, since the interparticle friction is expected to be much weaker in a suspension of spherical particles, it is now clear that the fibers should give a stronger MR response as compared to spheres. Another possible reason of an enhanced MR effect in fiber suspensions – a stronger magnetic permeability of the fiber suspension – is studied in the next section in conjunction with experiments and modeling of steady shear flows at shear stresses above the yield point.

1.2.3 Rheology: Hydrodynamic Interactions and Dynamic Yield Stress

Now, instead of a static shear deformation below the yield stress, we shall consider a steady shear flow of the fiber suspension generated by a continuous motion of the upper rheometer plate with velocity v. Consequently, the shear rate is equal to $\dot{\gamma} = v/b$, where b is the gap between the two plates. As previously, an external magnetic field $\mathbf{H_0}$ is applied perpendicularly to the planes.

In order to find a rheological law of the fiber suspension under steady shear flow, we introduce the following assumptions:

1. The fibers have a cylindrical shape and are characterized by a half-length l and a radius a. In the presence of field, the fibers attract to each other and form cylindrical aggregates with half-length L and radius A. We assume

that the fibers in the aggregates are all aligned parallel to each other and form therefore a closely packed bundle of cylindrical particles having an internal volume fraction of $\Phi_a = \pi^2/12$ [see Bideau *et al*[36]].

2. The aggregates are supposed to move affinely with the flow without rotation.

3. We do not take into account collisions, contact forces (compressive and friction forces) and hydrodynamic interactions between aggregates. Strictly speaking, this assumption is verified for diluted fiber suspensions with a volume fraction, $\Phi < (A/L)^2$, however, as we shall see, it still give a reasonable agreement with experiments until the volume fraction of $\Phi = 5\%$ for the experimental range of aspect ratios $7 < L/A < 13$.

4. We do take into account hydrodynamic interactions between the aggregates and the suspending liquid, which tend to align the aggregates in the flow direction by exerting a hydrodynamic torque on them. On the other hand, the external magnetic field applies a restoring torque on aggregates. So, at equilibrium, both torques are balanced and define an equilibrium angle θ_c of the aggregates' orientation with respect to the magnetic field.[15]

5. Under shear, the aggregates are subjected to tensile hydrodynamic forces, which break them at their weakest point – their central transverse section. On the other hand, the cohesive magnetic forces between fibers consolidate the aggregates. So, the equilibrium aggregate length (or rather aspect ratio L/A) is found from the balance of these forces. Such an approach has been employed in many calculations of the rheological properties of conventional MR fluids composed of spherical particles.[37–39] In our case of fiber suspensions, we use more rigourous expressions for the interparticle magnetic force [taking into account the magnetic saturation effects according to the model of Ginder *et al*.[40]] and for the hydrodynamic force and stress [using the slender body theory of Batchelor[41] with appropriate corrections accounting for a finite aggregate length].

6. At low shear rates or high magnetic fields, the aggregates become very long and, therefore, they may span the gap between the two planes (rheometer gap). In this case, the orientation of the aggregates and the shear stress developed in the suspension will be different from those found for unbounded shear flow. We assume that the aggregates are extensible: when they are inclined at an angle θ relative to the magnetic field direction, they are stretched and continue to touch the walls but slide over the walls without solid friction, such that their length is $L = b/2\cos\theta$. Alternatively, at high shear rates or low magnetic fields, all the aggregates are expected to be destroyed by the shear. In this case, the suspension is completely disaggregated and composed of isolated fibers whose aspect ratio, l/a, no longer depends on the shear rate. In conclusion, three distinct aggregation regimes are expected in a shear flow of magnetic fiber suspensions depending on the ratio of hydrodynamic to magnetic forces – the so-called Mason number, *Mn*: (1) aggregated state with confined aggregates at $Mn < Mn_c$; (2) aggregated state with free aggregates at $Mn_c < Mn < Mn_d$; and (3) disaggregated state at $Mn > Mn_d$, the expressions

for critical Mason numbers Mn_c and Mn_d being given in Gómez-Ramírez et al.[15]

To find the shear stress of the fiber suspension under steady shear flow, we use an expression similar to the one derived by Brenner[42] and Pokrovskiy[43] for a suspension of non-spherical force-free particles subjected to an external torque [magnetic torque T_m in our case, *cf.* eqn (1.2)]. Substituting the corresponding expressions for the aggregate orientation angle, θ_c, and the aspect ratio, L/A [*cf.* assumptions (4), (5)], as well as using the results of the slender body theory[41] for the rheological coefficients, we arrive to the final expression for the shear stress at the three considered aggregation regimes:

$$\sigma = \eta_0 \dot{\gamma}\left(1 + 2\frac{\Phi}{\Phi_a}\right) + \frac{2}{3}\frac{\Phi}{\Phi_a}\eta_0\dot{\gamma}\frac{[b/(2A_0\cos^{3/2}\theta_c)]^2}{\ln[b/(A_0\cos^{3/2}\theta_c)]}f_1^{\|}\sin^2\theta_c\cos^2\theta_c$$

$$+ \Phi\mu_0 H^2\frac{\chi_f^2(1-\Phi)}{2+\chi_f(1-\Phi)}\sin\theta_c\cos^3\theta_c, \quad Mn < Mn_c$$

(1.7a)

$$\sigma = \eta_0\dot{\gamma}\left(1 + 2\frac{\Phi}{\Phi_a}\right) + \Phi\mu_0 H^2\left\{\left(\frac{2M_S}{3H}\frac{\chi_f^2(1-\Phi)}{2+\chi_f(1-\Phi)}\frac{(f_1^{\|})^2}{f^{\perp}f_2^{\|}}\right)^{1/2}\sin^2\theta_c\cos^2\theta_c\right.$$

$$\left. + \frac{\chi_f^2(1-\Phi)}{2+\chi_f(1-\Phi)}\sin\theta_c\cos^3\theta_c\right\}, \quad Mn_c \le Mn \le Mn_d$$

(1.7b)

$$\sigma = \eta_0\dot{\gamma}(1+2\Phi) + \frac{2}{3}\Phi\eta_0\dot{\gamma}\frac{(l/a)^2}{\ln(2l/a)}f_1^{\|}\sin^2\theta_c\cos^2\theta_c$$

$$+ \Phi\mu_0 H^2\frac{\chi_f^2(1-\Phi)}{2+\chi_f(1-\Phi)}\sin\theta_c\cos^3\theta_c, \quad Mn > Mn_d$$

(1.7c)

Here η_0 is the viscosity of the suspending liquid; A_0 is radius of the unstrained fiber aggregates, which is found by the energy minimization; χ_f is the magnetic susceptibility of a single fiber; H is the internal magnetic field; M_S is the magnetization saturation of the cobalt fibers; and $f_1^{\|}, f_2^{\|}, f^{\perp}$ are numerical factors, coming from the slender body theory of Batchelor,[41] functions of L/A, taking into account the finiteness of the fiber aspect ratio [the expressions for these factors can be found in Gómez-Ramírez et al.[15]]. All the three expressions (1.7) for the shear stress contain three terms, the first of which corresponds to the solvent contribution, the second one comes from the longitudinal hydrodynamic stress generated by the aggregates, and the last one is connected to the magnetic torque exerted on aggregates.

The theoretical [eqn (1.7)] and experimental dependencies of the shear stress on the shear rate are shown in Figure 1.6 for the fiber suspension containing 5 vol% of particles.

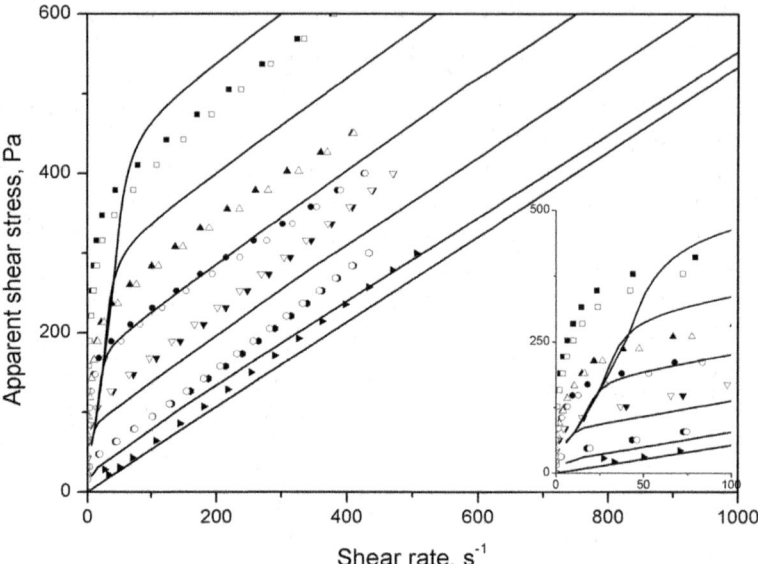

Figure 1.6   Flow curves for a fiber suspension with $\Phi = 0.05$ in the presence of a magnetic field. Lines correspond to the theory; the symbols correspond to experimental data obtained using a controlled-stress rheometer. Full and open symbols stand respectively for increasing and decreasing shear stress. Both theoretical and experimental curves correspond to magnetic fields of intensity H_0, from bottom to top: 0, 6.11, 12.2, 18.3, 24.4 and 30.6 kA m^{-1}. The inset shows the same flow curves at low shear rates (with kind permission from the *Journal of Rheology*).

We see that, both in experiments and in theory, the flow curves have two straight sections with different slopes, the left one with a steep slope and the right one with a less steep slope. Let us consider each part separately.

The left part of the flow curve corresponds to the state of confined aggregates. The aspect ratio of the aggregates is very high and they span the rheometer gap. Therefore they offer a high hydraulic resistance to the flow, which could explain the steepness of this part of the curves, corresponding to a high apparent viscosity at low shear rates. The theoretical zero-shear viscosity (the initial slope of the flow curves) can be easily found from eqn (1.7a) with an appropriate expression for the angle θ_c:

$$\eta_{\dot{\gamma} \to 0} \equiv (\tau / \dot{\gamma})_{\dot{\gamma} \to 0} = \eta_0 \left\{ 1 + \frac{\Phi}{\Phi_a} \left(2 + \frac{4}{3} \frac{[b / (2A_0)]^2}{\ln[b / A_0]} f^{\perp} \right) \right\} \qquad (1.8)$$

In the limit of small shear rates, the slopes of all the theoretical curves do not depend on the magnetic field intensity but depend on the rheometer gap b. For the fiber suspension with 5% volume fraction, the zero-shear viscosity predicted by eqn (1.8) is equal to $13\eta_0$, which is 12 times the differential viscosity (final slope) corresponding to high shear rates. As we can see in the inset of Figure 1.6, our theoretical model underestimates the zero-shear viscosity of the flow curve. This is probably because we have neglected hydrodynamic interactions between the aggregates and the walls. Recently, Berli and de Vicente[44] have proposed a structural viscosity model, which also predicts a large but finite zero-shear viscosity of MR fluids independently of the rheometer gap. Unfortunately, we cannot compare this theory with our experiments since the former employs an unknown parameter. Therefore, the questions of whether the zero-shear viscosity depends on the rheometer gap and what mechanism governs this quantity remain open.

The rounded part of the flow curves corresponds to the transition between the regime of confined aggregates to that of free aggregates, which happens at Mason number Mn_c. Note that the shear rate in the rheometer gap increases linearly with the radial distance from zero on the disk axis to a maximal value, $\dot{\gamma}_R$ at the disk edge. So, depending on the position in the gap, the zone of confined aggregates can coexist with the zone of free aggregates, and the transition between both regimes happens smoothly with increasing the shear rate $\dot{\gamma}_R$, which explains the rounded shape of the transition zone.

Starting from a shear rate $\dot{\gamma} \approx 50\ \mathrm{s}^{-1}$, the experimental and theoretical curves become linear and almost parallel to each other, which corresponds to the Bingham rheological law, $\sigma = \sigma_Y + \eta\dot{\gamma}$, with σ_Y being the dynamic yield stress and η the plastic viscosity. The dynamic yield stress is defined as a linear extrapolation of the flow curve to zero shear rate. We observe a reasonable quantitative correspondence between the theoretical and the experimental flow curves at $\dot{\gamma} > 50\ \mathrm{s}^{-1}$, without introducing any adjustable parameter. The Bingham behavior observed experimentally at $\dot{\gamma} > 50\ \mathrm{s}^{-1}$ is well predicted by our theory. In our theory, the stress σ_Y contains the hydrodynamic part, which is proportional to $(L/A)^2\eta_0\dot{\gamma}$ and the magnetic part, which does not depend on $\dot{\gamma}$. Due to the action of hydrodynamic tensile forces, the aggregate length appears to be proportional to $\dot{\gamma}^{-1/2}$, thus, the hydrodynamic part of the stress σ_Y and, consequently, the stress σ_Y itself do not depend on the shear rate. So, as in MR suspensions of spherical particles,[37,38] in our model this stress is assigned to the dynamic yield stress. Notice that, in both the theoretical and the experimental flow curves, the regime of isolated fibers is not distinguished from the regime of free aggregates. This is because the considered transition happens at relatively high shear rates (Mason numbers Mn_d), when the magnetic field does not play any significant role on the shear stress.

Notice finally, that the experimental flow curves show only a very narrow hysteresis, which could mean that the hydrodynamic forces dominate over the forces of solid friction, at least, for semi-dilute suspensions ($\Phi < 0.05$) and at Mason numbers $Mn > 1$. Therefore, assumption (3) of the absence of contact and friction forces seems to be justified for a steady flow at such parameters.

The most important parameter characterizing MR response of MR fluids in steady flows is the dynamic yield stress. Its theoretical value for the fiber suspension is easily obtained from eqn (1.7b) by putting the shear rate $\dot{\gamma}=0$ and the coefficients $f_1^{\|}=f_2^{\|}=f^{\perp}\approx1$:

$$\sigma_Y=\Phi\mu_0H^2\left\{\left(\frac{2M_S}{3H}\frac{\chi_f^2(1-\Phi)}{2+\chi_f(1-\Phi)}\right)^{1/2}\sin^2\theta_c\cos^2\theta_c+\frac{\chi_f^2(1-\Phi)}{2+\chi_f(1-\Phi)}\sin\theta_c\cos^3\theta_c\right\}$$

(1.9)

A similar expression has also been derived for the dynamic yield stress of the conventional MR fluids composed of spherical particles.[15] Let us now compare the rheology of suspensions of magnetic fibers with the rheology of suspensions of spherical particles, both made of the same material (cobalt) and at the same volume fraction 5 vol%. The dependencies of the apparent dynamic yield stress on the magnetic field intensity are shown in Figure 1.7 for both suspensions. Both experiments and theory show that the dynamic yield stress of the suspension of fibers is a few times higher than that of the suspension of spheres. Note that the magnetization of both spherical and fiber-like cobalt particles is similar, and both types of particles are micron-sized, so non-Brownian. Thus, the difference in the yield stress cannot be explained by different magnetic properties, nor by their Brownian motion, but rather comes from a shape-dependent demagnetizing field inside the particles. In more detail, the

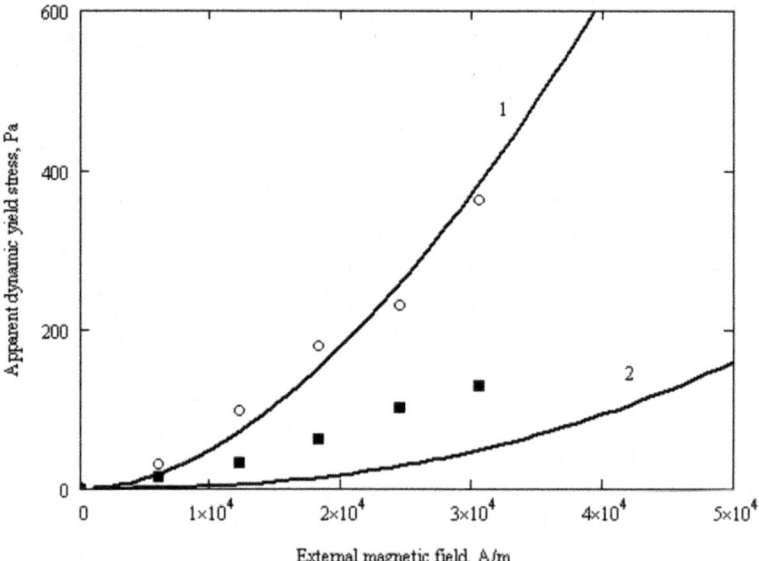

Figure 1.7 Dynamic yield stress for suspensions of cobalt particles as a function of the intensity of the external magnetic field, H_0. The volume fraction $\Phi=5\%$ in both cases. Circles – experiments for the suspension of cobalt fibers; squares – experiments for the suspension of spherical particles; 1 – theory for the fiber suspension; 2 – theory for the suspension of spherical particles (with kind permission from the *Journal of Rheology*).

magnetization M_a of a particle aggregate varies linearly with the magnetization M_p of a separate particle, and the latter is proportional to the magnetic field intensity H_p inside the particle: $M_a = \Phi_a M_p = \Phi_a \chi_p H_p$, where χ_p is the particle magnetic susceptibility. Because of the particle shape, the magnetic field H_p appears to be lower inside the spherical particles than inside the fiber-like particles, so the magnetization and the magnetic susceptibility of the aggregates composed of spherical particles is a few times lower than those of the aggregates of fibers. Since the yield stress is a growing function of the aggregate magnetic susceptibility, it appears to be larger for the fiber suspension.

Inspecting Figure 1.7, one can see that our theory predicts the dynamic yield stress for fiber suspension reasonably well, but it underestimates the dynamic yield stress for suspensions of spherical particles. This could be due to the underestimation of the magnetic susceptibility of the aggregates of spherical particles, as was shown by comparing the permeability of an elastomer containing chain like structures of magnetic particles to that of one having an isotropic distribution of particles.[45]

In addition to steady shear flows of MR fluids, oscillatory shear flows are even more frequently employed in MR devices, such as MR dampers and shock absorbers. The strain amplitude of such flows in real devices often overcomes by orders of magnitude the limit of the linear viscoelastic regime. Therefore, a large amplitude oscillatory shear (LAOS) response of magnetic fiber suspensions is of high practical interest and is considered in the next section 1.2.4.

1.2.4 Rheology: Non-linear Viscoelastic Response

The LAOS tests have been carried out with an MR fluid composed of cobalt fibers at a volume fraction $\Phi = 5\%$, using a controlled-stress rheometer under an external magnetic field applied perpendicularly to the rheometer plates. Experimental dependencies of the shear moduli, G_1' and G_1'' (the subscript "1" stands for the first harmonic of the strain response) on the stress amplitude, σ_0, are shown in Figure 1.8 for the excitation frequency, $f = 1\,\text{Hz}$ and for six values of the external magnetic field, H_0. In all cases, both moduli increase with the growth in the magnetic field intensity and decrease with the stress amplitude. In particular, a short linear viscoelastic plateau at $\sigma_0 \leq 1$ Pa is followed by a rapid decrease of the moduli until a second quasi-plateau, which is better distinguished for the loss modulus curves. After this second quasi-plateau, there is a second abrupt decrease of the moduli, at the end of which the storage modulus shows the third final plateau after some local minimum.

The first viscoelastic plateau, which appears only for the magnetic fields $H_0 \geq 12.2$ kA m^{-1}, corresponds to the strain amplitudes, γ_0, as low as 10^{-4}–10^{-3}. At such strains, the upper plate displacement during an oscillation cycle is as small as 20–200 nm, *i.e.* much smaller than the fiber's minor dimension – diameter $2a = 4.8$ μm. Thus, we cannot expect a homogenous deformation of the aggregates, but rather a rearrangement of fibers inside the aggregates accompanied by their microscopic displacement and/or by their elastic bending.

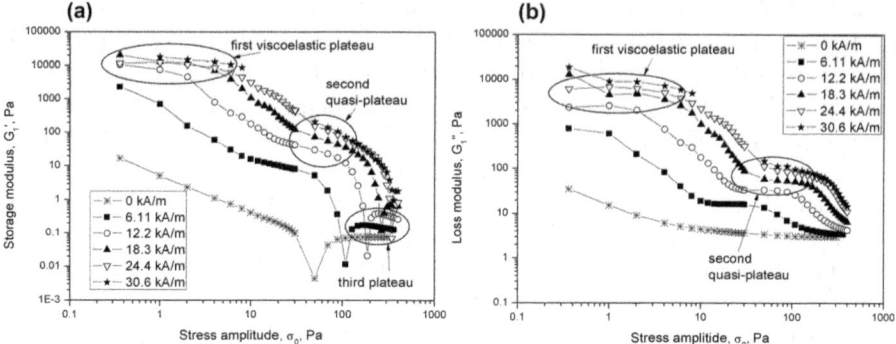

Figure 1.8   Experimental stress dependencies of the storage (a) and loss (b) moduli of the fiber suspension at the excitation frequency of 1 Hz (with kind permission from the *Journal of Non-Newtonian Fluid Mechanics*, Elsevier).

The latter could explain high values of the storage modulus at small amplitudes (more than 10 kPa at the particle volume fraction of 5%). The large values of the loss modulus could come from the non-affinity of the fiber displacement on microscopic scale, as pointed out by Klingenberg.[46] The first decrease of the storage moduli followed by a second quasi-plateau probably corresponds to a gradual transition from microscopic-to-macroscopic scale deformation of the suspension structure. At the end of this transition, the percolating aggregates are expected to be strained uniformly at small but measurable angles. Actually, at the magnetic field intensity, $H_0 = 30.6$ kA m^{-1}, the second quasi-plateau starts at $\gamma_0 \approx 0.1$ corresponding to the upper plate displacement of 20 µm, which is at least five times the fiber diameter. This second quasi-plateau is attributed to the second quasi-linear viscoelastic regime governed only by macroscopic deformations of the structure. After a second viscoelastic quasi-plateau (which extends from $\sigma_0 \approx 40$ Pa to $\sigma_0 \approx 100$ Pa), a more gradual decrease in shear moduli is caused first by an abrupt increase in oscillation amplitude of the aggregates and, second, by their rupture in response to tensile hydrodynamic forces.

To confirm the above interpretations of the experimental results, we develop a theoretical model, which considers a macroscopic deformation of structures and can only be applied for the stress amplitudes higher than those corresponding to the beginning of the second quasi-plateau.[47] Briefly, the theory supposes a coexistence of the aggregates spanning the rheometer gap (percolating aggregates), the aggregates attached by one of the ends to rheometer wall and the free branches attached by one of the ends to the percolating aggregates (the latter two are called pivoting aggregates). The percolating aggregates move affinely with the rheometer walls and contribute only to the storage modulus of the suspension, while the pivoting aggregates oscillate out of phase with the rheometer walls and contribute to both the storage and loss moduli. Starting from some critical stress amplitude, the percolating aggregates are detached from one of the walls (because of the structure instabilities revealed by simulations) and join to the class of pivoting aggregates. The stress *versus* shear rate relation is given by the sum of the contributions from percolating and pivoting aggregates, each of them weighed by a volume percentage of the corresponding aggregates. In simulations, we impose a harmonic stress signal, $\sigma(t) = \sigma_0 \cos(2\pi f t)$, with t being the time, and calculate a non-harmonic strain response, $\gamma(t)$, which is then expanded into Fourier series. The first harmonics of such an expansion gives us the shear moduli G_1', G_1'', whose theoretical stress-dependencies are compared with the experimental ones in Figure 1.9 for $H_0 = 30.6$ kA m^{-1} and the stress $\sigma_0 > 30$ Pa corresponding to the beginning of the second quasi-plateau.

The best correspondence between theory and experiments is achieved for the values of the volume percentage of pivoting aggregates equal to $\phi = 0.7$ (accordingly, the volume percentage of the percolating aggregates is 0.3). This parameter is kept the same throughout all our simulations. Nevertheless, in the broad range of ϕ ($0.5 < \phi < 1$), the calculated shear moduli differed by no more than two times from the values reported in Figure 1.9. As is seen from Figure 1.9, the storage modulus is subjected to a more drastic decrease than the loss modulus. The crossover of both moduli occurs at $\sigma_0 \approx 100$ Pa and is well captured by our model. In our calculations, we did not reproduce the small local minimum of the storage modulus at $\sigma_0 \approx 350$ Pa. A small increase of the storage modulus after this local minimum could occur because of the short-range hydrodynamic interactions and collisions between aggregates,

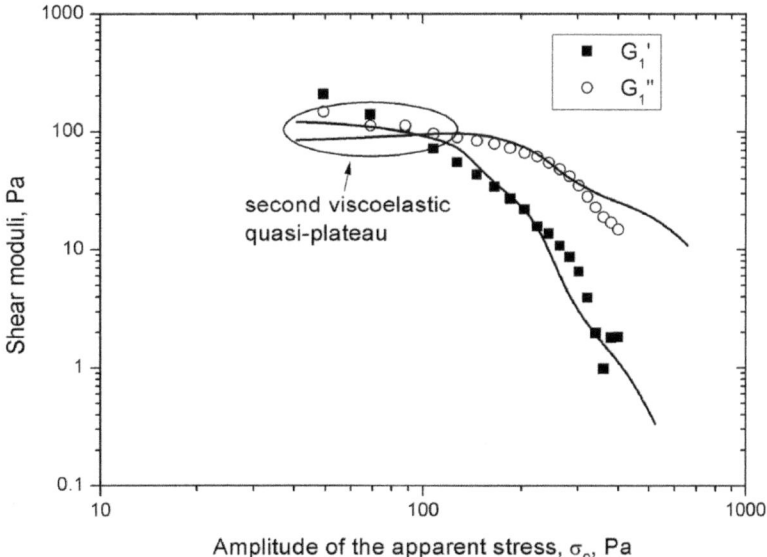

Figure 1.9 Comparison theory-experiments for the stress-dependence of the shear moduli at the magnetic field intensity, $H_0 = 30.6$ kA m^{-1} and frequency $f = 1$ Hz. The fit parameter of the model – the volume percentage of the pivoting aggregates – is chosen to be $\phi = 0.7$. Solid lines correspond to calculations and points to experimental results (with kind permission from the *Journal of Non-Newtonian Fluid Mechanics*, Elsevier).

which would restrict the aggregate motion to smaller amplitudes. Note that, apart from this local minimum, both experimental and theoretical curves of $G_1'(\sigma_0)$, $G_1''(\sigma_0)$ are relatively smooth in the whole range of the applied stresses; thus, the transitions between the different aggregation regimes are not clearly distinguishable in these curves. However, the transition between the regime of coexisting percolating and pivoting aggregates to the regime of purely pivoting aggregates requires a special attention. At $\sigma_0 > 140$ Pa, the solution for the strain $\gamma(t)$ becomes strongly asymmetric relative to the equilibrium position, $\gamma = 0$, which does not have any physical sense. The percolating clusters are considered to be unstable and they are supposed to break in the middle, remaining attached to one of the walls. So, they are transformed into pivoting aggregates, which, according to our calculations, are stable in the broad range of applied stress. Note that this instability is similar to the one, which defines the static yield stress through its maximum *versus* applied strain.[48]

Speaking about comparison between the viscoelastic response of fiber suspensions and the suspensions of spherical particles, the experimental results of de Vicente *et al.*[7,10] reveal higher shear moduli of the former at low-to-intermediate magnetic fields, while the difference becomes minor at high magnetic fields, at which the particle magnetization approaches saturation. Such a shape-induced enhancement of the shear moduli can be easily understood in terms of the shape-dependent demagnetization effect discussed in Section 1.2.3. At high magnetic fields, $H \sim M_S$, the demagnetization field vanishes inside both spherical and rod-like particles, and, in the absence of solid friction, the difference in the yield stresses and shear moduli of both suspensions should also vanish, as is apparently the case of smooth nano-sized iron or magnetite particles used by de Vicente *et al.*[7,10] Such an interpretation was originally given by these authors. A significant difference between the yield stresses in our suspensions at high magnetic fields [Section 1.2.2] was interpreted by a rather strong solid friction between our rough micron-sized fibers. In conclusion, a more detailed study is required to elucidate the effect of the particle surface state on the viscoelastic properties of MR suspensions (composed of spheres or fibers).

1.3 MR Fluid Flows in Longitudinal Fields

As already mentioned in the introduction, apart from adjusting the particle morphology, one can change the orientation of the magnetic field relative to the flow in order to improve the performance of a given MR smart device. In particular, the use of longitudinal magnetic fields instead of perpendicular ones allows a much more compact design of MR devices, without substantial loss of the MR effect – the yield stress in longitudinal fields appears to be of the same order of magnitude that the one in perpendicular fields, except for suspensions of weakly paramagnetic particles, whose viscosity decreases in longitudinal fields.[26] In this section, we consider shear and pipe flows of a conventional MR fluid in the presence of a longitudinal external magnetic field and give a qualitative explanation of the "longitudinal" MR effect followed by quantitative estimations of the dynamic yield stress.

Obviously, high mechanical stresses in longitudinal fields may only appear if the MR structures are misaligned relative to the fluid streamlines. For instance, if aggregate rotation is restricted to the shear plane, the aggregate shear stress scales as $\sigma \propto \eta_0 \dot{\gamma} r_e^2 < \cos^2 \theta \sin^2 \theta >$ [*cf.* Batchelor[49]], with θ being an angle between the aggregates and the streamlines, r_e the aggregate aspect ratio, and the angle brackets denote averaging over all possible orientations. Thus, even a small angle deviation of aggregate orientation from the flow direction may generate a non-negligible stress, if the aggregate aspect ratio is high. Contrarily to flow-aligned aggregates, a misaligned aggregate should have a large but finite length defined by the equilibrium between the tensile hydro-dynamic force and the magnetic cohesive force. Thus, the aggregate aspect ratio is expected to follow the same shear rate dependence as in the case of the perpendicular magnetic field, $r_e^2 \propto \dot{\gamma}^{-1}$ [*cf.* Section 1.2.3]. This condition, verified by our theory, could explain the appearance of the dynamic yield stress in longitudinal fields. The main question now is which mechanism can be responsible for aggregate misalignment. The main hypothesis of the present study is that the aggregates can deviate from their orientation along the streamlines because of magnetic dipole interactions with the neighboring aggregates. Since the aggregates are randomly spaced in the suspension, under shear flow, they will change their mutual positions and orientations in irregular way. Together with many-body interactions, this may cause a stochastic variation in dipolar forces and torques experienced by the aggregates and could produce some fluctuations in their orientations. This process can be regarded as a magnetically induced rotational diffusion of aggregates, by analogy with Brownian rotational diffusion[50] or flow-induced rotational diffusion of elongated particles caused by their collisions or short-range hydrodynamic interactions in sheared suspensions.[51,52] Stochastic torques coming from many-body magnetic interactions tend to randomize the aggre-gate orientation, while a shear flow and a restoring magnetic torque, exerted on aggregates by an external field, tend to align the aggregates with the flow. So, the fluctuations in aggregate orientation are not necessarily large and might not lead to collisions. In this case, we shall deal with a weak rotational diffusion caused solely by long-range dipole interactions. In support of this hypothesis, weak orientation fluctuations have recently been observed in experiments on kinetics of aggregation of diluted magnetic suspensions.[53]

Let us now estimate the dynamic yield stress of the MR fluid, whose ag-gregates are subjected to stochastic angular oscillations under a simple shear in the longitudinal field. The fluctuations of aggregate orientation can be seen as a random walk where the aggregates perform irregular jumps with a mean amplitude $\Delta\theta$ and a mean jump duration Δt. The intensity of such fluctuations is measured by a rotational diffusion constant, which, according to the random walk model, scales as [*cf.* Van de Ven[50]]:

$$D_r \sim \frac{(\Delta\theta)^2}{\Delta t} \approx \langle \omega^2 \rangle \Delta t \qquad (1.10)$$

where $\langle \omega^2 \rangle$ is the mean square angular velocity of the aggregates. The mean jump duration can be estimated by considering the mutual displacement of two neighboring aggregates (spaced laterally by a distance d) in shear flow: $\Delta t \sim 2L/(\dot{\gamma}d)$, where $2L$ is the aggregate length. The amplitude of orientational fluctuations is measured by a torque, T_{int}, created by many-body magnetic interactions with neighboring aggregates and called hereinafter interaction torque. The mean square angular velocity of the stochastic motion of aggregates can be estimated as $\langle \omega^2 \rangle \approx \langle T_{int}^2 \rangle / f_r^2$, with $\langle T_{int}^2 \rangle$ being the mean square interaction torque and f_r the rotational friction coefficient. Estimating both quantities $\langle T_{int}^2 \rangle$ and f_r and taking into account that the latter is proportional to the aggregate aspect ratio squared ($f_r \propto r_e^2 \propto \dot{\gamma}^{-1}$), we show that the rotational diffusion constant is linear in shear rate:[28]

$$D_r = C\dot{\gamma} \tag{1.11}$$

with C being a dimensionless factor proportional to the square of the particle volume fraction Φ. Note, that the same shear rate dependence was postulated by Folgar and Tucker[51] for the rotational diffusion of non-Brownian rod-like particles induced by their collisions in sheared suspensions. However, the physics is quite different because, in the latter situation, the random orientational walk was only dictated by the rate of collisions proportional to the shear rate, whereas, in our case, it is the interplay between long range dipolar forces and shear rate, which produces the same scaling.

In order to evaluate the suspension stress, we must first determine the orientation distribution of aggregates, or rather the second and the fourth statistical moments, $\langle e_i e_k \rangle$ and $\langle e_i e_k e_l e_m \rangle$, intervening into the expression for the stress tensor, where $\mathbf{e}$ is the unit vector along the aggregate major axis and e_i is its component along the axis Ox_i, $i = 1, 2, 3$. Here, the axis Ox_1 is aligned with the velocity direction, the axis Ox_2 corresponds to the velocity gradient direction and the axis Ox_3 to the vorticity direction. The quantities $\langle e_i e_k \rangle$ and $\langle e_i e_k e_l e_m \rangle$ can be determined by solving a set of equations describing temporal evolution of the statistical moments $\langle e_i e_k \rangle$ [these equations are not given here for brevity, the reader may consult the textbooks of Bird *et al.*[54] and Doi and Edwards[55] for more details] coupled with a certain closure relationship between the fourth and the second moments: $\langle e_i e_k e_l e_m \rangle = f(\langle e_i e_k \rangle)$. We choose a quadratic closure approximation, first postulated by Doi and Edwards,[55] $\langle e_i e_k e_l e_m \rangle \equiv \langle e_i e_k \rangle \langle e_l e_m \rangle$, which becomes exact in the limit of perfect alignment of aggregates and whose exactness decreases with decrease in degree of alignment of aggregates. In a steady shear flow considered here, the problem reduces to a system of four algebraic equations for the four unknown quantities, $\langle e_1 e_2 \rangle$, $\langle e_1^2 \rangle$, $\langle e_2^2 \rangle$, $\langle e_3^2 \rangle$, the first of which can be seen as a mean sine of the angle between the aggregates and the flow, and the last three quantities as mean square cosines of the angle that the aggregate makes with the flow, velocity gradient and vorticity, respectively. The aggregate aspect ratio r_e intervening into these equations is found from the balance of magnetic and hydrodynamic tensile forces acting on aggregates, in the same way as in the case of fiber suspensions [*cf.* Section 1.2.3]. In the wide range of magnetic fields and concentrations ($H < 15\,\text{kA m}^{-1}$ and $\Phi < 0.3$) the problem allows, within the 10% error, an approximate analytical solution, as follows:

$$\langle e_1 e_2 \rangle \approx \frac{1}{8} \alpha \left(\frac{\Phi \mu_0 \chi_a^2 H^2}{\Phi_a^2 f^m} \right)^2 \tag{1.12a}$$

$$\langle e_2^2 \rangle = \langle e_3^2 \rangle \approx \frac{1}{32} \alpha \left(\frac{\Phi \mu_0 \chi_a^2 H^2}{\Phi_a^2 f^m} \right)^4 \tag{1.12b}$$

$$\langle e_1^2 \rangle \approx 1 - \frac{1}{16} \alpha \left(\frac{\Phi \mu_0 \chi_a^2 H^2}{\Phi_a^2 f^m} \right)^4 \tag{1.12c}$$

where α is an adjustable parameter, $\Phi_a \approx \pi/6$ is the internal volume fraction of aggregates with the particles arranged into a simple cubic lattice, χ_a is the aggregate magnetic susceptibility, and f^m is the magnetic force (per unit particle cross-section) acting between two particles inside the aggregate. The last two quantities are functions of the external magnetic field H and calculated by numerical simulations of Maxwell magnetostatic equations.

Similarly to the case of fiber suspensions, the shear stress is calculated with the help of the well-known relationships developed for dilute and semi-dilute suspensions of anisotropic particles.[42,43] Replacing the aggregate aspect ratio by an appropriate relationship we obtain the final expression for the shear stress (12-component of the stress tensor):

$$\sigma = \underbrace{\eta_0 \dot{\gamma} \left(1 + 2 \frac{\Phi}{\Phi_a} \right)}_{\text{solvent stress}} + \underbrace{\Phi f^m \langle e_1 e_2 \rangle \langle e_1^2 \rangle}_{\substack{\text{hydrodynamic} \\ \text{aggregate stress}}} \quad \underbrace{-2 \frac{\Phi}{\Phi_a} \cdot \frac{\chi_a^2}{2 + \chi_a} \mu_0 H^2 \langle e_1 e_2 \rangle \langle e_2^2 \rangle}_{\text{magnetic stress}}$$

$$\underbrace{+ \frac{3}{2} \alpha \frac{\Phi^3}{\Phi_a^4} \frac{\left(\mu_0 \chi_a^2 H^2 \right)^2}{f^m} \frac{\langle e_1 e_2 \rangle^2}{\langle e_1^2 \rangle}}_{\text{diffusion stress}} \tag{1.13}$$

The first term in the right-hand side of this equation stands for the solvent contribution to the stress and the last three terms stand for the aggregate contribution. Among these three terms, the first one corresponds to the hydrodynamic part of the aggregate stress, the second one comes from the external magnetic torque (magnetic stress) and the last one arises from the random interaction torques inducing random fluctuations of aggregate orientations [defined by Leal and Hinch[56] as diffusion stress]. All the three contributions of the aggregate stress appear to be independent of the shear rate, so, their sum is considered as a dynamic yield stress. Furthermore, analysis shows that the magnetic stress gives a negligible contribution, so the final expression for the dynamic yield stress reads:

$$\sigma_Y \approx \underbrace{\Phi f^m \langle e_1 e_2 \rangle \langle e_1^2 \rangle}_{\text{hydrodynamic stress}} + \underbrace{\frac{3}{2} \alpha \frac{\Phi^3}{\Phi_a^4} \frac{\left(\mu_0 \chi_a^2 H^2\right)^2}{f^m} \frac{\langle e_1 e_2 \rangle^2}{\langle e_1^2 \rangle}}_{\text{diffusion stress}} \qquad (1.14)$$

To validate our theory, we have performed a detailed experimental study of magnetic suspension flow in the presence of a longitudinal magnetic field. Because of experimental constraints, experimental realization of simple shear flows with the magnetic field aligned with the fluid streamlines is quite problematic. Therefore, we had to use a pressure-driven flow through a cylindrical channel instead of the simple shear flow studied theoretically. The shear rate varies across the channel in capillary flows. However, using our model, we estimated that the shear rate variation along the aggregates was negligible, except for a narrow central flow region. So, we expect that the rheological behavior observed in the pressure-driven flow should be similar to that in the drag shear flow with a linear velocity profile. The experimental flow curves (not shown here for brevity) obtained for the capillary flow in the longitudinal magnetic field appear to be linear and can be interpolated by a linear rheological law, $\sigma = \sigma_Y + \eta \dot{\gamma}$, similar to the one predicted by our model.

Experimental and theoretical dependencies of the dynamic yield stress, σ_Y, on the magnetic field intensity are presented in Figure 1.10 for the magnetic suspensions of different volume fractions. The theoretical dependencies were fitted to experimental ones by using the least square method with a single free parameter, $\alpha \approx 1.5$. As is seen from Figure 1.10, the yield stress is an increasing function of the magnetic field intensity. The increasing field-dependence of the yield stress can be easily understood by the two mechanisms, as follows. First, the magnetic interactions between aggregates increase with the increasing magnetic field. This leads to larger fluctuations of aggregate orientation and therefore to a larger viscous dissipation. This mechanism appears in eqn (1.14) for the yield stress through the statistical moment $\langle e_1 e_2 \rangle$, which is a growing function of the magnetic field intensity [eqn (1.12a)]. Second, magnetic interactions between particles, composing the aggregates also increase with a growing magnetic field. The aggregates become more resistive against destructive shear forces, their length increases with the field, so they generate higher stresses.

The concentration dependence of the dynamic yield stress is presented in Figure 1.11 for the magnetic field strength $H = 15$ kA m^{-1}. The theoretical values of the yield stress were calculated using the appropriate value of the free parameter, $\alpha = 1.5$. Again, we obtain a reasonably good correspondence with experiments. As is seen from Figure 1.10, both theory and experiments show that the yield stress increases with the particle volume fraction more strongly than linearly. Such nonlinear behavior could be easily explained by concentration-enhanced interactions between aggregates. In more detail, the mean distance between aggregates and, consequently, the magnetic interaction

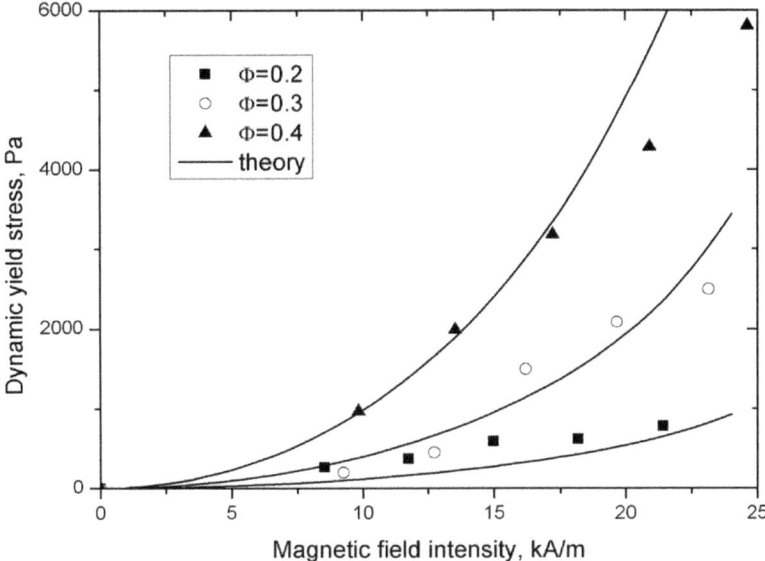

Figure 1.10 Theoretical and experimental dependencies of the dynamic yield stress of the MR fluid (composed of spherical particles) on the magnetic field strength at different particle volume fractions. Points correspond to experimental data and lines to the theory. The magnetic field is parallel to the flow direction (with kind permission from the *Journal of Rheology*).

torque increase with the particle volume fraction. Therefore, the aggregates will be subjected to stronger fluctuations of their orientation and will generate a stronger viscous dissipation. More quantitatively, the hydrodynamic stress – the most important contribution to the yield stress – is equal to $\Phi f^m \langle e_1 e_2 \rangle \langle e_1^2 \rangle$, with, according to eqn (1.12a) and (1.12c), $\langle e_1^2 \rangle \sim 1$ and $\langle e_1 e_2 \rangle \propto \Phi^2$. Therefore, the hydrodynamic stress, and consequently the yield stress, varies as $\sigma_Y \propto \Phi^3$. To the best of our knowledge, such a strong concentration behavior has not been observed in magnetic fields perpendicular to the flow. This is likely because, in perpendicular fields, the MR response of the suspension is mostly governed by the magnetic interaction between aggregates and external field (which results in a high restoring torque), whereas, in longitudinal fields, the MR effect is produced by relatively strong dipolar interactions between aggregates (Figure 1.11).

1.4 Concluding Remarks

In this chapter, we have reviewed recent experimental and theoretical results on the magnetorheology of fiber suspensions in magnetic fields perpendicular to the shear as well as of suspensions of spherical particles in longitudinal magnetic fields. Both these problems, weakly related to each other at first sight, reveal essentially similar physics. Upon magnetic field application, both

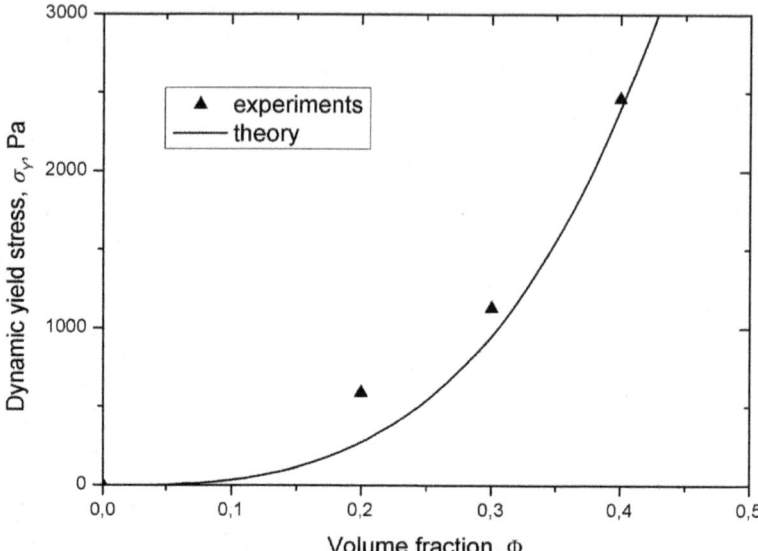

Figure 1.11 Theoretical and experimental dependencies of the dynamic yield stress of the MR fluid (composed of spherical particles) on the particle volume fraction for the magnetic field strength $H = 15$ kA m^{-1}. The magnetic field is parallel to the flow direction (with kind permission from the *Journal of Rheology*).

spherical particles and fibers form strongly elongated aggregates exhibiting a similar behavior in shear flows. The two types of aggregates tend to align with the streamlines, they are progressively destroyed by the shear forces at increasing shear rates and resist these forces thanks to magnetic interactions between their constitutive particles. The differences lie in (1) a stronger magnetic permeability of the aggregates of fibers; (2) a presumably stronger solid friction between fibers; and (3) eventually a more sparse structure of the aggregates of fibers. Such differences between the aggregates composed of fibers and spheres lead to a non-similar MR response of both suspensions: in most of the cases, a few times enhancement of the yield stress and shear moduli was reported for fiber suspensions by different research groups, at least for magnetic fields below the magnetization saturation limit of particles.[7,8,11,17]

The particle shape effect on the magnetic permeability enhancement is commonly recognized to be the major effect contributing to a stronger MR response of fiber suspensions, at least in steady shear flows at Mason numbers, $Mn > 1$ and at low-to-moderate magnetic fields, $H < 50$ kA m^{-1} [*cf.* de Vicente *et al.*,[7] Gómez-Ramírez *et al.*[15]]. As the magnetic field increases, demagnetizing fields inside particles decrease and their magnetization approaches a saturation value, which is almost the same for both spheres and fibers. So, at high magnetic fields, the scenario of enhanced permeability does not explain anymore the difference in the rheological behavior of fiber suspensions and suspensions of spheres. Note that experimental results obtained by different groups are not similar for high magnetic fields. De Vicente *et al.*[10] reported only a slight difference in shear moduli of both suspensions, which is consistent with the scenario of enhanced magnetic permeability, while López-López *et al.*[14] still found a few times increase of the yield stress of fiber suspensions as compared to conventional MR fluids. In the former case, the rod-like particles seem to have a relatively smooth surface, so the solid friction is likely minimized providing that these sub-micron-sized particles are subjected to a weak Brownian motion that can separate them and reduce direct contacts between them. In the latter case, the particle surface seems to be relatively rough [Figure 1.1] and the suspension microstructure appears to be quite intricate and entangled [Figure 1.2], which supports the existence of solid friction between fibers. Interestingly, the contribution to the yield stress coming from friction forces is strongly dependent on the applied magnetic field. This is easily explained by the fact that interfiber friction forces are proportional to the normal contact forces and the latter appear as the result of the restoring magnetic torque acting on fibers. The fibers tend to align with the field and press the neighboring fibers with a certain force proportional to the magnetic torque. Of course, the friction scenario, employed by Vereda *et al.*[16] and Kuzhir *et al.*,[17] may only be applied to a quasi-static deformation of the fiber suspension below the yield point, for which the hydrodynamic lubrication is unable to avoid direct solid contacts between fibers – the case considered in Section 1.2.2. In these regimes, the interparticle friction induces non-perfect alignment of the fibers with the magnetic field, and the fiber orientation distribution may exhibit a hysteresis with respect to increasing and decreasing magnetic fields. Such behavior will result in a relatively strong hysteresis of magnetization curves, whose atypical shape could not be explained by the remnant magnetization of fibers.[11] However, in steady shear flows or large amplitude oscillatory flows at high Mason numbers, $Mn > 1$, the cases considered in Section 1.2.3 and 1.2.4, the direct interparticle contacts seem to be absent, even for the particles having a rough surface. The absence (or smallness) of flow curve hysteresis confirms the absence of solid friction in this case. The stronger dynamic yield stress and shear moduli of fiber suspensions are again explained by the scenario of enhanced magnetic permeability. In conclusion, the overview of the existing experimental results and models shows that the systematic information about the magnetic field effect and particle morphology effects is still missing for fiber suspensions. New studies covering a broad range of parameters (fiber aspect ratio, particle surface roughness, rheometer gap, magnetic field intensity, *etc.*) are highly desirable.

In all the theories considered above for fiber suspensions, the interactions between aggregates were ignored. If, neglecting these interactions, one can still explain the MR response of MR suspensions (composed of either fibers or spheres) in perpendicular fields, this is not the case for longitudinal fields. We explain a high level of stress generated in longitudinal magnetic fields by many-body magnetic interactions between aggregates, which induce misalignments of particle aggregates from the streamlines and result in stochastic oscillations of their orientation. Random fluctuations in aggregate orientation are mimicked by an effective rotational diffusion process. The rotary diffusivity, D_r, is estimated using a random walk model and is found to be proportional to the mean square interaction torque, $\langle T_{int}^2 \rangle$ – a net magnetic torque exerted to a given aggregate by all the neighboring aggregates. The theory predicts that the diffusion constant is linear in shear rate, $D_r = C\dot{\gamma}$. Using a mathematical apparatus inherent to the rotational diffusion process, we found the orientation distribution of aggregates and a suspension shear stress, which follows the Bingham law, $\sigma = \sigma_Y + \eta\dot{\gamma}$, observed in experiments. Both experiments and theory suggest a strong concentration dependence of the yield stress ($\sigma_Y \propto \Phi^3$ at $\Phi < 0.3$ and $H < 15$ kA m^{-1}) which is attributed to a strong concentration dependence of the rotary diffusivity. In fact, an increase in the particle volume fraction diminishes the mean lateral spacing between aggregates and, consequently, enhances magnetic interactions between them. Even though the present theory describes experimental data reasonably well, we cannot state with confidence that the stochastic misalignments of aggregates from the flow direction is the unique mechanism of the longitudinal MR effect. Unfortunately, we do not have reliable experimental evidence of such a mechanism. We could expect that at high concentrations, the particle structure will be dendrite-like, as the one shown in Figure 1.2c, with part of the particle aggregates more or less aligned with the flow and another part constituted by chains bridging the former aggregates. Such bridging chains are transverse to the flow and could provide a major contribution to the stress. They are expected to break and reform periodically in the flow, and their orientation distribution and length could, in principle, be found by an approach similar to the one developed in Section 1.3 for misaligned aggregates. Finally, a synergy between the rotational diffusion concept [Section 1.3] and a representation of the MR structure by a cross-linked network is expected to give a more realistic description of the MR fluid behavior in longitudinal fields.

References

1. J. M. Ginder, *MRS Bull.*, 1998, **23**, 26–29.
2. G. Bossis, O. Volkova, S. Lacis and A. Meunier, Magnetorheology: Fluids, structures and rheology, *Lect. Notes Phys.*, 2002, **594**, 186–230.
3. J. D. Carlson, D. M. Catanzarite and K. A. St. Clair, *Int. J. Mod. Phys. B*, 1996, **10**, 2857.
4. W. I. Kordonski and S. D. Jacobs, *Int. J. Mod. Phys. B*, 1996, **10**, 2837.
5. R. C. Bell, E. D. Miller, J. O. Karli, A. N. Vavreck and D. T. Zimmerman, *Int. J. Mod. Phys. B*, 2007, **21**, 5018.
6. M. T. López-López, G. Vertelov, P. Kuzhir, G. Bossis and J. D. G. Durán, *J. Mater. Chem.*, 2007, **17**, 3839.
7. J. de Vicente, J. P. Segovia-Guitérrez, E. Anablo-Reyes, F. Vereda and R. Hidalgo-Alvarez, *J. Chem. Phys.*, 2009, **131**, 194902.

8. R. C. Bell, J. O. Karli, A. N. Vavreck, D. T. Zimmerman, G. T. Ngatu and N. M. Wereley, *Smart Mater. Struct.*, 2008, **17**, 015028.
9. F. Vereda, J. de Vicente and R. Hidalgo-Álvarez, *Langmuir*, 2007, **23**, 3581–3589.
10. J. de Vicente, F. Vereda and J.-P. Segovia-Gutiérez, *J. Rheol.*, 2010, **54**, 1337–1362.
11. A. Gómez-Ramírez, M. T. López-López, J. D. G. Durán and F. González-Caballero, *Soft Matter*, 2009, **5**, 3888–3895.
12. G. T. Ngatu, N. M. Wereley, J. O. Karli and R. C. Bell, *Smart Mater. Struct.*, 2008, **17**, 045022.
13. R. C. Bell, D. Zimmerman and N. M. Wereley, Impact of Nanowires on the Properties of Magnetorheological Fluids and Elastomer Composites, in *Electrodeposited Nanowires and their Applications*, ed. N. Lupu, Intech Publishers, Vienna, Austria, 2010, ch. 8, pp. 189–212.
14. M. T. López-López, P. Kuzhir and G. Bossis, *J. Rheol.*, 2009, **53**, 115–126.
15. A. Gómez-Ramírez, P. Kuzhir, M. T. López-López, G. Bossis, A. Meunier and J. D. G. Durán, *J. Rheol.*, 2011, **55**, 43–67.
16. F. Vereda, J. de Vicente and R. Hidalgo-Álvarez, *Chem. Phys. Chem.*, 2009, **10**, 1165–1179.
17. P. Kuzhir, M. T. López-López and G. Bossis, *J. Rheol.*, 2009, **53**, 127–151.
18. K. Asano, H. Suto and K. Yatsuzuka, *J. Electrostat.*, 1997, **40–41**, 573–578.
19. Y. Otsubo, *Colloids Surf., A*, 1999, **153**, 459–466.
20. K. Tsuda, Ya. Takeda, H. Ogura and Ya. Otsubo, *Colloids Surf., A*, 2007, **299**, 262–267.
21. M. M. Ramos-Tejada, M. J. Espin, R. Perea and A. V. Delgado, *J. Non-Newtonian Fluid Mech.*, 2009, **159**, 34–40.
22. R. C. Kanu and M. T. Shaw, *J. Rheol.*, 1992, 42 657–660.
23. A. Kawai, U. Kunio and I. Fumikazu, *Int. J. Mod. Phys. B*, 2002, **16**, 2548–2554.
24. Ya. K. Kor and H. See, *Rheol. Acta*, 2010, **49**, 741–756.
25. Z. P. Shulman and W. I. Kordonsky, *Magnetorheological effect*, Nauka i Tehnika, Minsk, 1992 (in Russian).
26. R. Tao and K. Huang, *Phys. Rev. E*, 2011, **84**, 011905.
27. P. Kuzhir, G. Bossis, V. Bashtovoi and O. Volkova, *J. Rheol.*, 2003, **47**, 1385–1398.
28. P. Kuzhir, C. Magnet, G. Bossis and A. Meunier, *J. Rheol.*, 2011, **55**, 1297–1318.
29. S. G. Mason, *TAPPI J.*, 1950, **33**, 440–444.
30. C. F. Schmid, L. H. Switzer and D. J. Klingenberg, *J. Rheol.*, 2000, **44**, 781–809.
31. L. H. Switzer and D. J. Klingenberg, *Int. J. Multiph. Flow*, 2004, **30**, 67–87.
32. B. N. J. Persson, *Sliding Friction. Physical Principles and Applications*, Springer-Verlag, Berlin, 2000.
33. R. G. Larson, *The Structure and Rheology of Complex Fluids*, Oxford University Press, New York, 1999.
34. S. Toll and J.-A. E. Manson, *J. Rheol.*, 1994, **38**, 985–997.

35. C. Servais, J. A.-E. Manson and S. Toll, *J. Rheol.*, 1999, **43**, 991–1004.
36. D. Bideau, J.-P. Troadec and L. Oger, *C. R. Seances Acad. Sci., Ser. 2*, 1983, **297**, 319–322.
37. Z. P. Shulman, V. I. Kordonsky, E. A. Zaltsgendler, I. V. Prokhorov, B. M. Khusid and S. A. Demchuk, *Int J. Multiphase Flow*, 1986, **12**, 935–955.
38. J. E. Martin and R. A. Anderson, *J. Chem. Phys.*, 1996, **104**, 4814–4827.
39. O. Volkova, G. Bossis, M. Guyot, V. Bashtovoi and A. Reks, *J. Rheol.*, 2000, **44**, 91–104.
40. J. M. Ginder, L. C. Davis and L. D. Elie, *Int. J. Mod. Phys. B*, 1996, **10**, 3293–3303.
41. G. K. Batchelor, *J. Fluid. Mech.*, 1970, **44**, 419–440.
42. H. Brenner, *Int. J. Multiphase Flow*, 1974, **1**, 195–341.
43. V. N. Pokrovskiy, *Statistical Mechanics of Diluted Suspensions*, Nauka, Moscow, 1978.
44. C. L. A. Berli and J. de Vicente, *Appl. Phys. Lett.*, 2012, **101**, 021903.
45. J. de Vicente, G. Bossis, S. Lacis and M. Guyot, *J. Magn. Magn. Mater.*, 2002, **251**, 100–108.
46. D. J. Klingenberg, *J. Rheol.*, 1993, **37**, 199–214.
47. P. Kuzhir, A. Gómez-Ramírez, M. T. López-López, G. Bossis and A. Yu. Zubarev, *J. Non-Newtonian Fluid. Mech.*, 2011, **166**, 373–385.
48. G. Bossis, E. Lemaire, O. Volkova and H. Clercx, *J. Rheol.*, 1997, **41**, 687–704.
49. G. K. Batchelor, *J. Fluid. Mech.*, 1971, **46**, 813–829.
50. G. M. Van de Ven, *Colloidal Hydrodynamics*, Academic Press Limited, London, 1989.
51. F. Folgar and C. L. Tucker, *J. Reinforced Plast. Composites*, 1984, **3**, 98–119.
52. J. Férec, G. Ausias, M. C. Heuzey and P. J. Carreau, *J. Rheol.*, 2009, **53**, 49–72.
53. G. Bossis, P. Lançon, A. Meunier, L. Iskakova, V. Kostenko and A. Zubarev, *Physica A*, 2013, **392**, 1567–1576.
54. R. B. Bird, O. Hassager, R. C. Armstrong and C. F. Curtiss, *Dynamics of Polymeric Liquids Volume II. Kinetic Theory*, John Wiley and Sons, New York, 1977.
55. M. Doi and S. F. Edwards, *The Theory of Polymer Dynamics*, Oxford Press, New York, 1986.
56. L. G. Leal and E. J. Hinch, *J. Fluid Mech.*, 1972, **55**, 745–765.

CHAPTER 2

Magnetorheology of Fe Nanofibers Dispersed in a Carrier Fluid

R. C. BELL,*[a] D. T. ZIMMERMAN[b] AND N. M. WERELEY[c]

[a] Department of Chemistry, The Pennsylvania State University, Altoona College, Altoona, PA 16601, USA; [b] Department of Physics, The Pennsylvania State University, Altoona College, Altoona, PA 16601, USA; [c] Department of Aerospace Engineering, University of Maryland, College Park, MD 20742, USA
*Email: RCB155@psu.edu

2.1 Introduction

Conventional magnetorheological (MR) fluids are composed of roughly spherical, micron-sized ferromagnetic particles suspended in a carrier fluid such as water, silicone, or hydrocarbon oil depending on the desired properties.[1,2] By application and removal of an externally applied magnetic field, this smart material has the unique ability to rapidly change its effective viscosity in a matter of milliseconds and in a nearly reversible manner. In the absence of a magnetic field (off-state), MR fluids are viscous liquid/particle suspensions in which the viscosity can range from 0.1–3 Pa s depending upon the particle loading, shape, and size. Upon application of an external magnetic field (on-state), the particles acquire a magnetic polarization and attract one another, forming chain-like structures that join to form columnar structures parallel to the applied field. These columnar structures span the opposing surfaces of the

RSC Smart Materials No. 6
Magnetorheology: Advances and Applications
Edited by Norman Wereley
© The Royal Society of Chemistry 2014
Published by the Royal Society of Chemistry, www.rsc.org

device parallel to the field lines resulting in a material that behaves as a Bingham plastic fluid, displaying an increased viscosity and apparent yield stress under shear. In the presence of the magnetic field, the suspension is converted to a semi-solid with a substantial change in the viscosity of up to 10^6 Pa s and a concomitant field-induced yield stress of up to 100 kPa (in suspensions that contain close to 40 vol% ferromagnetic particles).[1,3] The viscosity and yield stress of the suspension is scalable with the magnitude of the applied magnetic field until magnetic saturation of the particles is reached.[4] Exceeding the yield stress of the fluid causes the fibril chains and columnar structures to continuously break and re-form, resulting in a post-yield viscosity.

There are a wide variety of applications ideally suited to exploit this material's uniquely large change in the field-dependent yield stress and apparent viscosity. The fast response and controllability of MR fluids[5] make them ideally suited for implementation in semi-active smart vibration–absorption systems,[6] primary vehicle suspension systems,[7] adaptive crew seats for vibration,[8–10] landing gear for aircraft,[11–13] shock isolation,[14,15] actuators,[16] base isolation systems for earthquake damping,[17] clutches,[18] and optical polishing instruments.[19] The use of MR fluids in more diverse applications have been limited due to a number of critical issues including the high manufacturing costs,[5] settling of the particles in the absence of continual mixing due to the difference in densities between the particles and the carrier fluid[20] and particle wear[5] that tends to reduce the efficiency of the suspension over time and eventually leads to device failure.

The rheological properties of MR fluids depend on many factors including the strength of the applied magnetic field, type and amount of additives within the suspension, and the properties of the particles, such as loading (vol% or wt%), size, composition, magnetic properties, and, as discussed in this chapter, the geometry of the suspended particles. Particle loading is the most important factor affecting the achievable yield strength of MR fluids. Increasing the concentration of particles in the suspension results in an increase in the achievable yield strength, but this also results in an increase in the off-state viscosity. Ideal MR fluids have low viscosities (<3 Pa s) giving them the largest possible range of controllable viscosities and allowing for easy flow within a device. Conventional MR fluids utilize high particle loading (30 to 40 vol%) to achieve the high yield stresses required for most applications. As the size of the particles increase, the shear strength of the fluid also increases;[1,21] however, spherical particles larger than about 10 μm tend to settle rapidly. Thus, particles in the range of 1–10 μm are ideal for MR fluids.

A disadvantage of conventional MR fluids is that the particles are prone to settle under no- or low-magnetic field conditions in the absence of constant or frequent mixing.[20] This is due to the intrinsic density difference between the iron particles and the carrier fluid.[22] The iron particles sediment and agglomerate at the bottom of the container, rendering the device ineffective. Once the particles sediment, the remnant magnetism and hydrostatic inter-actions of particles results in undesired tightly-bound particle clusters making re-dispersion extremely difficult.[22–24] Since the yield stress of MR fluids is directly related to the volume fraction of particles in the suspension, the behavior of the fluid in this inhomogeneous state is less predictable until the suspension is remixed. Many approaches to reduce sedimentation while maintaining the desired magnetorheological properties of MR fluids have been studied, including the addition of thixotropic additives,[24–26] nanoscale additives,[27] nanometre-sized organo-clay additives[28,29] and polyvinyl butyral (PVB) coating of the particles.[30,31] Addition of nano-particles to the suspension effectively mitigates sedimentation;[32–38] however, increasing the nanoparticle concentration causes the apparent yield stress of the MR fluids to decline by as much as 50%.[39] The use of suspensions containing strictly nanoparticles can remain suspended in the fluid indefinitely due to Brownian motion by thermal convection, but the yield strength of the MR fluid is greatly reduced.[40,41] Sedimentation of the particles in conventional MR fluids is a key hurdle to overcome before more diverse applications can be realized.[42]

Numerous studies have shown that the yield stress and settling properties of magnetorheological fluids depend on particle morphology.[42–52] The use of elongated particles in MR fluids has distinct advantages over suspensions that contain only spherical particles. The maximum achievable yield stress can be twice that (or greater) of conventional fluids (at the same particle solids loading). Studies by Chin *et al.* showed that the addition of elongated ferromagnetic particles Co-γ-Fe_2O_3 and CrO_2 to MR suspensions composed of carbonyl iron particles reduced sedimentation and slightly increased the yield stress.[42] MR fluids utilizing cobalt microfibers produced by the agglomeration of spherical particles in the presence of a magnetic field during the formation process, displayed enhanced sedimentation properties and an increase in yield stress for suspensions.[45,48,50–52] For example, one of these studies examined MR fluids containing up to 7 vol% fibers that were 60 μm long with 4.8 μm diameters. When compared to sphere-based suspensions containing 1.34 μm diameter cobalt particles at the same loading,[50] the yield stress of the fiber-based fluids were up to 3 times greater than that of the sphere-based suspensions. It was suggested that the enhancement of the yield stress was due to inter-fiber solid friction. However, the fibers used in these studies had extremely rough surfaces and non-uniform dimensions; in addition, the difference in the size of the spherical particles compared to fibers made a direct comparison between the properties of the two types of suspensions difficult to interpret. All of these studies suffered from broad size distribution, inconsistency in particle shape, surface roughness, and particle fragility, making it difficult to form conclusions or infer the underlying physics.

The following chapter addresses the advantages and shortfalls of altering the particle shape of the suspended ferromagnetic particles in MR fluids. Nanofiber-based MR fluids have shown promising results for not only reducing or preventing settling, but also display increased yield stress of the suspension.[44,46,47,53] Unlike the ferromagnetic oxide particles used in previous studies[42] and fibrils formed from spherical particles adjoined in the presence of a magnetic field,[45,48,50–52] nanofibers have well defined shapes with smooth edges and controllable length distributions. This allows for a more direct comparison between systematic experimental results and theoretical studies;

this will undoubtedly lead to a greater understanding of the physical properties of MR fluids and to better design control of future MR fluids.

2.2 MR Fluid Synthesis and Testing

MR fluids consisting of two types of ferromagnetic particles were investigated. To represent conventional MR fluids, spherical iron particles having 1–3 μm or 6–10 μm diameters (Alfa Aesar) were used. Nanofiber-based fluids were synthesized using fibers generated by template-based electrodeposition techniques. These particles were then suspended in a silicone carrier fluid both separately or mixed in various proportions.

2.2.1 Nanofiber Generation and Characterization

The nanofibers were generated *via* template-based electrodeposition using commercially available, anodized alumina membranes (Whatman) as templates. The electrolyte solution for iron nanofibers consists of 0.9 M $FeSO_4$, 0.03 M $FeCl_2$, 0.10 M NH_4Cl, 0.01 M $C_6H_8O_6$, and 0.5 M H_3BO_3 at a pH of 3 adjusted using H_2SO_4. A 99.99% iron foil was used as the working electrode. Cobalt nanofibers were deposited from a solution containing 0.2 M $CoSO_4$, 0.02 M $CoCl_2$, and 0.3 M H_3BO_3 adjusted to a pH of 3.5 adjusted using H_2SO_4. A 99.97% cobalt foil was used as the working electrode. The working electrodes were suspended approximately 2 cm from the alumina template in the electrolytic solutions. Fibers were electrodeposited using a current density ranging from 4.8–5.5 mA cm^{-2} under ambient conditions without agitation. The diameter of the fibers was fixed by the diameter of the channels within the membrane and the lengths of the fibers were controlled by the current and deposition time. The nanofibers were recovered by dissolving the fiber-filled templates in a 1 M sodium hydroxide solution.

Particle morphology including dimensions and size-distribution information for both the spherical particles and nanofibers were obtained using Hitachi S570 (LaB_6 filament) and Hitachi S800 (field emission) scanning electron microscopes (SEM). Both microscopes are equipped with Orion PCI digital imaging systems. Hundreds of images were examined to give an average diameter and length with a standard deviation of particle size. The particle shape and size distribution of the spheres and fibers can be seen in the SEM micrographs in Figure 2.1.

The stoichiometric purity of all fiber compositions was determined using induction-coupled plasma spectrometry (ICP). The ICP data indicates that the fibers consisted of iron (or cobalt) with a purity greater than 99.4% and, in particular, do not contain significant amounts (at the ppb range limit of the ICP) of gallium, indium, copper, aluminum, or other elements that come in contact with the fibers during fabrication. Oxygen (in the form of oxides) is not detectable in ICP analysis; therefore, the presence of oxygen was determined by XRD.

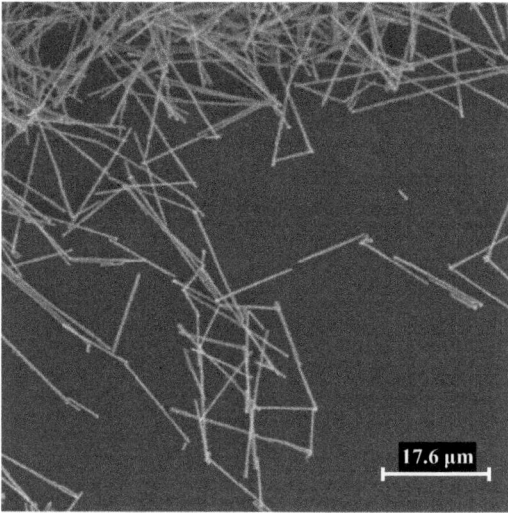

17.6 μm

Figure 2.1 SEM micrograph of iron nanofibers.

The bulk crystal structure and approximate grain size of the crystallites forming the spheres and nanofibers was determined using a Scintag X2 automated diffractometer equipped with a Peltier solid-state detector. X-Ray diffraction (XRD) patterns of the free fibers indicate that iron fibers grow in a bcc polycrystalline structure[46,47] and that cobalt grows in a hexagonal polycrystalline structure.[53] Using Scherrer's formula,[54] the approximate crystallite size for as-deposited fibers were determined to be 14.8 nm and 42.2 nm for iron and cobalt nanofibers, respectively. The XRD revealed the susceptibility of the iron nanofibers to oxidation (some samples display a composition as high as 28% iron oxide) while those of cobalt displayed only minor peaks (or none at all) for their corresponding oxides. Only minor metal oxide peaks were observed for the spherical iron particles.

2.2.2 MR Fluid Composition

The magnetorheological fluids were prepared by thoroughly mixing the ferromagnetic particles in a 0.18 Pa s viscosity silicone oil (GE SF96-200). Lecithin was added (2 wt% as per total particle mass) to the oil and thoroughly mixed prior to the addition of particles. The lecithin acts as a surfactant in order to produce stable dispersions. Magnetorheological fluids used for the percolation studies (Section 2.4) were prepared using a 0.45 Pa s viscosity silicone oil. The more viscous fluid was used for these low vol% suspensions to avoid sedimentation and to prevent the oil from being expelled from the rheometer.

2.2.3 Rheological Fluid Testing

The rheological measurements were made on an Anton-Paar Physica MCR300 parallel plate rheometer equipped with a MRD180 cell. A standard gap of

1 mm was maintained between the plates and a nominal 0.3 mL sample of fluid was placed between them. For the percolation studies, a gap of 0.5 mm was maintained between the plates and a nominal 0.15 mL sample of fluid was used along with a viscous carrier fluid (silicone oil, 0.45 Pa s) to avoid expulsion of the low vol% suspensions from between the plates of the rheometer at high shear rates. A Hall probe (F.W. Bell FH301) was placed within the gap to calibrate the input current of the electromagnet of the rheometer in terms of the magnetic flux density within fluid-containing gap. A magnetic flux density of up to 1.0 T with the magnetic field oriented perpendicular to the parallel plates was used in these studies. The temperature of all samples was maintained at 25 °C *via* a closed-cycle cooling system. To avoid sedimentation, the tests were carried out as soon as the thoroughly mixed fluid was inserted between the disks.

Rotational tests were carried out using the parallel plate rheometer to establish the steady-state flow curves (shear stress, τ, *versus* shear rate, $\dot{\gamma}$) at specified magnetic fields as shown in Figure 2.2. The Bingham-Plastic (BP) constitutive model[55] was then fitted to the flow curves to determine the apparent yield stress and viscosity of the samples. The BP model for the viscoplastic flow with yield stress is given by

$$\tau = \tau_y + \eta\dot{\gamma} \quad (\dot{\gamma} > 0) \tag{2.1}$$

where τ_y is the apparent yield stress, η is the post-yield viscosity and $\dot{\gamma}$ is the shear rate. The BP model predicts that in the pre-yield region, where stress is below the apparent yield stress, the fluid exhibits a rigid behavior. The values of τ_y and η were determined by fitting with a weighted least-squares error minimization for each fluid tested at all values of the applied field. By using the

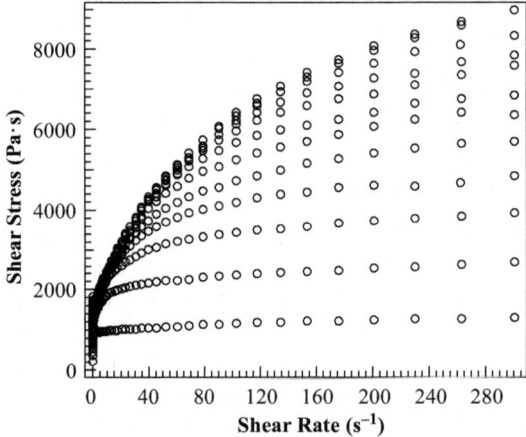

Figure 2.2 Experimental flow curves for fluids containing 6 vol% 7.6 μm Fe nanofibers. The applied magnetic field was varied from 0.0 T (bottom) to 0.72 T (top). The applied field is increased in 0.05 T increments from 0.0 T to 0.50 T; the last two curves are at 0.60 T and 0.72 T. Note the curves begin to overlap as the magnetic saturation of the particles is reached.

measured shear rates as the weighting factor, the model provided a better fit to the high shear rate data.

2.2.4 Sedimentation Testing

The inherent density difference between the carrier fluid and magnetic particles ultimately results in sedimentation of the particles resulting in a volume of supernatant fluid (the clarified fluid above the sediment mud line). As noted previously, this is detrimental to applications that require the MR fluid to remain ideal for long periods of time in which the particles settle (in the absence of constant mixing or large forces), thus rendering the MR fluid useless until which time the particles could be remixed. To quantify the sedimentation properties of fluids, both the degree to which the particles settled, the percentage of sedimentation, and the sedimentation rate were examined. The percentage of sedimentation is given by[46]

$$\text{percent sedimentation} = \left(\frac{\text{volume of supernatent fluid}}{\text{total volume of fluid}}\right) \times 100\% \quad (2.2)$$

The percentage of sedimentation quantifies the degree to which the particles settle and is an indication of how tightly-packed the particles become once settled; this is directly related to the ease with which the suspensions can be re-dispersed.

The sedimentation rate (speed at which particles settle) within the suspensions was determined by exploiting the magnetic properties of the ferro-magnetic particles in the suspension. During settling, particles sediment towards the bottom of the container, creating a volume of supernatant fluid (the clarified carrier fluid above the sediment mud line), as seen in Figure 2.1. To quantitatively compare the sedimentation rate of conventional sphere-based MR fluids with those containing nanofibers, an inductance-based solenoid sensor[39,56] was used to track the mud line location of the settling fluid as shown in Figure 2.3. The sedimentation rate is defined as the velocity at which the mud line descends due to particle settling. As the MR fluid sediments, the mud line travels downwards, until all the iron particles fully deposit at the bottom of the container and little or no further compacting is possible. Since the magnetic permeability of MR fluids is highly dependent on the volume fraction of ferromagnetic particles in the suspension, the inductive sensor solenoid is kept within this region. The permeability of the MR fluid within the sensor region is related to the sensor inductance L as follows

$$L = \frac{N^2 A \mu_0}{l} \mu_r \quad (2.3)$$

where N is the number of turns, A is cross-sectional area of the wire of the solenoid, l is solenoid length, μ_0 is magnetic permeability of vacuum and μ_r is relative permeability of the enclosed MR fluid. Thus, by measuring the rate of change of magnetic inductance of the sensor, the rate of change of the mud line

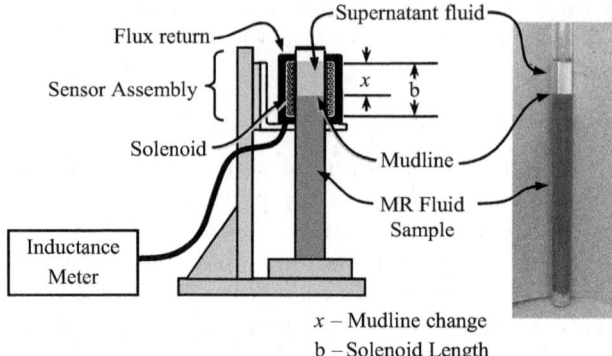

x – Mudline change
b – Solenoid Length

Figure 2.3   Inductance measuring instrument for determining settling rate of MR
fluids is comprised of the sensor assembly, an inductance meter, and a
stand to mount the sensor assembly and sample.

position can be estimated as it traverses through the sensor. The inductance
measuring instrument is comprised of an inductance meter, the sensor assembly
and a stand to mount the apparatus as shown in Figure 2.3.[39] Using this instrument, the change in sedimentation rate as a function of change in fluid
composition was examined.

2.3   Iron Nanofiber-Based Magnetorheological Fluids

To understand the shape-dependence on the rheological properties of MR
fluids, suspensions containing strictly nanofibers were compared to conventional MR fluids.[44,46] Iron nanofiber-based MR fluids with two
distinct length distributions of 5.4 ± 5.2 µm and 7.6 ± 5.1 µm with diameters of
260 ± 30 nm were compared with conventional sphere-based MR fluids that
consisted of 1–3 µm diameter spherical iron particles.[46] These studies displayed
the distinct differences between conventional MR fluids employing spherical
particles and nanofiber-based fluids with the same ferromagnetic composition
and particle loadings.

2.3.1   Magnetorheological Properties of Nanofiber-Based MR
Fluids

The dynamic yield-stress displays a strong dependence on the applied field for
all the fluids studied as shown in Figure 2.4. At a saturated magnetic flux
density, fluids containing the 5.4 µm iron nanofibers in silicone oil displayed
maximum yield stresses of 0.65, 2.23, and 4.76 kPa for the 2, 4, and 6 vol%
fluids, respectively. For suspensions utilizing a longer average fiber length of
7.6 µm, the yield stress increased to 8.23 kPa for the 6 vol% fluid.

As seen in Figure 2.4, the nanofiber-based fluids exhibit a more sensitive
response demonstrating a higher yield stress at low magnetic field strengths as
compared to the sphere-based fluids. This enhanced response is likely attributed

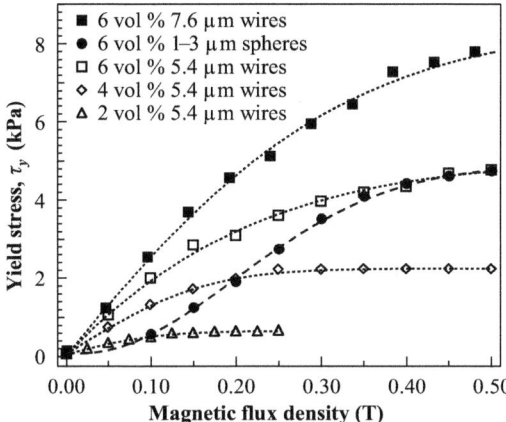

Figure 2.4   Plots of the yield stress *versus* magnetic field for MR fluids containing 5.4 μm and 7.6 μm iron nanofibers and for conventional MR fluids with 1–3 μm iron spheres. The dotted lines are simply used as a visual aid. Adapted from ref. 46.

to magnetic shape-anisotropy of the fibers. Even at low field strengths, the nanofibers experience a torque within the applied magnetic field forcing them to align with their long axis parallel to field and they experience a greater magnetization due to the shape anisotropy.[57,58]

The yield stress (τ_y) for spherical-based fluids display a characteristic response at intermediate fields below saturation as given by[59]

$$\tau_y \propto \phi\mu_0 M_S^{1/2} H^{3/2} \tag{2.4}$$

where μ_0 is the permeability of free space, ϕ is the volume fraction, M_S is the magnetic saturation of the material, and H is the strength of the applied magnetic field. In a separate study, the yield strength of two spherical-based MR fluids with median particle diameters of 2 and 8 μm at applied magnetic fields between 0.2 to 0.6 T displayed similar behavior as given in eqn (2.4).[1] However, the authors found that the data was better fitted using a second order polynomial. They concluded that the difference in the yield stress must be attributed to the difference in the magnetic saturation of the two particle sizes. The spherical-based suspensions used in this study displayed similar behavior. However, the trend in the yield stress for the nanofiber-based suspensions is found to be proportional to the square root of the applied field,

$$\tau_y \propto \phi\mu_0 H^{1/2} \tag{2.5}$$

At present, it is still uncertain how the magnetic saturation and aspect ratio of the particles quantitatively affect the yield stress of the nanofiber-based suspensions; studies varying the size and composition of the nanofibers with well-defined magnetic properties will be required. The magnetic saturation of the nanofibers is a function of their length and diameter, with longer fibers (with fixed diameter) having a higher magnetic saturation.[60] The substantial difference in the yield stresses of the fluids containing the two fiber lengths observed in these studies cannot be accounted for by the difference in the magnetic saturation alone. Therefore, other factors such as magnetostatic coupling between neighboring fibers,[60] inter-fiber friction, and structural differences in the two types of suspensions in an applied magnetic field must also be considered.

It is essential to consider the structure of the suspension both in the presence (on-state structure) and absence (off-state structure) of a magnetic field to be able to actually model the properties of MR fluids. In order to directly examine the differences in the on-state structures of the different types of fluids, the particles were suspended in a drop of water and a planar magnetic field was applied across the droplet.[44] The droplet was then allowed to evaporate while the applied field was maintained. Figure 2.5 shows SEM micrographs of spheres and nanofibers in a 0.26 T planar magnetic field. This image indicates that the spheres form similar structures to those supported in epoxy or paraffin wax.[61,62] These images clearly show that the fibers form columnar structures similar to that of spherical particles, but that there is far greater inter-particle contact for the fiber-based fluids.

Many factors must be considered to understand the increase in the yield stress of the nanofiber-based fluids as compared to the sphere-based fluids. As noted previously, for sphere-based fluids, yield stress scales with particle size.[1] If the increase in yield stress depended on the volume of the individual particles alone, the fluids containing the 1–3 μm spheres would be expected to exhibit much higher yield stresses than the fiber-based fluids. The volume of a 2 μm diameter sphere is 4.19 μm^3 and the volumes of 5.4 and 7.6 μm long fibers with 260 nm diameters have volumes 0.29 and 0.40 μm^3, respectively. This means that the mass of individual spherical particles are at least an order of magnitude greater. The more sensitive response at lower magnetic fields is likely due to the smaller demagnetization factor of fibers as compared to spherical particles.

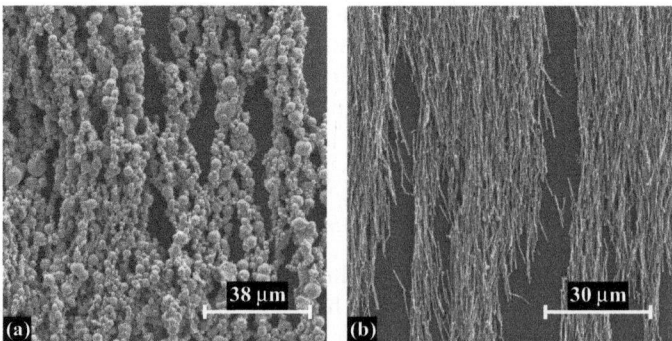

Figure 2.5   SEM micrograph showing the columnar structures formed in the presence of a planar 0.26 T magnetic field for (a) 6–10 μm diameter spherical iron particles and (b) 7.6 μm iron nanofibers with 260 nm diameters.

Suspensions containing the same volume fraction of iron particles, regardless of particle shape, would be expected to have similar global magnetic permeability values. However, the local packing density of the nanofibers is probably higher than that of the pure spheres,[46] resulting in a higher local permeability and thus stronger columnar structures within the nanofiber suspension. The packing efficiency of particles depends greatly on the packing geometry and particle size distribution. The most efficient packing geometry is achieved by close-packing in which a central particle has 12 equidistant neighbors. For spheres of identical diameter, the particles will occupy 74% of the total suspension volume. The packing efficiency drops to 52% for a primitive cubic structure in which each sphere is placed around a central sphere with only six equidistant neighbors. A large size distribution can greatly enhance the packing efficiency of spherical particles in excess of 90%.[63] The packing efficiency for nanofibers is expected to be slightly higher than that of spheres, but these too vary widely depending on the distribution of aspect ratios.[64] As for the nanofiber fluids containing the two different length fibers, the packing density distributions should be similar and, therefore, the yield stress of these two fluids would be expected to be similar. This is not the case, as it is apparent from Figure 2.4 that the longer nanofibers display a much greater yield stress than the shorter fibers at the same loading, therefore, the packing efficiency of the fluids may only play a minor role in the overall yield stress of the fluid, at least for different fiber-based suspensions.

Another important factor to consider when developing new types of MR fluids is the off-state viscosity. In many applications, the MR fluid must be able to freely flow in the off-state. In addition, a lower off-state viscosity would allow for a larger range of viscosities. The pure silicone oil used in these studies[46] had a viscosity of 0.41 Pa s and the off-state viscosity for the 6 vol% fluids containing spheres was 0.66 Pa s. The 6 vol% nanofiber-based fluids demonstrate a slight increase in off-state viscosity with a value of 0.68 Pa s, which drops to 0.50 Pa s for the 2 vol% fluids. However, the nanofiber-based fluids display a much higher degree of viscoelasticity, as seen in the sedimentation studies discussed in the next section.

2.3.2 Sedimentation Properties of Nanofiber-Based MR Fluids

Fluids containing strictly nanofibers display greatly reduced sedimentation compared to conventional fluids containing only spherical particles. For fluids with 6 vol% nanofibers, there is no sign of settling after the suspension was allowed to sit undisturbed for several months, while fluids containing 6 vol% spheres displayed 79% settling in just 48 hours as determined from eqn (2.2).[46] Note that the exact vol% required to prevent settling depends on the length of the fibers where shorter fibers will require higher loadings while longer fibers will require less. Note that the rate of sedimentation of the fibers compared to the spheres was not examined in this study, only the percentage of sedimentation.

The difference in the sedimentation and the viscoelastic properties of the two types of suspensions is a result of several factors, including nanofiber entanglement, increased surface area of the fibers as compared to spheres resulting in greater hydrostatic interactions, and magnetic interactions between nanofibers, particularly those due to remnant magnetization. Entanglement occurs in concentrated suspensions of long slender particles, as the rotational, end-over-end motion, as well as the translational motion perpendicular to the fiber axis of each fiber is severely restricted.[65] The total surface area of the particles in fluids with the same volume fraction is substantially different; fluids containing 5.4 μm long nanofibers with 260 nm diameters have more than 5 times the total surface area compared to 2 μm diameter spheres (this difference changes little with fiber length at the same volume fraction of particles). These factors suggest that the settling rate hindering factor is greatly increased in concentrated suspensions due to particle–particle and particle–fluid interactions. However, this does not greatly affect the off-state viscosity of the fiber-containing suspensions under shear because the fibers tend to line up under shear and the viscosity tends to be close to that of the carrier fluid at low particle concentrations.[66]

As demonstrated in these studies, MR fluids utilizing non-spherical ferromagnetic particles exhibit enhanced rheological properties as well as greater stability against sedimentation.[46,67] However, a major disadvantage of suspensions containing only nanofibers is that the maximum volume fraction of particles (<10 vol%) is much less than the desired 30–40 vol%, achievable with spherical particles in conventional suspensions, necessary for high yield-stress applications. The low loading is due to their high wetted area and fiber-to-fiber interactions, which also results in an increase in the static viscosity of the suspension.[68] As the concentration of particles is further increased, there is no longer enough fluid to lubricate the relative motion of particles, thus increasing static viscosity to infinity.[69] For these reasons, suspensions that contain strictly nanofibers would be more useful in applications where sedimentation would be extremely detrimental and where low yield-stress is sufficient.

2.4 Percolation Behavior of Cobalt Nanofiber-Based MR Fluids

Composite materials of two or more distinct components are often best described by an effective medium model in which the macroscopic properties of a medium are based on the known properties and the relative fractions of its components. The ability to predict the effective-medium physical properties of composite materials has been a challenge for researchers for many years.[70] However, most of the current theories have difficulty in describing percolating systems where abrupt phase changes occur at a critical volume fraction of the component of interest. This phase change is known as a percolation transition which has its origins in the random-walk theory of fluid flow within porous media.[71] Most studies over the past few decades have focused on the electronic and thermodynamic transport while many fewer studies have examined the elastic and mechanical properties of materials near the critical point.[72,73] Formulating an effective-medium model is complicated due to the great variation of the physical properties of the constituents. Being able to understand and control the effective-medium physical properties of MR fluids would allow for more diverse MR fluids to be manufactured.

Several percolation models have been developed using both discrete-lattice networks and continuum networks.[74] These models attempt to determine the scaling behavior of the system by using different critical exponents to describe the nature of the transition as the critical volume fraction, ϕ_c, is approached. Elastic percolation networks (EPNs) consist of a two-component mixture with bonds that have a finite elastic modulus together with completely non-rigid bonds. This would be the case for MR fluids when the loading (in terms of volume fraction, ϕ) of ferromagnetic particles is greater than the critical volume fraction, $\phi > \phi_c$. The general elastic modulus, G, of such systems is believed to vanish as ϕ_c is approached from above,[75–77]

$$G \sim (\phi - \phi_c)^f \qquad (2.6)$$

Below the critical point, $\phi < \phi_c$, the composite is predicted to behave as a super-elastic percolation network (SEPN) consisting of a mixture of perfectly rigid bonds and bonds with finite elastic constants. Below the critical point, the general elastic moduli is expected to diverge as,

$$G \sim (\phi_c - \phi)^{-c} \qquad (2.7)$$

Studies on the rheology of magnetic suspensions exhibiting elastic behavior in the region of $\phi > \phi_c$, suggest that f depends on loading, increasing from 1.0 to 2.26 with increasing particle concentration.[78] However, most experimental measurements and theoretical models indicate that f-values are in the range of 3.5–4.0, which is attributed to the higher tensor order of EPNs.[75–77] Numerical simulations of SEPN behavior in two dimensions indicate that the super-elastic critical exponent c should be in the range of 1–1.3 for suspensions exhibiting elastic behavior.[79,80]

To investigate the rheological behavior of nanofiber-based MR fluids, and in particular, the long-range correlations in elastic materials and their contribution to the critical behavior, the volume fraction of 330 nm diameter cobalt-nanofibers with lengths of 8–12 μm in MR fluids were systematically varied to measure the changes in the dynamic yield stress at various applied magnetic field strengths.[53] A percolation transition was observed for the dynamic yield-stress of the MR fluids; moreover, it was found that the location of the critical point as well as the behavior of the fluid near the transition is dependent on the magnitude of the applied field.

MR fluids consisting of cobalt nanofibers with diameters of 330 nm and lengths of 8–12 μm were used in these studies. The steady-state flow curves were obtained for samples having cobalt-nanofiber volume fractions in the range $\phi = 0$–0.06 (note volume fraction rather than vol% were used in these studies).

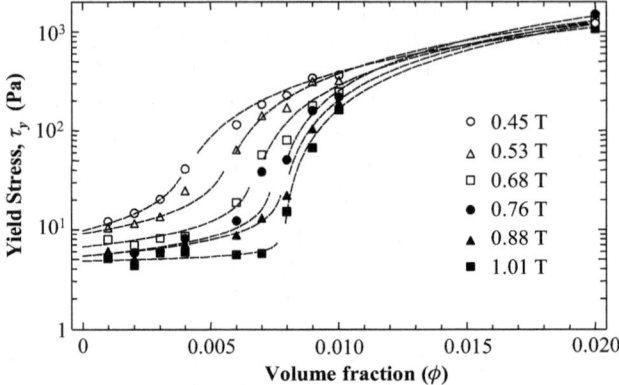

Figure 2.6   The apparent yield stress of cobalt-nanofiber MR fluids as a function of nanofiber volume fraction for varying applied magnetic fields. The solid curves represent the best fit to the power-law dependence given in eqn (2.8) and (2.9).
Adapted from ref. 53.

Flow curves were obtained at a constant applied magnetic field and were incrementally increased over the range of 0–1.0 T. The data were fitted with a Bingham-plastic constitutive model as given by eqn (2.1) to determine the apparent yield stress and the post-yield viscosity as described earlier.

The yield stress *versus* volume fraction at increasing magnetic field strengths (data points) is shown in Figure 2.6. The dependence of the onset of the critical volume fraction transition on the applied field is apparent from this graph. The data was characterized with a normalized, two-parameter fit based on eqn (2.8) and (2.9) using values of the yield stress extrapolated to $\phi = 0$ ($\tau_{low} \sim 5$–12 Pa) and far beyond ϕ_c ($\tau_{high} \sim 10^5$ Pa).

$$\tau_y = \tau_{low}\left(\frac{\phi_c - \phi}{\phi_c}\right)^{-c}, \; \phi < \phi_c \tag{2.8}$$

$$\tau_y = \tau_{high}\left(\frac{\phi - \phi_c}{1 - \phi}\right)^{f}, \; \phi > \phi_c \tag{2.9}$$

From the fitted curves, it is observed that ϕ_c increases with increasing magnetic field strength, while c decreases and f is relatively constant, independent of the applied field, as seen in Table 2.1. The values of f for all field-strengths range from 1.0 to 1.2 and are similar to the values of the elastic exponent ($f \sim 1.1$–1.3) found in two-dimensional, static-matrix numerical studies.[76,81–83] This suggests that an MR fluid system subjected to a relatively uniform magnetic field behaves two-dimensionally; this is most likely due to the fact that the applied field defines a preferred spatial direction for fibril formation. The values of the critical volume fraction are reasonable for nanofibers having a high aspect ratio (~ 30), but the dependence of c and ϕ_c on the strength of the magnetic field is difficult to explain in terms of the known scaling models.

Table 2.1 The fitted percolation parameters of cobalt-nanofiber MR fluids for varying applied magnetic field. Adapted from ref. 53.

Applied magnetic field/T	Critical volume fraction (ϕ_c)	Super-elastic critical exponent (c)	Elastic exponent (f)
0.45	0.0038 ± 0.0007	0.4–0.6	1.1–1.2
0.53	0.0055 ± 0.0005	0.4–0.7	1.0–1.1
0.68	0.0065 ± 0.0004	0.2–0.4	1.1–1.2
0.76	0.0067 ± 0.0004	0.36–0.42	1.1–1.2
0.88	0.0077 ± 0.0002	0.32–0.38	1.0–1.2
1.01	0.0078 ± 0.0002	0.09–0.13	1.1–1.2

Unlike the present study, involving a dynamic-matrix fluidic system with both short-range contact and long-range magnetic interactions, the vast majority of other studies have considered static systems with short-range interactions or those involving direct particle–particle contacts. In conductivity networks, ϕ_c is expected to decrease as the metal to insulator particle-size ratio decreases,[84] and at the same time, increase with aspect ratio, due to the effect of excluded volume.[85] Yet, neither of these is a variable in the MR fluid system under study. However, the trend in ϕ_c may be explained by the degree of alignment of the nanofibers. For percolating systems containing high-aspect-ratio fillers, calculations have shown that the critical volume fraction increases as the particles become more aligned; the increased degree of alignment is certainly consistent with higher-aspect-ratio ferromagnetic fibrils being formed as the strength of the magnetic field is increased.[86] The super-elastic critical exponent, c, presumably related to the growth of elastic clusters below the transition (ferromagnetic chains in this case), varies considerably with the magnetic field. The variation in c may be ascribed to the changes in the random reinforcement of the MR fluid by chain growth that does not span the sample below the critical volume fraction.[77] The dependence of c and ϕ_c on the applied field is unexpected and is most likely attributed to the dynamic nature of the measurement technique and the inhomogeneities of magnetic field in the rheometer sample space. The importance of this issue cannot be underestimated in light of the numerous applications that exploit the MR effect.

2.5 Dimorphic Magnetorheological Fluids

MR fluids utilizing non-spherical ferromagnetic particles exhibit enhanced rheological properties as well as greater stability.[46,67] However, the use of strictly nanofibers limits the loading of MR fluids to volume fractions of 10 vol% (48 wt%) or less, thus producing small yield stresses that are largely insufficient for most dynamic applications.

To enhance the sedimentation properties of MR fluids while still maintaining the concentration required of conventional MR fluids, dimorphic fluids were created.[47] Dimorphic MR fluids are suspensions composed of ferromagnetic particles with different shapes. This type of suspension was developed to harness the enhanced properties of the nanofiber-based MR fluids and still achieve the particle loadings required by conventional MR fluids. Dimorphic MR fluids were expected to display low sedimentation rates, but with sufficiently high particle-loading to achieve high yield stress. The rheology and sedimentation stability of these dimorphic fluids, in which a given percentage of spheres within a conventional MR fluid were substituted with nanofibers were investigated. For these studies, iron spheres with diameters of 1–3 μm and iron nanofibers with diameters of 230 nm and lengths ranging 7.6 ± 5.1 μm suspended in a 0.18 Pa s silicone oil were used. Various MR fluid samples with total particle loadings of 50, 60, and 80 wt% combined with nanofiber substitution from 0 to 8 wt% in 2 wt% increments were examined.

2.5.1   Rheological Properties of Dimorphic MR Fluids

The yield-stress behavior as a function of magnetic field of the dimorphic MR fluids containing 80, 60 and 50 wt% total loading is shown in Figure 2.7. At low field strengths, a small increase in yield stress with increasing fiber substitution was observed, consistent with the enhanced response observed in pure nanofiber MR fluids at lower field strengths.[46] At moderate field strengths, there were no differences in performance at any substitution percentage. The

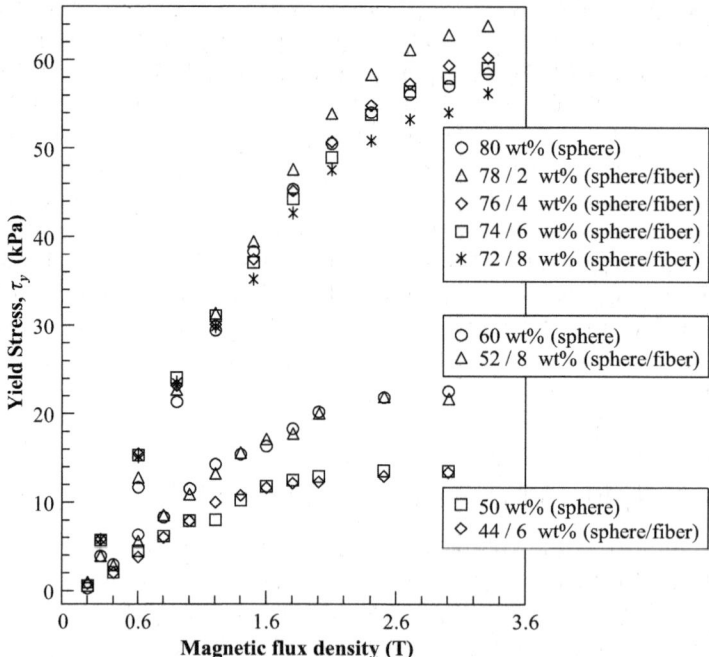

Figure 2.7   Yield stress *versus* magnetic field for conventional sphere-based fluids at 50, 60, and 80 wt% and for dimorphoric fluids of the same total particle loading.
Adapted from ref. 47.

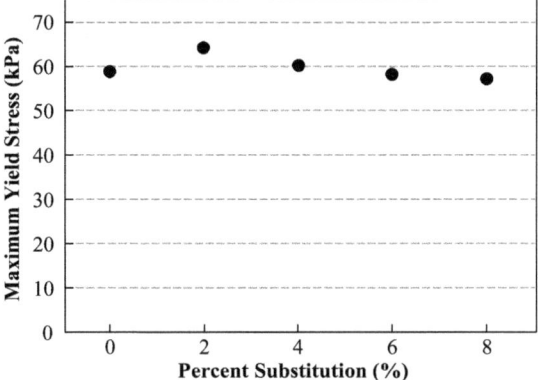

Figure 2.8 The effect on yield stress of substitution of spherical particles with fibers in dimorphic MR fluids.

maximum apparent yield stress for the 80 wt% total loading samples as a function of magnetic flux density is shown in Figure 2.8. The 2 wt% sample displayed a 10% increase in the maximum yield stress as compared to conventional MR fluids. However, further increasing the percentage of nanofiber substitution resulted in a steady drop in the yield stress as compared to purely sphere-based suspensions. For 8 wt% substituted dimorphic fluids a ~5% *decrease* from the conventional MR fluid was observed. In all, there is little variation in yield stress among the various dimorphic fluids samples and that of conventional MR fluids; that is, the presence of nanofibers at this level of concentration does not degrade the apparent yield strength of the MR fluids as is the case for many others additives aimed at reducing sedimentation.

The behavior of the dimorphic MR fluids as compared to the conventional fluids can be hypothetically explained by considering the structures formed in the presence of the magnetic field. To examine the particle orientation within the fluids, samples were prepared by suspending the particles in ethanol at representative loadings which were then deposited onto a silicon chip. The ethanol was then allowed to evaporate locking in the structure of the suspended particles. The dimorphic MR fluid particles under no magnetic field indicate that the nanofibers are uniformly dispersed and randomly oriented throughout the fluid as seen in the SEM micrograph shown in Figure 2.9a. This same random orientation was observed for all fluid compositions. Application of a 0.26 T planar magnetic field to the sample before and during the evaporation of the ethanol allows the nanofibers and microspheres to align along the field lines. The SEM micrograph of a dimorph MR fluid containing 78 wt% spheres 2 wt% fibers, shown in Figure 2.9b, indicates that at low nanofiber loadings the microspheres still form chained clusters and that the fibers are arranged very sparsely in the voids between and alongside these chain structures. Figure 2.9c represents a dimorphic MR fluid containing 72 wt% spheres 8 wt% fibers; as the nanofiber substitution (concentration) increases, the fibers not only fill the gaps between the spheres in the chain structure, but begin to interfere with the

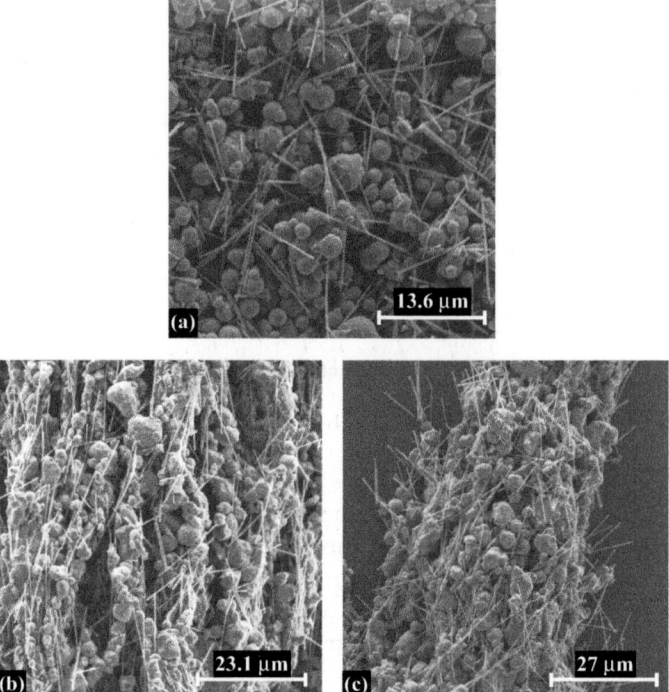

Figure 2.9   SEM images of various representative dimorphic fluids. (a) Sphere/
nanofiber mixture with no external field. Images of dimorphic mixtures
in a 0.26 T planar external field with (b) a low fiber substitution concentration and with (c) a high fiber substitution concentration.

chain structure formed by the spherical particles as a larger number of nanofibers become lodged between the spheres (long axis perpendicular to the field
lines). It is possible that this interference in the chain structure of the spheres is
the cause of the decrease in maximum achievable yield stress. Further experimental and theoretical studies will need to be performed to confirm this
hypothesis.

   The off-state viscosity of the dimorphic MR fluids increased with increased
concentration of nanofibers in the suspension, as shown in Table 2.2. For example, the off-state viscosity for fluids with an 80 wt% loading containing
strictly spherical particles is 1.3 Pa s, while that for the 8 wt% nanofiber substituted increased to 4.4 Pa s at the same total loading. As was the case for
suspensions containing strictly nanofibers, the increase in the viscosity is likely
due to their high wetted area and fiber–fiber interactions.[68]

2.5.2   Sedimentation Properties of Dimorphic MR Fluids

The greatest improvement demonstrated by the dimorphic MR fluids was seen
in the sedimentation stability for all ranges of particle loading and substitution.

Table 2.2 Off-state viscosity, percent sedimentation, and sedimentation rates for selected conventional and dimorphic MR samples. Adapted from ref. 47.

Composition/wt%		Off-state	Percent	Sedimentation
Spheres	Nanofibers	viscosity/Pas	sedimentation [%]	rate/μm s^{-1}
50	–	–	72.9	1.9
44	6	–	23.4	0.036
60	–	–	66.7	0.86
52	8	–	14.0	0.017
80	–	1.3	14.8	0.0254
78	2	2.1	8.8	0.0122
76	4	3.5	5.7	0.0021
74	6	3.5	–	0.0053
72	8	4.4	–	–

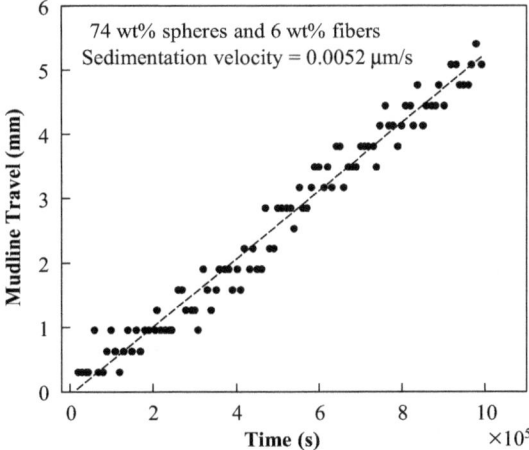

Figure 2.10 Plot of sedimentation rate for dimorphic MR fluid consisting of 74 wt% spheres and 6 wt% nanofibers.

Both the sedimentation rate and percentage of sedimentation were studied; however, due to the length of time required for the sedimentation tests at one month per sample, and one sample at a time, sedimentation studies were performed on a select set of samples. The 80 total wt% samples were examined in more detail as these are the loadings that are more commonly employed in industrial applications. Only one conventional and one dimorphic sample were tested for fluids in the moderate particle loading ranges of 50 and 60 wt%.

The sedimentation rate and percent sedimentation was determined as outlined in Section 2.2.4. The sedimentation plot as obtained from inductance measurements for dimorphic MR fluid containing 74 wt% spheres and 6 wt% fibers is shown in Figure 2.10. A linear least-squares fitting was used to determine the sedimentation rate. Sedimentation tests of the moderately loaded fluids showed an improved (order of magnitude lower) sedimentation rate for both the 50 wt% and 60 wt% conventional fluids, as compared to the 6 wt% and 8 wt% substituted dimorphic fluids, respectively. Dimorphic fluids with a total loading of 80 wt% also showed a significant sedimentation rate decrease. The rate of 2 wt% substituted fluids decrease by a factor of two while the rate of 4 wt% substituted dimorphic fluids displayed an order of magnitude decrease.

The second method used to characterize stability was to determine percent sedimentation of the fluids as given by eqn (2.2). This ratio was determined after the fluids were allowed to settle undisturbed, for a minimum of 1 month. These results are shown in Table 2.2. The difference in the degree of sedimentation between conventional and dimorphic fluids for the same total particle loadings was decreased by 50% in 50 wt% suspensions at a 6 wt% substitution. The percent sedimentation decreased by 80% in 60 wt% loaded fluids at an 8 wt% substitution, while the 80 wt% suspensions showed a 40% decrease in total sedimentation in 2 wt% substituted fluids and 60% in 4 wt% substituted fluids. It is seen that all dimorphic fluids have a lower percent sedimentation, implying a smaller magnitude of sedimentation than conventional microsphere-based MR fluids. This reduction in percent sedimentation results in a more porous agglomeration of particles at the bottom of the container, making re-dispersion much easier and providing a homogenous mixture more quickly.

Referring to the SEM image in Figure 2.9a, the dimorphic MR fluid under no magnetic field indicates that the fibers are uniformly dispersed and randomly oriented throughout the fluid. The fibers arrange themselves randomly between the spherical particles. The interaction of nanofibers with neighboring particles appears to cause entanglement with the spherical particles due to their limited rotational and translational motions. These types of inter-particle interactions would increase considerably as the nanofibers percentage increases. Therefore, the presence of the nanofibers tends to increase the off-state viscosity of the dimorphic suspensions generating fluids with an increased sedimentation hindrance factor. All these effects are expected to account for the reduction in the average particle settling rate, and considerably reducing overall sedimentation making re-dispersion much easier.

2.6 Summary

Nanofibers are ideal model particles for studying the shape-dependent properties of suspended particles in magnetorheological fluids. There are several advantages of nanofiber-based MR suspensions as compared to those that contain strictly spherical particles. The nanofiber-based suspensions perform with greater sensitivity at low magnetic fields, have relatively higher yield stresses, and allow for more sensitive control of the fluid properties. The sedimentation stability of the suspensions is also enhanced compared to conventional MR fluids with a greatly reduced sedimentation rate. The percent sedimentation is also greatly reduced allowing for the suspension to be more easily re-dispersed after prolonged periods of non-use. A disadvantage of the suspensions containing strictly nanofibers is the much-reduced limit on the volume fraction of particles, far below the desired 30–40 vol% achievable with spherical particles. This results in an achievable yield stress that is far below that required of most commercial applications. For these reasons, suspensions that contain strictly nanofibers would be more useful in applications where sedimentation would be extremely detrimental and low yield stresses are sufficient.

We observe a type of elastic percolation transition in the apparent yield stress of magnetized, cobalt-nanofiber MR fluids as a function of nanofiber volume fraction and applied magnetic field, H. A transition from minimal to greatly enhanced apparent yield stress occurs at variable critical point, ϕ_c, the location of which depends on H. Likewise, the behavior of the transition near ϕ_c, is characterized by critical exponents. For the region above ϕ_c, the elastic critical exponent, f, appears to be independent of the applied magnetic field, having a value near that seen in 2-D conductor networks. This suggests that an on-state MR-fluid exhibits 2-D behavior, due to the preferred spatial direction for fibril formation defined by a relatively uniform magnetic field. Describing the region below ϕ_c, the super-elastic exponent, c, is found to decrease with increasing magnetic field and is smaller than that seen in analogous 2-D or 3-D networks. While the values of ϕ_c are reasonable for nanofibers having a high aspect ratio (~ 30), the dependence of ϕ_c on the magnetic field is unexpected and likely attributed to the dynamic nature of the measurement technique and the inhomogeneities of the applied magnetic field inside the rheometer sample space. Unlike the present study, involving a dynamic fluid system with both short-range-contact and long-range-magnetic interactions, the vast majority of other studies have considered static systems with short-range interactions or those involving direct particle–particle contact. Future experiments that vary the nanofiber aspect ratio, use other materials (*e.g.*, iron and nickel), and employ quasistatic, rather than dynamic measurements, should help separate these issues from the underlying physical mechanisms, and clarify our interpretation of the results.

To overcome the shortfall of particle loading of the MR fluids containing only nanofibers, we generated dimorphic MR fluids, containing both spherical and nanofiber particles, allowing for maximum particle loading of 30–40 vol% of suspended particles. These fluids display both enhanced sedimentation properties and yield stresses comparable to conventional MR fluids. Suspensions containing low volume fractions of nanofibers display yield stresses about 10% higher than spherical-based fluids at the same volume fraction. As the fiber concentration is increased, the maximum achievable yield stress is seen to drop, but is still comparable to the conventional MR fluids, but with greatly enhanced sedimentation properties. The major advantage of dimorphic MR fluids over conventional MR fluids are clearly displayed with the greatly reduced percent sedimentation and sedimentation velocities. The substitution of the spherical particles with the fibers significantly reduced the rate of particle settling, enabling the MR fluid to maintain a uniform dispersion without marked sedimentation for an extended period of time. Even after the onset of sedimentation, the presence of the nanofibers tended to produce more porous particle sediment which suggests ease of re-dispersion. As the percent of nanofiber substitution is increased, both the percent sedimentation and rate of sedimentation decrease. However, the higher the percentage of nanofiber substitution results in an increase in the off-state viscosity. All of these observations can be explained by observing the chain structure of the particles both in the presence and absence of an applied magnetic field. The off-state viscosity is increased due to the entanglement of the nanofibers, which increases with increasing fiber loadings. As for the yield stress, the nanofibers appear to structurally support the chains that are primarily formed by spheres at low substitution concentration; however, as fiber substitution is increased, a threshold is reached where the fibers begin to interfere with the columnar structures formed by the spheres resulting in a lower yield stress.

Acknowledgements

The authors acknowledge funding support from the National Science Foundation (NSF-CBET-0755696), The Pennsylvania State University, and Altoona College. This publication was supported by the Pennsylvania State University Materials Research Institute Nano Fabrication Network, the National Science Foundation Cooperative Agreement No. 0335765, and the National Nanotechnology Infrastructure Network with Cornell University. Additional support was provided by a DARPA SBIR Phase 2 Contract No. W31P4Q-06-C-0400 (N. M. Wereley).

References

1. S. Genç and P. P. Phulé, *Smart Mater. Struct.*, 2002, **11**, 140.
2. D. J. Klingenberg, *AIChE J.*, 2001, **47**, 246.
3. J. M. He and J. Huang, *Int. J. Mod. Phys. B*, 2005, **19**, 593.
4. T. B. Jones and B. Saha, *J. Appl. Phys.*, 1990, **68**, 404.
5. J. D. Carlson, *Int. J. Veh. Des.*, 2003, **33**, 207.
6. Y. T. Choi, N. M. Wereley and Y. S. Jeon, *AIAA J. Aircr.*, 2005, **42**, 1244.
7. H. Sahin, Y. Liu, X. Wang, F. Gordaninejad, C. Evrensel and A. Fuchs, *J. Intell. Mater. Syst. Struct.*, 2007, **18**, 1161.
8. S. B. Choi, M. H. Nam and B. K. Lee, *J. Intell. Mater. Syst. Struct.*, 2000, **11**, 936.
9. S. J. McManus, K. A. St. Clair, P. E. Boileau, J. Boutin and S. Rakheja, *J. Sound Vib.*, 2002, **253**, 313.
10. G. J. Hiemenz, W. Hu and N. M. Wereley, *AIAA J. Aircr.*, 2008, **45**, 945.
11. Y. T. Choi and N. M. Wereley, *AIAA J. Aircr.*, 2003, **40**, 432.
12. D. C. Batterbee, N. D. Sims, R. Stanway and Z. Wolejsza, *Smart Mater. Struct.*, 2007, **16**, 2429.
13. D. C. Batterbee, N. D. Sims, R. Stanway and M. Rennison, *Smart Mater. Struct.*, 2007, **16**, 2441.
14. Y. T. Choi and N. M. Wereley, *J. Auto. Eng.*, 2005, **219**, 741.

15. G. J. Hiemenz, Y. T. Choi and N. M. Wereley, *AIAA J. Aircr.*, 2007, **44**, 1031.
16. L. Zipser, L. Richter and U. Lange, *Sens. Actuators, A*, 2001, **92**, 318.
17. J. C. Ramallo, E. A. Johnson and B. F. Spencer, Jr., *J. Eng. Mech., ASCE*, 2002, **128**, 1088.
18. N. C. Kavlicoglu, B. M. Kavlicoglu, Y. Liu, C. A. Evrensel, A. Fuchs, G. Korol and F. Gordaninejad, *Smart Mater. Struct.*, 2007, **16**, 149.
19. C. Miao, J. C. Lambropoulos and S. D. Jacobs, *Appl. Opt.*, 2010, **49**, 1951.
20. L. S. Chen and D. Y. Chen, *Rev. Scien. Instrum.*, 2003, **74**, 3566.
21. B. J. de Gans, N. J. Duin, D. van den Ende and J. Mellema, *J. Chem. Phys.*, 2000, **113**, 2032.
22. P. P. Phulé, M. Mihalcin and S. Genç, *J. Mater. Res.*, 1999, **14**, 3037.
23. P. P. Phulé and J. M. Ginder, *MRS Bulletin*, 1998, **23**, 19.
24. M. T. López-López, A. Zugaldia, F. González-Caballero and J. D. G. Durán, *J. Rheol.*, 2006, **50**, 543.
25. K. D. Weiss, D. A. Nixon, J. D. Carlson and A. J. Margida, *Thixotropic Magnetorheological Materials*, US Patent No. 5 645 752, 1997.
26. J. de Vicente, M. López-López and F. González-Caballero, *J. Rheol.*, 2003, **47**, 1093.
27. P. P. Phulé and J. M. Ginder, *Int. J. Mod. Phys. B*, 1999, **13**, 2019.
28. S. T. Lim, M. S. Cho, H. J. Choia and M. S. Jhon, *Int. J. Mod. Phys. B*, 2005, **19**, 1142.
29. S. T. Lim, H. J. Choia and M. S. Jhon, *IEEE Trans. Magn.*, 2005, **41**, 3745.
30. I. B. Jang, H. B. Kim, J. Y. Lee, J. L. You, H. J. Choia and M. S. Jhon, *J. Appl. Phys.*, 2005, **97**, 10Q912.
31. J. L. You, B. J. Park, H. J. Choi, S. B. Choi and M. S. Jhon, *Int. J. Mod. Phys. B*, 2007, **21**, 4996.
32. E. Lemaire, A. Meunier, G. Bossis, J. Liu, D. W. Felt, P. Bashtovoi and N. Matoussevitch, *J. Rheol.*, 1995, **39**, 1011.
33. C. W. Wu and H. Conrad, *J. Appl. Phys.*, 1998, **83**, 3880.
34. J. Trihan, J. H. Yoo, N. M. Wereley, S. Kotha, A. Suggs, R. Radhakrishnan, T. Sudarshan and B. J. Love, *Proc. SPIE*, 2003, **5052**, 175.
35. P. Poddar, J. L. Wilson, H. Srikanth, J. H. Yoo, N. M. Wereley, S. Kotha, L. Barghouty and R. Radhakrishnan, *J. Nanosci. Nanotechnol.*, 2004, **4**, 192.
36. A. Chaudhuri, G. Wang, N. M. Wereley, V. Tasovksi and R. Radhakrishnan, *Int. J. Mod. Phys. B*, 2005, **19**, 1374.
37. N. M. Wereley, A. Chaudhuri, J. H. Yoo, S. John, S. Kotha, A. Suggs, R. Radhakrishnan, B. J. Love and T. S. Sudarshan, *J. Intell. Mater. Syst. Struct.*, 2006, **17**, 393.
38. K. H. Song, B. J. Park and H. J. Choi, *IEEE Trans. Magn.*, 2009, **45**, 4045.
39. G. Ngatu and N. Wereley, *IEEE Trans. Magn.*, 43, 2007, 2474.
40. N. Rosenfeld, N. M. Wereley, R. Radakrishnan and T. S. Sudarshan, *Int. J. Mod. Phys. B*, 2002, **16**, 2392.

41. A. Chaudhuri, N. M. Wereley, S. Kotha, R. Radhakrishnan and T. S. Sudarshan, *J. Magn. Magn. Mater.*, 2005, **293**, 206.
42. B. D. Chin, J. H. Park, M. H. Kwon and O. O. Park, *Rheol. Acta*, 2001, **40**, 211.
43. H. Nishiyama, K. Katagiri, K. Hamada, K. Kikuchi, K. Hata, P. Sang-Kyu and M. Nakano, *Int. J. Mod. Phys. B*, 2005, **19**, 1437.
44. R. C. Bell, E. D. Miller, J. O. Karli, A. N. Vavreck and D. T. Zimmerman, *Int. J. Mod. Phys. B*, 2007, **21**, 5018.
45. M. T. López-López, G. Vertelov, G. Bossis, P. Kuzhir and J. D. G. Durán, *J. Mater. Chem.*, 2007, **17**, 3839.
46. R. C. Bell, J. O . Karli, A. N. Vavreck, D. T. Zimmerman, G. T. Ngatu and N. M. Wereley, *Smart Mater. Struct.*, 2008, **17**, 015028.
47. G. T. Ngatu, N. M. Wereley, J. O. Karli and R. C. Bell, *Smart Mater. Struct.*, 2008, **17**, 045022.
48. A. Gómez-Ramírez, M. T. López-López, J. D. G. Durán and F. González-Caballero, *Soft Matter*, 2009, **5**, 3888.
49. P. Kuzhir, M. T. López-López and G. Bossis, *J. Rheol.*, 2009, **53**, 127.
50. M. T. López-López, P. Kuzhir and G. Bossis, *J. Rheol.*, 2009, **53**, 115.
51. J. de Vicente, F. Vereda, J. P. Segovia-Guitérrez, M. D. Morales and R. Hidalgo-Álvarez, *J. Rheol.*, 2010, **54**, 1337.
52. P. Kuzhir, A. Gómez-Ramírez, M. T. López-López, G. Bossis and A. Y. Zubarev, *J. Non-Newtonian Fluid Mech.*, 2011, **166**, 373.
53. D. T. Zimmerman, R. C. Bell, J. A. Filer II, J. O. Karli and N. M. Wereley, *Appl. Phys. Lett.*, 2009, **95**, 014102.
54. B. D. Cullity, *Elements of X-Ray Diffraction*, Addison-Wesley, Reading, MA, 1978, pp. 281–284.
55. M. Jolly, J. Carlson and J. Bender, *J. Intel. Mater. Syst. Str.*, 1999, **10**, 5.
56. S. R. Gorodkin, W. I. Kordonski, E. V. Medvedeva, Z. A. Novikova, A. B. Shorey and S. D. Jacobs, *Rev. Sci. Instrum.*, 2000, **71**, 2476.
57. Y. Qi, L. Zhang and W. Wen, *J. Phys. D: Appl. Phys.*, 2003, **36**, L10.
58. L. Sun, Y. Hao, C. L. Chien and P. C. Searson, *IBM J. Res. Dev.*, 2005, **49**, 79.
59. J. M. Ginder, L. C. Davis and L. D. Elie, *Rheology of Magnetorheological Fluids: Models and Measurements, 5th Int. Conf. on ER Fluids and MR Suspensions*, ed. W. Bullough, World Scientific, Singapore, 1995, pp. 504–514.
60. G. C. Han, B. Y. Zong and Y. H. Wu, *IEEE Trans. Magn.*, 2002, **38**, 2562.
61. J. E. Martin, E. Venturini, J. Odinek and R. A. Anderson, *Phys. Rev. E*, 2000, **61**, 2818.
62. S. Henley and F. E. Filisko, *Int. J. Mod. Phys. B*, 2002, **16**, 2286.
63. A. B. Yu and N. Standish, *Ind. Eng. Chem. Res.*, 1991, **30**, 1372.
64. R. P. Zou, X. Y. Lin, A. B. Yu and P. Wong, *J. Am. Ceram. Soc.*, 1997, **80**, 646.
65. M. Doi and S. F. Edwards, *J. Chem. Soc., Faraday Trans. 2*, 1978, **74**, 560.
66. M. Bercea and P. Navard, *Macromolecules*, 2000, **33**, 6011.

67. M. L. Levin, D. É. Polesskii and I. V. Prokhorov, *J. Eng. Phys. Thermophys.*, 1997, **70**, 769.
68. J. Ferguson and Z. Kemblowski, *Applied Fluid Rheology*, 1991, Elsevier Applied Science, London.
69. A. Metzner, *J. Rheol.*, 1985, **29**, 739.
70. J. C. M. Garnett, *Phil. Trans. R. Soc. Lond.*, 1904, **203**, 385.
71. S. R. Broadbent and J. M. Hammersley, *Proc. Cambridge Phil. Soc.*, 1957, **53**, 629.
72. M. B. Isichenko, *Rev. Mod. Phys.*, 1992, **64**, 961.
73. C. Chiteme and D. S. McLachlan, *Phys. Rev. B*, 2003, **67**, 024206.
74. D. Stauffer and A. Aharony, *Introduction to Percolation Theory*, CRC Press, New York, 1994, 1–114 and references therein.
75. S. Feng, B. I. Halperin and P. N. Sen, *Phys. Rev. B*, 1987, **35**, 197.
76. M. Zhou and P. Sheng, *Phys. Rev. Lett.*, 1993, **71**, 4358 and references therein.
77. M. Sahimi, *Chem. Eng. J.*, 1996, **64**, 21.
78. H. Kanai, R. C. Navarrete, C. W. Macosko and L. E. Scriven, *Rheol. Acta.*, 1992, **31**, 333.
79. D. J. Bergman, *Phys. Rev. B*, 1986, **33**, 2013.
80. S. Feng, *Phys. Rev. B*, 1985, **32**, 510.
81. M. Sahimi and J. D. Goddard, *Phys. Rev. B*, 1985, **32**, 1869.
82. M. Plischke and B. Joós, *Phys. Rev. Lett.*, 1998, **80**, 4907.
83. O. Farago and Y. Kantor, *Phys. Rev. Lett.*, 2000, **85**, 2533.
84. R. P. Kusy, *J. Appl. Phys.*, 1978, **48**, 5301.
85. I. Balberg, C. H. Anderson, S. Alexander and N. Wagner, *Phys. Rev. B*, 1984, **30**, 3933.
86. A. Celzard, E. McRae, C. Deleuze, M. Dufort, G. Furdin and J. F. Marêché, *Phys. Rev. B*, 1996, **53**, 6209.

CHAPTER 3

Magnetoelasticity

M. ZRINYI

Laboratory of Nanochemistry, Department of Biophysics and Radiation Biology, Faculty of Medicine, Semmelweis University, H-1089 Budapest Nagyvarad ter 4, Hungary
Email: mikloszrinyi@gmail.com

3.1 Introduction

Materials producing strain in magnetic field are known as magnetoelastic or magnetostrictive materials. Most of the magnetostrictive materials are solids which develop mechanical deformation when subjected to an external magnetic field. These effects are fundamental to all ferro- and ferrimagnetic materials[1]. Although there are several magnetostrictive effects, *e.g.* volume and torsional changes, we shall consider only linear magnetostriction that is changes in the length.

Until recently the observed magnetostrictive deformations are quite small. For iron, nickel and cobalt the strain is of the order of 10^{-5}. The largest magnetostriction has been found in the rare-earth metals and compounds. Terbium-dysprosium-iron alloys offer strains up to 0.002. Although the magnetostrictive dimensional changes are small, they are driven by strong interatomic forces, therefore traditional magnetostrictive materials find application where very strong forces, but not large movements are required.

Solid magnetic materials can be characterized as densely packed magnetic domains. If finely divided filler particles having strong magnetic properties are dispersed in liquids or in highly elastic matrix, then we have magnetic soft matters (not to be confused with soft magnets!) like magnetic fluids (ferrofluids), magnetic gels (ferrogels) and magnetic elastomers (magnetoelasts).[2–5]

RSC Smart Materials No. 6
Magnetorheology: Advances and Applications
Edited by Norman Wereley
© The Royal Society of Chemistry 2014
Published by the Royal Society of Chemistry, www.rsc.org

As opposed to solid magnetic materials, magnetic soft matters can be considered as a collection of diluted magnetic domains. As a result due to the larger average distance between domains, the domain interactions are weak. On the other hand the interstitial space between the particles is mobile. This means that despite the weaker domain–domain interactions, the mobile matrix enhances the magnetostrictive effects. The consequence of the soft matrix is the motility of the magnetic fluid or the giant linear magnetostriction of magnetic polymers under imposition of a non-unifrom magnetic field. In a non-uniform field, the field gradient induces flow or deformation.

Magnetic field sensitive polymers are a special type of magnetic materials, where finely divided filler particles having strong magnetic properties are dispersed in a highly elastic polymeric matrix (Figure 3.1). They can be characterised by a giant magnetostriction with strains up to 0.5, but weak forces of entropic origin.[5]

It is worth mentioning that composite materials consisting of rather rigid polymeric material filled with magnetic particles have been known for a long time. These composites are successfully used as permanent magnets and magnetic cores, connecting and fixing elements in many areas. These traditional magnetic polymers have low flexibility and practically do not change their size, shape and elastic properties in the presence of an external magnetic field. The new generation of magnetic elastomers represents a new type of composite consisting of small (mainly nanosized) magnetic particles, having superparamagnetic behaviour, dispersed in a high elastic polymeric matrix. If there are strong attractive interactions between the network polymers and the surface of the solid particles, this provides a coupling between forces acting on the solid particles and conformational change of the surrounding macromolecules.

The following chapters are devoted to discussing the preparation of magnetic field responsive polymer systems, which include polymer gels and liquid free elastomers. This is followed by a discussion of magnetoelasticity, which describes the effect of magnetic field on the mechanical behaviour.

3.2 Preparation of Magnetic Field Responsive Polymer Gels and Elastomers

Magnetic- and magnetorheological fluids contain small, dispersed particles in the size range from nanometres to micrometres.[2,3,6] Since polymer gels contain

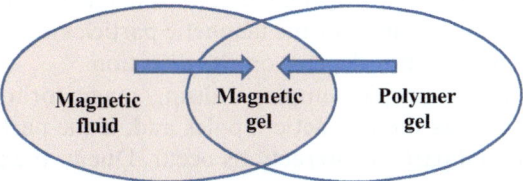

Figure 3.1 A magnetic gel combines the elasticity of a rubberlike network and the magnetic behavior of a magnetic fluid.

substantial amount of liquid as swelling agent, it is possible to fabricate magnetic field sensitive gels (sometimes called as ferrogels) by using a polymer network swollen by magnetic fluid. Preparation of ferrogels does not require a special polymer or a special type of magnetic material. As a polymer network one may use every flexible chain molecules which can be cross-linked. The magnetic filler particles can be prepared from ferro- or ferrimagnetic materials. An important requirement is the strong adsorptive interaction between the solid particles and polymer. The magnetic particles are located in the swelling liquid and attached to the flexible network chains. Strong attractive interactions between particles and polymer chains inhibit particle mobility that is, the fine particles can neither flow away nor rotate. Preparation of a magnetic gel is similar to that of other filler loaded elastomeric networks.[7] One can precipitate well-dispersed particles in the polymeric material dissolved. The *"in situ"* precipitation can be made before, during and after the cross-linking reaction. According to another method, the preparation of magnetic particles is made separately, and the cross-linking reaction takes place after the polymer solution and the magnetic sol are mixed together.[8–10] One can prepare magnetic hydrogels as well as polymer networks swollen by organic ferrofluid. During the preparation it is possible to vary the initial polymer concentration, the degree of cross-linking, size and quality of the dispersed magnetic material, as well as the amount of magnetic particles incorporated into the gel. The preparation procedure consists of three main steps. The first step is to mix the filler particles with the reaction mixture, which contains the polymers, the cross-linking agent and the catalyst. The second step is to stabilize the system in order to avoid aggregation and sedimentation of the solid particles. Then it is followed by the crosslinking reaction, which fixes the spatial arrangement of the filler paarticles. One can prepare isotropic magnetic polymers with randomly distributed filler particles and anisotrpic samples containing aligned particle shains. A detailed description of the preparation process can be found in several papers. Zrinyi *et al.* prepared superparamagnetic magnetite loaded polyvinyl alcohol hydrogels,[8–14] poly(*N*-isopropylacryilamide) hydrogels[15] and carbonyl iron and magnetite loaded polydimethylsiloxane gels.[16] Mitsumata and Ohori have prepared carbonyl iron loaded polyurethane elastomers.[17] Bajpai and Gupta reported maghemite and magnetite filled polyvinyl alcohol based hydrogels.[18] Choubey and Bajpai have investigated chemically crosslinked gelatin floaded with iron oxide.[19] Ghost and Cai have prepared poly(*N*-isopropylacryilamide) hydrogels filled with ferromagnetic magnetite nanoparticles.[20] Abramchuk *et al.* investigated several magnetic elastomers.[21,22] In these cited works the distribution of the magnetic particles is uniform, therefore the material properties do not depend on the direction

One can prepare anisotropic samples utilising magnetorheological effects. The imposed field orients the magnetic dipoles and, if the particles are spaced closely enough, mutual particle interactions occur. Due to the attractive forces a pearl chain structure develops.

The same phenomenon occurs if the liquid is replaced by the monomeric mixture of the polymer. If the polymerization reaction is performed under a uniform external field, then due to the mutual interaction between particles, a pearl chain structure develops. The chemical cross-linking locks in the chainlike structure, aligned along the direction of the field. The resulting sample becomes highly anisotropic. A schematic representation of the synthesis is shown in Figure 3.2. As a representative example, the mixture of the silicone prepolymer and magnetite particles confined in a mould was subjected to $B = 400$ mT uniform magnetic field between the poles of a large electromagnets.[23] It took a few seconds to induce the pearl chain structuring of the filler particles before the reaction is completed. As a result, particles aggregate aligned parallel to the field direction and are built into the network. After the cross-linking polymerization was completed, the gels were removed from the moulds.

Depending on the concentration of the magnetic particles as well as on the applied magnetic field, columnar structures of the magnetic particles built in the elastic matrix can be varied in a wide range. Figure 3.3 shows a PDMS elastomer containing dense column of pearl chains of magnetic particles.

It is worth mentioning that not only the magnetorheological effect, but also the elecrorheological effect can be applied to create particle alignment.[24] Since electric and magnetic fields do not interfere each other, one can use both fields simultaneously to prepare highly anisotropic samples.[25] Careful evaporation of

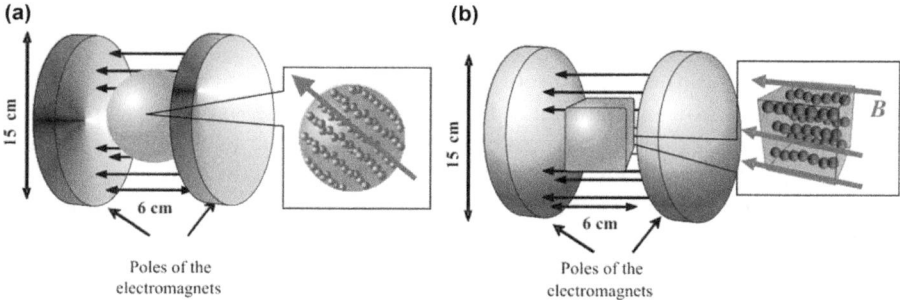

Figure 3.2 Magnetorheological effect assisted preparation of uniaxially ordered spherical (a) and cube (b) shaped polymer composites under a uniform magnetic field.

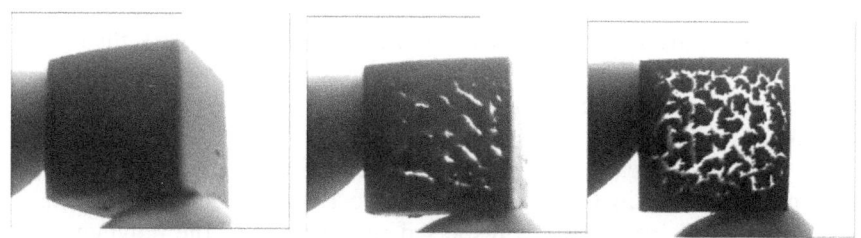

Figure 3.3 Columnar structure of iron particles built in a PDMS elastomer. In order to visualize the columnar structure of magnetic particles in the PDMS composite the sample is rotated. The angle of rotation has been decreased from left to right.

the liquid in the magnetic gel results in a magnetic elastomer which is composed of a flexible polymer network and dispersed magnetic particles.

3.3   Elastic Properties of Magnetic Gels and Elastomers

In the absence of an external magnetic field a ferrogel presents a mechanical behaviour very close to that of a swollen filler-loaded network. Since a typical magnetic gel can be considered as a dilute magnetic system in the absence of an external magnetic field, we may neglect the influence of magnetic interactions on the modulus. Thus the stress–strain dependence of a unidirectional deformed gel sample can be expressed on the basis of statistical theories of rubber elasticity:[26–28]

$$\sigma_n = G(\lambda - \lambda^{-2}) \tag{3.1}$$

where the nominal stress, σ_n, is defined as the ratio of the equilibrium elastic force and the undeformed cross-sectional area of the sample. The deformation ratio, λ, is the length, h, (in the direction of force) divided by the corresponding undeformed length, h_o. In eqn (3.1), which is known as the neo-Hookean law, G stands for the elastic modulus. In a dilute filler loaded elastomer, the dependence of elastic modulus on the volume fraction of the filler particles, φ_m can be expressed as:[29]

$$G = G_0(1 + k_E \varphi_m) \tag{3.2}$$

where G_o denotes the modulus of gel without filler particles and k_E is the Einstein–Smallwood parameter. For non-interacting spherical particles $k_E = 2.5$. Eqn (3.2) describes the reinforcement effect due to the filler–polymer interactions. The modulus of a ferrogel can be varied by the cross-linking density through G_o and by the concentration of colloidal particles *via* φ_m. It must be mentioned, that in many cases eqn (3.1) can not be used to fit the experimental data. Deviation from the Gaussian theory of rubber elasticity may be due to finite chain extensibility, entanglement effects and filler chain interactions. For filler loaded elastomers this non-ideal rubber elastic behaviour is often represented by the Mooney–Rivlin equation:[26–28]

$$\frac{\sigma_n}{\lambda - \lambda^{-2}} = C_1 + C_2\lambda^{-1} \tag{3.3}$$

where C_1 and C_2 are structural constants, and for a given network the modulus is $G = C_1 + C_2$.

Stress–strain measurements for isotropic PDMS magnetoelasts are shown in Figure 3.4.

These figures indicate that within the experimental accuracy the unidirectional stress–strain behavior of magnetic PDMS elastomers having different amounts of iron particles dispersed randomly into the polymer matrix obeys the neo-Hookean law. Deviation from the ideal behavior, characterized by the second Mooney–Rivlin constant, C_2 can be neglected. This is not the case for anisotropic samples where there is a slight deviation from eqn (3.1) as shown in Figure 3.5.

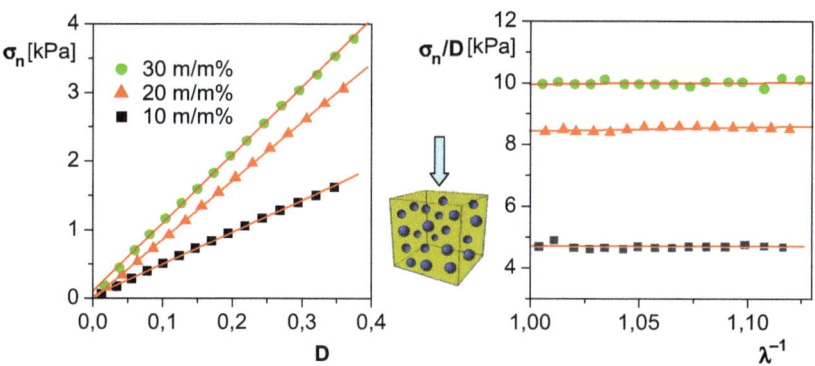

Figure 3.4   Stress–strain dependence of iron loaded PDMS network plotted according to neo-Hookean law (a), and Mooney–Rivlin representation (b). The arrow indicates the direction of mechanical stress. The concentration of randomly distributed iron particles are given in the figure, where D is defined as $D = -(\lambda - \lambda^{-2})$.

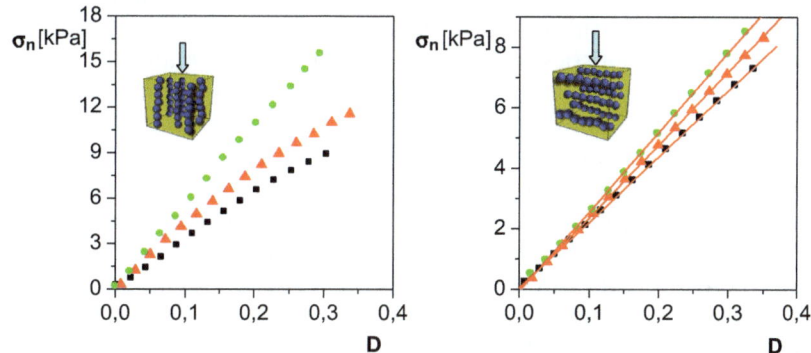

Figure 3.5   Stress–strain dependence of anisotropic iron loaded PDMS network plotted according to the neo-Hookean law. The concentration of aligned iron particles are given in the Figure 3.4. In the figure D is defined as $D = -(\lambda - \lambda^{-2})$.

There are two different situations when comparing the direction of mechanical stress to the particles alignment. The direction of the compression can either be perpendicular or parallel to the pearl chain structure locked into the PDMS network. When the direction of the compressive force is perpendicular to the pearl chain structure, we have found a slight deviation from the ideal behavior, as shown in Figure 3.5.

3.4   Magnetic Properties of Ferrogels

In the ferrogel, the finely distributed nano-sized magnetic particles are located in the swelling liquid and are attached to the flexible network chains by adhesive forces. The solid nanoparticles are the elementary carriers of a magnetic moment. These fine particles have properties that are significantly different from those of the corresponding bulk materials.[2,30–33] The primary reasons for these are the monodomain structure and the large fraction of surface atoms which have significant influence on the magnetic properties. If the concentration of magnetic particles is below the percolation threshold, the gel contains a collection of single domain particles.

In the absence of an applied field, the magnetic moments are randomly fluctuating, and thus the gel has no net magnetisation. As soon as an external field is applied, the magnetic moments tend to align with the field to produce a bulk magnetic moment. With low field strengths, the tendency of the dipole moments to align with the applied field is partially overcome by thermal fluctuation, such as the molecules of paramagnetic gas. As the field strength increases, all the particles eventually align their moments along the field direction, and as a result, the magnetisation saturates. The single domain particles are therefore the most efficient magnets. If the field is turned off, the magnetic dipole moments quickly randomize and thus the bulk magnetisation is again reduced to zero. On the basis of relaxation times one can define a blocking temperature, T_B. Below this temperature, $T < T_B$, the magnetic material exhibits ferromagnetic characteristics with hysteresis and remnant magnetisation. It is worth mentioning that the blocking temperature is proportional to the volume of the magnetic particles, which means even a modest increase in particle size can result in a significant increase in the value of T_B. With increasing temperature or decreasing particle size, the ferromagnetic characteristics decrease and vanish. Above T_B all apparent ferromagnetic characteristics disappear, and the magnetic nanoparticles exhibit superparamagnetic behaviour with a giant magnetic moment. This can be about 10^4-times larger than the individual atomic moment. Consequently the particle size plays an essential role in the magnetic behaviour. For the magnetic particles in the nanosize regime the superparamagnetic behaviour dominates, whereas for larger particles the ferromagnetic characteristics can be observed as shown in Figure 3.6.

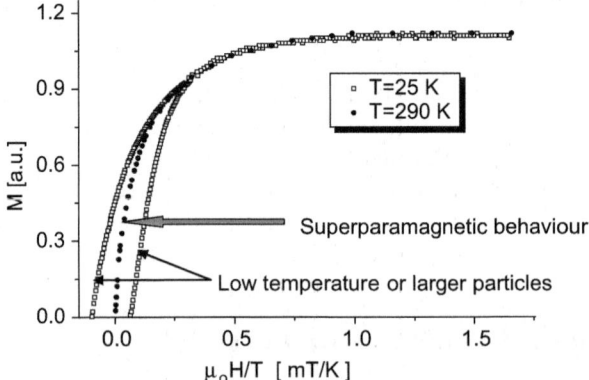

Figure 3.6   The influence of temperature or particle size on the magnetisation curve.

If we suppose that monodisperse nanosized (superparamagnetic) particles are randomly distributed in the polymer network, and the magnetisation of individual particle in gel equals to the saturation magnetisation of the pure and bulk corresponding material, M_s. In this case the magnetisation of ferrogel, M similarly to that of ferrofluids can be described by the Langevin function:[2,6,30–32]

$$M = \varphi_m M_s L(\xi) = \varphi_m M_s \left(\coth \xi - \frac{1}{\xi} \right) \qquad (3.4)$$

where φ_m stands for the volume fraction of the magnetic particles in the whole gel, and the parameter of the Langevin function, $L(\xi)$ is defined as

$$\xi = \frac{mH}{k_B T} \qquad (3.5)$$

where H represents the strength of an external magnetic field, m is the magnetic moment of subdomain particles, k_B denotes the Boltzmann constant and T stands for the temperature.

In the following sub-chapters we will separately treat the effect of uniform- and non-uniform magnetic field distribution on the magnetoelastic behavior of polymers.

3.5 Coupled Elastic and Magnetic Behaviour in Non-uniform Magnetic Field

When a magnetic gel or elastomer is placed in a gradient of a magnetic field, forces act on the superparamagnetic particles and the magnetic interactions are enhanced. The stronger field attracts the particles and due to their small size and strong interactions with molecules of dispersing liquid and polymer chains they all move together. Because of the cross-linking bridges in the network, changes in molecular conformation can accumulate and lead to macroscopic shape changes which go together with motion. The magnetic field drives and controls the motion and the final shape is due to the balance of magnetic and elastic interactions. The force density on a piece of magnetic gel can be written as:

$$f_m = \mu_0 (M \nabla) H \qquad (3.6)$$

where μ_0 means the magnetic permeability of vacuum, M represents the magnetization and ∇H takes into account the gradient of magnetic field, H. The orientation of f_m is parallel to the direction of magnetic field. In a non-accelerating system the force density manifests itself as a stress distribution which must be balanced by the network elasticity. A completely balanced set of forces is in this respect equivalent to no external force at all. However they affect the gel internally, tending to change its shape or size or both. The magnetic force density vector varies from point to point in accordance with the position dependence of product $(M \nabla) H$ resulting in non-homogeneous deformation.[35] The description of the macroscopic deformations of ferrogels requires a special treatment due to the complex nature of the mechanism of magnetic field induced deformations. The non-linear character of both elastic and magnetic interactions results in some novel features of the deformation process. In the theory of non-linear elasticity, a material body is regarded as a set of elastically joined material points.[34] The forces are considered to act on the material points and the condition of mechanical equilibrium is characterised by the balance between external and elastic forces.

Since the field, and the force alike, is non-uniform, different material points experience forces of different strength and direction, which leads to a non-homogeneous deformation often with a complex deformation pattern. As an illustrative example of the complex deformation that ferrogels can achieve in magnetic field we present the two-dimensional deformation of a ferrogel sheet in Figure 3.7. A magnetic gel layer was placed in between two glass plates so that no perpendicular deformations were allowed. The magnetic field was induced by a couple of permanent magnets placed around the gel. The gel sheet is shown in the undeformed (A) and deformed (B) and (C) states, respectively. The complexity and non-homogeneity of the deformation is well demonstrated on the distortion of the grid that was painted to the surface of the gel.

Due to the relatively complex form of the Langevin-type magnetization and the magnetic force density – even in a simple magnetic field distribution – it is not possible to achieve an analytical solution of the driving equations. In order to demonstrate the characteristics of the magnetic field induced deformations of a magnetic continuum it is worth examining a simple one-dimensional situation. Let us consider a long and thin ferrogel cylinder that is suspended in water vertically. The magnetic field is induced by a solenoid-based electromagnet placed under the gel. The axis of the gel cylinder (z) is parallel with the magnetic field and its gradient. In this case, the deformation of the gel is uniaxial and can be considered as one-dimensional. The governing equation for this situation describing the displacement of each point of the gel along the z axis is the following second-order, non-linear ordinary differential equation.[36–38]

$$G\left(\frac{\mathrm{d}^2 u_z(Z)}{\mathrm{d}Z^2} + \frac{2}{(\mathrm{d}u_z(Z)/\mathrm{d}Z)^3}\frac{\mathrm{d}^2 u_z(Z)}{\mathrm{d}Z^2}\right) + M(u_z(Z))\frac{\mathrm{d}H(u_z(Z))}{\mathrm{d}Z} = 0 \qquad (3.7)$$

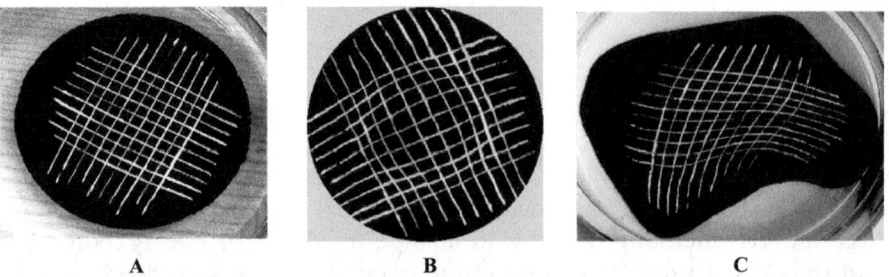

A   B   C

Figure 3.7   Illustration of the complex non-homogeneous deformation of a ferrogel sheet. Undeformed (A) and deformed (B) and (C) states. The magnetic field was induced by permanent magnets placed around the gel.[36–40]

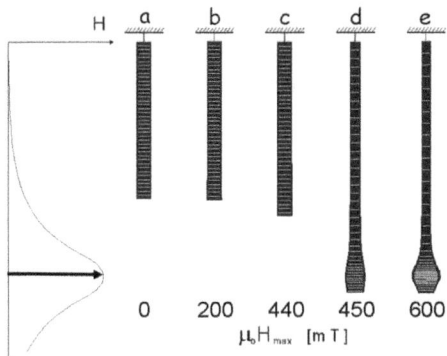

Figure 3.8 Schematic representation of the uniaxial deformation of a ferrogel cylinder calculated numerically from eqn (3.7). The external magnetic field
distribution is shown in the left side of the figure. The arrow indicates the
maximum magnetic field intensity produced by the electromagnet. The gel
on the left-hand side is undeformed ($H_{max} = 0$).[36–40]

where $u_z(Z)$ represents the displacement given in the reference – undeformed –
configuration, G is the shear modulus of the gel, M denotes the magnetization,
and H stands for the magnetic field strength. As boundary conditions, the
displacements and/or the surface tractions must be prescribed on the two ends
of the gel cylinder. In our particular case the position of the top surface of the
gel cylinder was fixed by a rigid, non-magnetic copper thread, while the bottom
surface was free and unloaded. Based on eqn (3.7) we have calculated the
unidirectional deformation of a ferrogel cylinder. This is shown in Figure 3.8.
On the left-hand side of the figure the magnetic field strength along the z axis is
plotted. The distribution of the field we employed in the calculations was
similar to that in real experiments. The shape of the gel is plotted against the
maximum magnetic field intensity, H_{max} produced by the electromagnet.

As one can see, the gel eventually elongates as the magnetic field intensity
increases. It may be seen that at small magnetic field intensities, which means a
small magnetic field gradient, the elongation is rather weak. However at higher
field intensity a comparatively large, abrupt elongation occurs. This non-
continuous change in the size appears within an infinitesimal change in the
magnetic field distribution. The white lines on the gel body demonstrate the
non-homogeneity of the deformation. Differents distance between adjacent
lines indicates different degrees of deformation. The high degree of non-
homogeneity is clearly seen. At lower field intensities (gels b and c) the upper
part of the gel elongates to a greater extent than the lower part, whereas at
higher field intensities (gels d and e) the lower part of the gel contracts while the
upper part elongates.

The analytical determination of the displacement field is a rather difficult
task, therefore numerical methods like finite element method (FEM) can provide a useful tool to study three-dimensional non-linear deformations. An illustrative example of three-dimensional deformation of a ferrogel block is

shown in Figure 3.9. The magnetic field is uniaxial and has a Gaussian distribution along the gel axis as shown on the top of the figure. We represented the deformation of the block at two different field intensities. Looking at the set of pictures one may see the association with the motion of a living worm.

Depending on the magnetic field distribution, not only elongation, but also contraction can be realised in an inhomogeneous magnetic field. This is demonstrated by Figure 3.10, where a ferrogel is shown in different types of

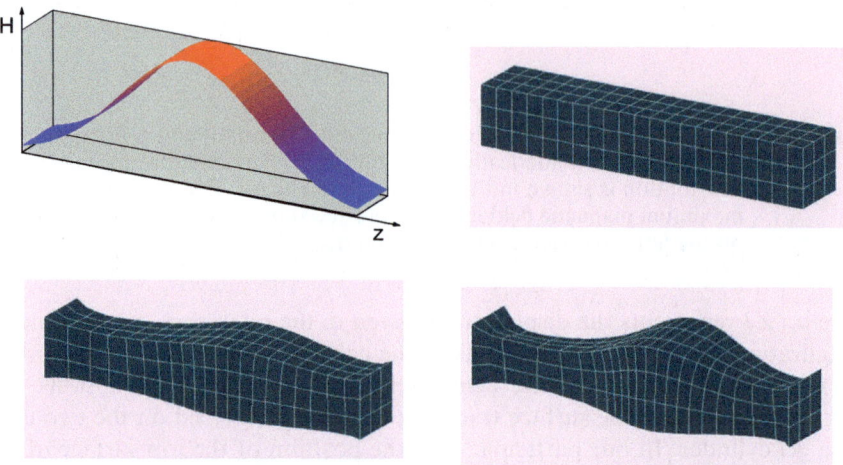

Figure 3.9   FEM calculation of the deformation of a ferrogel block in a uniaxial magnetic field at different field intensities. The magnetic field strength has a Gaussian distribution along the block, as shown in the figure.[36-40]

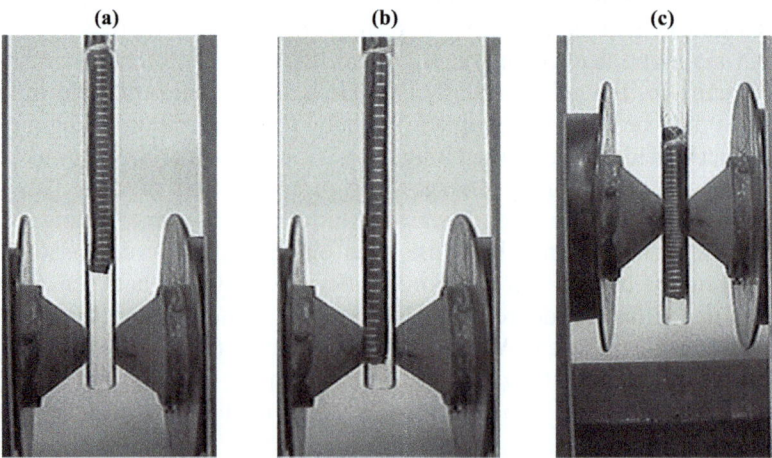

Figure 3.10   Shape distortion of ferrogels due to non-uniform magnetic fields:[40] (a) no external magnetic field; (b) the maximal field strength is located under the lower end of the gel; (c) the maximal field strength is focused in the middle of the gel along its axis.

magnetic fields produced by two planparallel poles of electromagnets. The position of the poles of electromagnets can be varied along the direction of z. We present here three different situations: (a) no magnetic field is applied, (b) the axis of magnetic poles is below the lower end of the gel, (c) the axis of the poles is in the middle of the gel along z. In the case of (a) no deformation occurs. In the presence of an applied magnetic field, a gradient develops parallel to the gel axis and results in elongation in case (b) and contraction at arrangements denoted by (c).

An approximate magnetoelastic equation for uniaxial deformation was developed.[8–12]

$$\lambda_H^3 - \beta\left(H_b^2 - H_t^2\right)\lambda_H - 1 = 0 \qquad (3.8)$$

where λ_H represents the deformation ratio due to magnetic interactions only (no load), H_b and H_t denote the magnetic field strength at the bottom and the top of a gel cylinder, β is a constant defined as

$$\beta = \frac{\mu_0 \chi}{2G} \qquad (3.9)$$

where χ stands for the magnetic susceptibility of the magnetic gel.

Eqn (3.8) was derived by considering homogeneous deformation and a linear relationship between magnetisation and magnetic field strength. This equation says if we suspend a magnetic gel in a non-homogeneous magnetic field in such a way that $H_b > H_t$ then elongation occurs. In the opposite case, $H_b < H_t$, contraction occurs. It is rather difficult to achieve an analytical solution of eqn (3.8) since during deformation the position of the bottom of the gel keeps on changing, therefore H_b depends on λ_H.

3.6 Coupled Elastic and Magnetic Behaviour in a Uniform Magnetic Field

As we have discussed earlier, similar to conventional magnetostrictive materials, ferrogels also elongate in uniform fields. Recently, there has been considerable interest in soft materials with uniform magnetic field induced deformation.[6,33,41–54]

Evolution of the shape of ferrofluid magnetic drops in the presence of a magnetic field has also been reported.[51,52]

When an immiscible ferrofluid drop is subjected to a magnetic field, the drop deforms into a prolate ellipsoid, whose axis of rotation is along the field direction. Interpretation of these phenomena is based on the competition of field energy and interfacial energy. In contrast to ferrofluids the magnetic gels are kept together by elastic polymer chains. As a consequence it may be assumed that beside the magnetic- and elastic energies, the surface energy of a gel droplet can be neglected.

More recently the dynamic response of a ferrogel under the sudden change of uniform magnetic field has been reported.[46] It was found that the time dependence of the elastic shear modulus causes elongation to increase with time and the rapid excitation causes the ferrogel sphere to vibrate. The phenomena were described by the theory of Raikher and Stolbov.[46,55,56]

We have studied the effect of a uniform magnetic field on the shape of carbonil iron loaded poly(dimethyl siloxane) (PDMS) gels.[54] In these works it was assumed that the shape of a ferrogel is controlled by a competition between magnetic- and elastic energy. For a constant volume of a ferrogel bead and a given magnetic field, the shape of the gel can be obtained by minimalization of the total energy with respect to the aspect ratio (a/R).[55,56] This process yields:

$$\frac{a}{R} = 1 + \kappa_a \frac{\chi_m^2 H^2}{G} \tag{3.10}$$

$$\frac{b}{R} = 1 + \kappa_b \frac{\chi_m^2 H^2}{G} \tag{3.11}$$

where κ_a as well as κ_b are the magneto-deformational susceptibilities parallel and perpendicular to the magnetic field. For an incompressible material, $\kappa_a = 1/15$ and $\kappa_b = -1/30$. Eqn (3.10) and (3.11) predict that the sphere elongates along the magnetic field lines and the measure of elongation scales with the square of magnetic field intensity. It is also seen that the higher the elastic modulus, the weaker the magnetodeformation effect. The experimental results preformed on carbonyl iron loaded poly(dimethyl siloxane) gels support these ideas, as shown in Figure 3.10.

It has only recently been accepted that the elastic properties of magnetic elastomers can be increased rapidly and continuously by application of external magnetic field.[50–56] The elastic modulus of magnetic composites was measured under uniform magnetic field at 293 K. We were able to measure the stress–strain dependence under magnetic induction, B between 0 and 400 mT. Figure 3.11. shows the effect of a uniform magnetic field on the modulus of

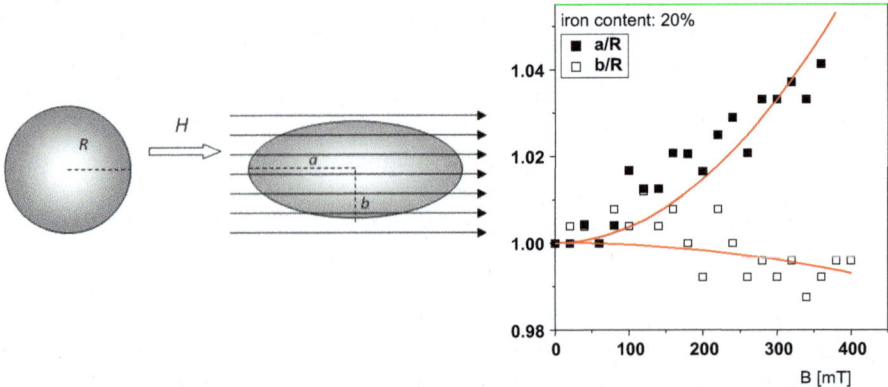

Figure 3.11 The dependence of aspect ratio on the external magnetic field intensity. The solid lines represent the square law dependence of aspect ratio on field intensities.[54]

PDMS samples containing randomly distributed magnetite particles. There are two basic experimental situations: The compressive force (F_x) and the direction of magnetic field can be either parallel or perpendicular, as shown in Figure 3.11. This finding is in accordance with our previous result where a similar temporary reinforcement effect was reported for magnetite (Fe_3O_4) loaded poly(vinyl alcohol) hydrogels.[14] It is also seen on the figures that the concentration of the filler particles increases the field free modulus, G_o. A larger amount of magnetite results in a larger elastic elastic modulus, in accordance with eqn (3.2).

The elastic modulus of the anisotropic magnetoelasts can be determined by five different experimental set-ups, as shown in Figure 3.12. On the basis of Figure 3.11 and 3.12, it may be concluded that the spatial orientation of the force, the field and the particle arrangement play a decisive role in the temporary reinforcement effect. A weak effect has been found when the field is perpendicular to the particles alignment (Figure 3.13).

If the columnar arrangements of the particles are parallel to the direction of the magnetic field, the elastic modulus increases significantly. At small field intensities of up to 30 mT a slight increase has been observed. Above 30 mT the modulus increases significantly. At higher field intensities (from 200 mT) the elastic modulus tends to level off. It is also seen that by increasing the concentration of the iron particles in the polymer matrix the elastic modulus, G_o, also increases. The most significant reinforcement effect was found when the magnetic field, force and the particle alignment are parallel. The properties of ferrogels in homogeneous magnetic fields have been studied using a simple microscopic model and Monte Carlo simulations by Wood and Camp.[57] The main phenomena of interest concerns the anisotropy and enhancement of the elastic moduli that result from applying uniform magnetic fields before and after the magnetic grains are locked in to the polymer-gel matrix by

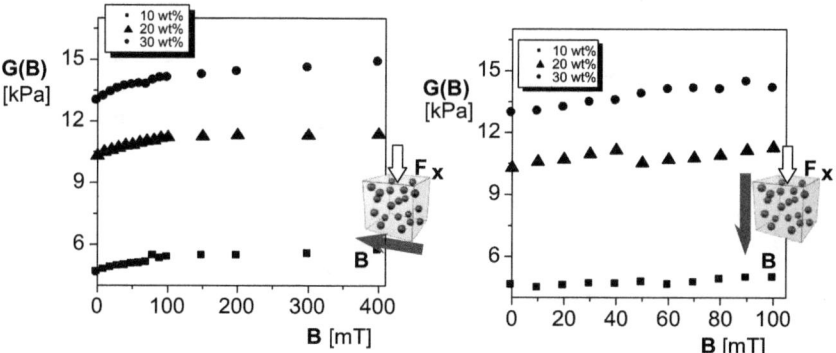

Figure 3.12 Dependence of the magnetic field intensity on the elastic modulus for PDMS containing different amount of randomly distributed magnetite particles.[47] The concentration of the filler particles are indicated in the figure. White arrows indicate the direction of the force; black arrows show the direction of magnetic field.

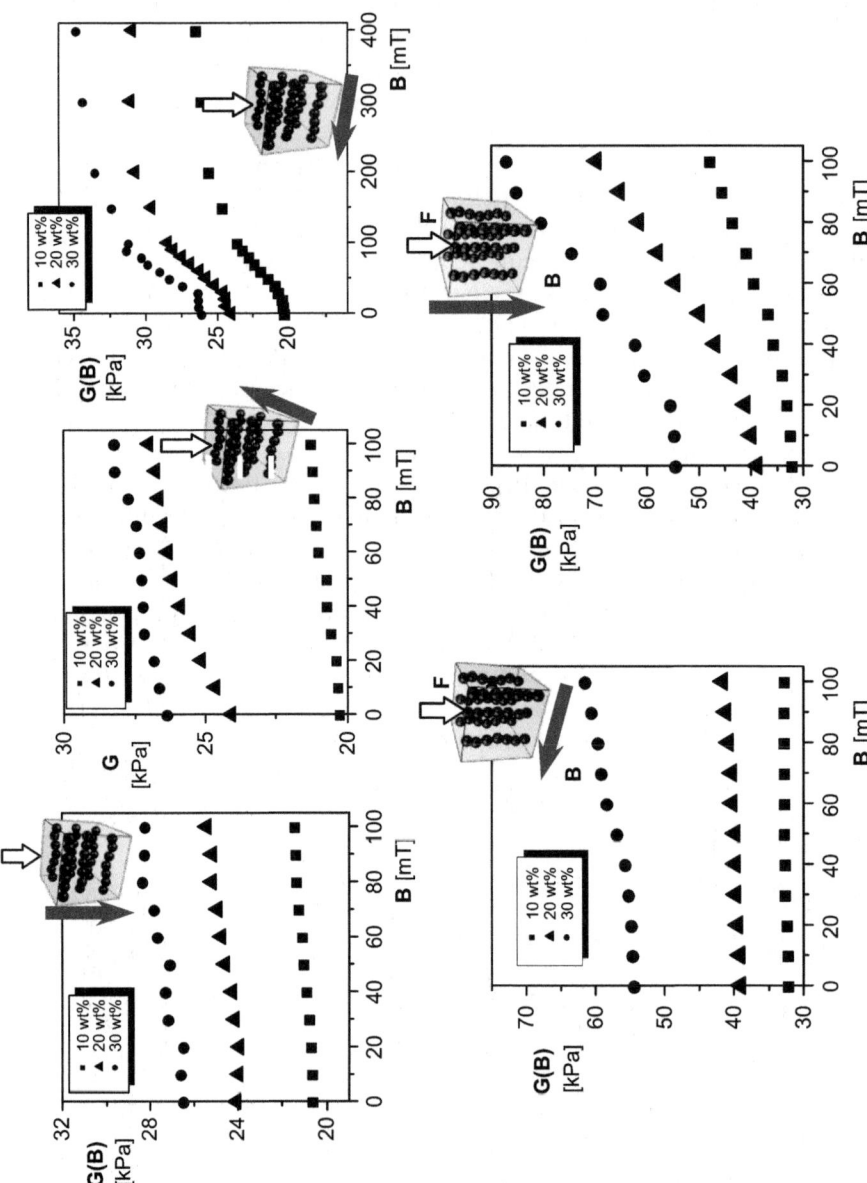

Figure 3.13 Effect of the magnetic field intensity on the elastic modulus of anisotropic samples. [46] The iron content of the PDMS elastomers is indicated in the figure. The white and black arrows show the direction of the force and the uniform magnetic field, respectively.

cross-linking reactions. It has been demonstrated using MC simulations that the model ferrogel mimics experimental trends quite reliably, while providing specific insights on the microscopic arrangements of the magnetic grains.

3.7 Summary

Magnetic field sensitive polymer gels and elastomers have been prepared by introducing magnetic particles of colloidal size into chemically cross-linked polymer networks. Randomly distributed particles as well as uniaxial aligned particles have been built into high elastic matrices. The magnetic and magneto-elastic properties of these materials have been studied. The magnetic properties of ferrogels containing magnetic nanoparticles can be described by a Langevin function indicating a strong analogy between magnetic properties of ferrogels and that of ferrofluids. Peculiar magneto-elastic behaviour has been observed in non-uniform magnetic fields. Shape distortion occurs instantaneously and disappears abruptly when an external magnetic field is produced or removed. The description of the macroscopic deformations of ferrogels requires a special treatment due to the complex nature of the mechanism of magnetic field induced deformations. The non-linear character of both elastic and magnetic interactions results in non-homogeneous deformation – that is the displacement of material points during the deformation varies point by point. The peculiar magneto-elastic properties may be used to create a wide range of motions and to control the shape change and movement, which are smooth and gentle similar to that observed in muscle.

The elastic modulus of magnetoelasts could be increased by a uniform external magnetic field. We have shown that uniaxial field structured composites exhibit a much larger increase in modulus than random particle dispersions. These magnetic field sensitive materials with tuneable elastic properties may find usage in elastomer bearings and vibration absorbers.

Acknowledgements

The financial support of OTKA Foundation (Grant number: NK 101704) is gratefully acknowledged.

References

1. J. B. Restorf, *Encyclopedia of Applied Physics*, 1994, **9**, 229.
2. R. E. Rosenweig, *Ferrohydrodynamics*, 1985, Cambridge University Press.
3. R. Massart, Dubois, V. Cabuil and E. Hasmonay, *J. Magn. Magn. Mater.*, 1995, **149**, 1–5.
4. A. Hajdu, E. Tombacz, E. Illes, D. Bica and L. Vekas, *Progr Colloid Polym Sci., Progr Colloid Polym Sci.*, 2008, **135**, 29–37.
5. M. Zrinyi, *Colloid Polym. Sci.*, 2000, **27**(2), 98–103.

6. M. Nakano and K. Koyama, *Electro-rheological Fluids, Magneto-rheological Suspensions and their Applications*, World Scientific Publishing Co. Pte. Ltd., 1997.
7. J. E. Mark, *Br. Polym. J.*, 1985, **17**, 144.
8. L. Barsi, A. Büki, D. Szabó and M. Zrínyi, *Progr. Colloid Polym. Sci.*, 1996, **102**, 57.
9. M. Zrínyi, L. Barsi and A. Büki, *J. Chem. Phys.*, 1996, **104**, 20.
10. M. Zrínyi, L. Barsi, D. Szabó and H. G. Kilian, *J. Chem. Phys.*, 1997, **106**, 5685.
11. M. Zrínyi, L. Barsi and A. Büki, *Polym. Gels Networks*, 1997, **5**, 415–427.
12. M. Zrinyi, *Trends Polym. Sci.*, 1997, **5**(9), 280–285.
13. D. Szabó, G. Szeghy and M. Zrinyi, *Macromolecules*, 1998, **31**, 6541–6548.
14. T. Mitsumata, K. Ikeda, J. P. Gong, Y. Osada, D. Szabó and M. Zrínyi, *J. Appl. Phys.*, 1999, **85**, 1–5.
15. M. Xulu, G. Filipcsei and M. Zrinyi, *Macromolecules*, 2000, **33**(5), 1716–1719.
16. Z. S. Varga, G. Filipcsei and M. Zrínyi, *Polymer*, 2006, **47**(1), 227–233.
17. T. Mitsumata and S. Ohori, *Polym. Chem*, 2011, **2**, 1063–1067.
18. A. K. Bajpai and R. Gupta, *J. Mater. Sci.*, 2011, **22**, 357–369.
19. J. Choubey and A. K. Bajpai, *J. Mater. Sci, Mater. Med.*, 2010, **21**, 1573–1568.
20. S. Ghost and T. Cai, *J. Phys. D.*, 2010, **43**, 1–10.
21. S. Abramchuk, E. Kramarenko, G. Stepanov, L. V. Nikitin, G. Filipcsei, A. R. Khokhlov and M. Zrinyi, *Polym. Adv. Technol.*, 2007, **18**, 883–890.
22. S. Abramchuk, E. Kramarenko, G. Stepanov, L. V. Nikitin, G. Filipcsei, A. R. Khokhlov and M. Zrinyi, *Polym. Adv. Technol.*, 2007, **18**, 513–518.
23. Z. S. Varga, G. Filipcsei and M. Zrínyi, *Polymer*, 2005, **46**(18), 7779–7787.
24. R. Thao and. G. D. Roy, *Electrorheological Fluids*, World Scientific Publishing Co. Pte. Ltd., 1994.
25. M. Zrínyi, Szabó, G. Filipcsei and J. Fehér, *Polymer Gels and Networks*, ed. Osada and Khokhlov, Marcel Dekker Inc., NY, 309–355.
26. L. R. G. Treolar, *The Physics of Rubber Elasticity*, Clarendon Press, Oxford, 1975.
27. K. Dusek and W. Prins, *Adv. Polym. Sci.*, 1969, **6**, 1.
28. J. E. Mark and B. Erman, *Rubberlike Elasticity: a Molecular Primer*, John Wiley Sons New York, Chichester, Brisbance, Toronto, Singapore, 1988.
29. H.-G. Kilian, *Colloid Polym. Sci.*, 1987, **265**, 410.
30. J. L. Dormann and D. Fiorani, ed. *Magnetic Properties of Fine Particles*, North-Holland, Amsterdam, 1992.
31. G. C. Hadijipanayis and R. W. Siegel, ed. *Nanophase Materials*, Kluwer Academic Publishers, 1994.
32. B. Berkovski and V. Bashtovoy, ed. *Magnetic Fluids and Applications Handbook*, Begell House, Inc., New York, Wallingford, 1996.
33. D. Szabo, I. Czako-Nagy, M. Zrinyi and A. Vertes, *J. Colloid Interface Sci.*, 2000, **221**, 166–172.

34. R. W. Odgen, *Non-linear Elastic Deformations*, Ellis Horwood Limited, Chichester, 1984.
35. D. Szabó, Mágneses polimer gélek, PhD thesis, Budapest University of Technology and Economics, Budapest, Hungary, 1999.
36. Electric and Magnetic Field -Sensitive Smart Polymer Gels, in *Polymer Gels and Networks*, ed. Y. Osada and A. Khokhlov, Marcel Dekker, Inc., New York, 2001, p. 309–355.
37. M. Zrínyi, D. Szabó and L. Barsi, Magnetic Field Sensitive Polymeric Actuators in *Polymer Sensors and Actuators*, ed. Y. Osada and D. E. Rossi, Springer Verlag Berlin Heidelberg, 1999, pp. 385–408.
38. M. Zrinyi, Magnetic polymer gels as intelligent artificial muscles, in *Intelligent Materials*, RSC Publishing, 2007, ch. 11, pp. 282–299.
39. G. Filipcsei, I. Csetneki, A. Szilágyi and M. Zrínyi M, Magnetic field-responsive smart polymer Composites (review), in *Advances in Polymer Science, Oligomers, Polymer Composites, Molecular Imprinting*, Springer-Verlag Berlin Heidelberg, 2007, pp. 137–189.
40. D. S. Wood and P. J. Camp, *Phys. Rev. E*, 2011, **83**, 011402.
41. J. D. Carlson and M. R. Jolly, *Mechatronics*, 2000, **10**, 555–569.
42. J. M. Ginder and L. C. Davis, *Appl. Phys. Lett.*, 1994, **65**, 3410.
43. J. M. Ginder, M. E. Nichols, L. D. Elie and J. L. Tardiff, *Proc. SPIE*, 1999, **3675**, 131–138.
44. J. M. Ginder, S. M. Clark, W. F. Schlotter and E. Nichols, *Int. J. Modern. Phys. B*, 2002, **16**(17–18), 2412–2418.
45. C. Gollwitzer, A. Turanov, M. Krekhova, G. Lattermann, I. Rehberg and R. Richter, *J. Chem. Phys.*, 2008, **128**(16), 164709.
46. Z. Varga, G. Filipcsei and M. Zrínyi, *Polymer*, 2006, **47**(1), 227–233.
47. L. Nikitin, G. Stepanov, L. Mironova and A. Samus, *J. Magn. Magn. Mater.*, 2003, **258–259**, 468.
48. L. Nikitin, L. Mironova, K. Kornev and G. Stepanov, *Polym. Sci., A*, 2004, **46**(3), 301.
49. R. Patel, R. V. Upadhyay and R. V. Mehta, *J. Phys.: Condens. Matter*, 2008, **20**, 204116.
50. J. C. Bacri and D. Salin, *J. Phys. Lett.*, 1982, **43**, L179.
51. J. C. Bacri and D. Salin, *J. Phys. Lett.*, 1982, **44**, 415.
52. C. Flament, S. Lacis, J. Bacri, A. Cebers, S. Neveu and R. Perzynski, *Phys. Rev. E*, 1996, **53**, 4801.
53. G. Filipcsei and M. Zrínyi, *J. Phys.: Condens. Matter*, 2010, **22**, 276001.
54. Y. L. Raikher and O. V. Stolbov, *J. Magn. Magn. Mater.*, 2003, **258–259**, 477.
55. Y. L. Raikher and O. V. Stolbov, *J. Magn. Magn. Mater.*, 2005, **289**, 62–65.
56. D. S. Wood and P. J. Camp, *Phys. Rev. E*, 2011, **83**, 011402.
57. P. J. Camp, *Magnetohydrodynamics*, 2011, **47**, 123.

CHAPTER 4

MR Fluids at the Extremes: High-Energy and Low-Temperature Performance of LORD® MR Fluids and Devices

DANIEL E. BARBER

Open Technology Innovation, LORD Corporation, 111 LORD Drive, Cary, NC 27511, USA
Email: daniel_barber@lord.com

4.1 Introduction

Magneto-Rheological (MR) fluids and devices were first successfully commercialized in 1998 by LORD Corporation with the introduction of the Motion Master™ seat suspension system, comprised of a small MR damper in combination with a position sensor and controlling electronics. As described by Carlson,[1,2] the key requirement for successful commercialization of these devices was the development of an MR fluid formulation that could meet the durability requirements of the application. Early MR fluid formulations suffered from a characteristic degradation phenomenon known as *in-use thickening*, or IUT, which is a substantial increase in the off-state viscosity of the fluid as a result of use in a damper or clutch device. The increased viscosity causes the device to have an off-state force that is too high for the specified application. The IUT problem was addressed using patented and/or proprietary improvements to the MR formulation,[3] and the improved LORD® MR

RSC Smart Materials No. 6
Magnetorheology: Advances and Applications
Edited by Norman Wereley
© The Royal Society of Chemistry 2014
Published by the Royal Society of Chemistry, www.rsc.org

fluids allowed for the successful introduction of MR technology into many other applications, including automotive shock absorbers and struts,[4,5] cab dampers for agricultural vehicles,[6] and suspension dampers for military and other large off-road vehicles.[7,8]

IUT in early LORD MR fluids was attributed to the formation of fine iron oxide particles produced by mechanical degradation of the carbonyl iron particle surface. This IUT could be simulated by adding up to 1% fine ferrite particles to a fresh MR fluid and observing a substantial increase in off-state viscosity.[1] In automotive dampers, a similar thickening of the MR fluid was described and was attributed to several possible mechanisms, including mechanical degradation of the soft reduced carbonyl iron particles under the high side load conditions of the test; adherence of fumed silica particles (used as a thickener in the tested formulations) to the deformed soft iron particles; and iron-catalyzed polymerization of the hydrocarbon base oil used in the fluid.[9] Iyengar and Foister claimed that these problems were overcome using mechanically hard non-reduced carbonyl iron in combination with appropriate base oils, anti-settling agents, and anti-wear and friction-reducing additives.

The purpose of this chapter is to explore recent work on the effects of extreme environments on LORD MR fluid properties. Most of the work discussed relates to high-energy applications that place the greatest demands on fluid durability, but there will also be some discussion of low-temperature fluid performance. Because the composition and detailed properties of current LORD MR fluids are proprietary, only *changes* in composition and properties relative to a fresh, unused fluid will be discussed. These studies will demonstrate the wide operating range of LORD MR fluid and of properly designed devices containing it.

4.2 High-Energy Applications

4.2.1 General Fluid and Device Considerations

As with any system using MR technology, the lifetime of the system depends on both fluid properties and elements of the device design. Carlson previously defined the *lifetime dissipated energy* (LDE) of an MR fluid as the total mechanical energy dissipated by a device per unit volume of MR fluid and claimed that commercial LORD MR fluids could achieve LDE values of approximately 10^7 J cm^{-3}.[1,2] The LDE was intended as an *ad hoc* measure useful for estimating durability and device feasibility, but there are several difficulties with using LDE to assess the actual in-use lifetime of a device. One main difficulty is determination of the actual amount of energy dissipated during use, either in a real application or in a more complex simulated life test. In these situations, the MR damper cycles between its off-state and various degrees of on-state depending upon the excitation profile experienced by the damper. Thus, the amount of energy dissipated over time is a complex function varying from low (off-state) to high (maximum on-state) values as opposed to an easily-determined value that could be obtained with a constant-current sinusoidal damper test.

Another major difficulty with applying the LDE is that certain mechanical design parameters can cause premature failure of a fluid that would normally have an LDE in the predicted range of 10^7 J cm^{-3}. For example, dampers with inadequate seals or a rod finish that is not sufficiently smooth will result in leakage of base oils, leading to premature IUT. Design of and choice of materials for the damper piston and damper wall can also have a significant effect on both device and fluid durability.[10]

The manner in which the damper is used can also have a significant effect. Certain vehicle designs can create a high force perpendicular to the stroke axis of the damper, called the *side load*. Side loads can also be inadvertently or deliberately introduced during a simulated life test. High side loads can cause mechanical wear of the device and also of the iron particles in the MR fluid, leading to premature failure of the fluid by IUT.[9]

Because of these complexities, values of LDE for the lifetime durability tests discussed below have not been determined. The durability tests are generally complex, proprietary tests specified by the automotive industry or military requirements, and lifetimes will generally be discussed as a percentage, with 100% life test corresponding to the minimum test duration (time and/or number of test cycles) specified by the particular test.

4.2.2   MR Fluids from Automotive Damper Tests

LORD MR fluids have been used in automotive MR dampers since their introduction by Delphi Corporation for the 2002 Corvette. In the decade since, use of MR dampers has expanded to include more than a dozen vehicle platforms, including the Corvette, Cadillac and Buick by General Motors Corporation, the TT, R8, and A3 models by Audi, the MDX and ZDX by Acura,[11] and most recently a next-generation damper for the Range Rover Evoque.[12] Thousands of MR dampers from these vehicles have survived typical vehicle lifetimes greater than 150,000 km with no field failures.

Typical industry durability life tests include both standard sinusoidal tests and more complex vehicle simulation tests. Both tests place more demand on the damper and the MR fluid than will typically be experienced by automotive dampers. Current LORD MR fluid and damper designs routinely survive 200% or more of the required lifetimes in these tests, accounting for the excellent performance of MR dampers in the field.

Recently, Barber and Faulk investigated changes in the dynamic and static rheology of LORD MR fluids obtained from dampers subjected to one of these standard industry durability life tests.[13] The MR fluid was a proprietary oil-based formulation containing 20–30% iron particles by volume and various additives to optimize durability and dispersibility of the fluid. The damper durability test protocol was a standard automotive durability life test involving a proprietary sine-on-sine excitation profile under variable on- and off-state conditions and at a controlled temperature. Dampers were removed from the test rig after completing 20%, 45%, and 100% of life test, and used MR fluids were extracted from the dampers. All dampers were operating normally when they were removed from the life test.

Upon standing for many hours, the used fluids seemed to become quite thick and pasty, whereas fresh MR fluids maintain good flow indefinitely. Shaking the used samples vigorously by hand or on a commercial paint shaker was sufficient to return the samples to a more fluid state. This off-state gel formation was examined by characterization of the physical, rheological, and chemical changes occurring in the MR fluid during use, as described below.

4.2.2.1   Density, Particle Size and Dynamic Rheology

The relative densities and dynamic rheological properties are summarized in Table 4.1. Used fluid densities differed from that of a fresh fluid by only 1–3%, indicating negligible loss of oils and no change in the iron volume percentage of the fluid. The relative particle size distributions of the used fluids, shown in Figure 4.1, were nearly superimposed on that of a fresh fluid, indicating that no significant particle wear occurred during the life test. (The small particle distribution at very low particle size in the 100% used fluid is a measurement artifact and can occasionally also be seen in fresh MR fluids.) These observations, combined with the observed normal operation of the dampers when they were removed from test, indicate that the thickening of the fluid upon standing is not the result of IUT.

Plots of shear stress *versus* shear rate from which dynamic viscosity and yield stress were determined are shown in Figure 4.2. The fluid samples were shaken and degassed prior to loading them into the Couette cell, and the dynamic viscosity was tested at a controlled temperature of $40 \pm 0.5\,^{\circ}\mathrm{C}$ using the following protocol: temperature equilibration and a conditioning step (5 minutes at $100\ \mathrm{s}^{-1}$) followed by shear rate ramps from 0–$1200\ \mathrm{s}^{-1}$ over 10 minutes (up curve) and from 1200–$0\ \mathrm{s}^{-1}$ over 10 minutes (down curve). All samples displayed nearly Bingham rheology, and the overlap of the up and down curves for each sample indicated that none of the fluids were thixotropic. The reported dynamic viscosity is the slope of the best-fit line to the down curve from 800–$1200\ \mathrm{s}^{-1}$, and the dynamic yield stress is the y-intercept of the data.

The relative dynamic viscosity and yield stress values determined from these curves are included in Table 4.1. Dynamic viscosities for tested samples were 8–13% higher *versus* the fresh fluid, and off-state yield stresses were up to eight

Table 4.1   Summary of densities and dynamic rheological properties of tested fluids.

Test Duration (% of Life Test)	Relative Density	Relative Dynamic Viscosity	Relative Dynamic Yield Stress
0	1	1	1
20%	1.01	1.08	4.0
45%	1.02	1.13	8.0
100%	1.03	1.08	2.9

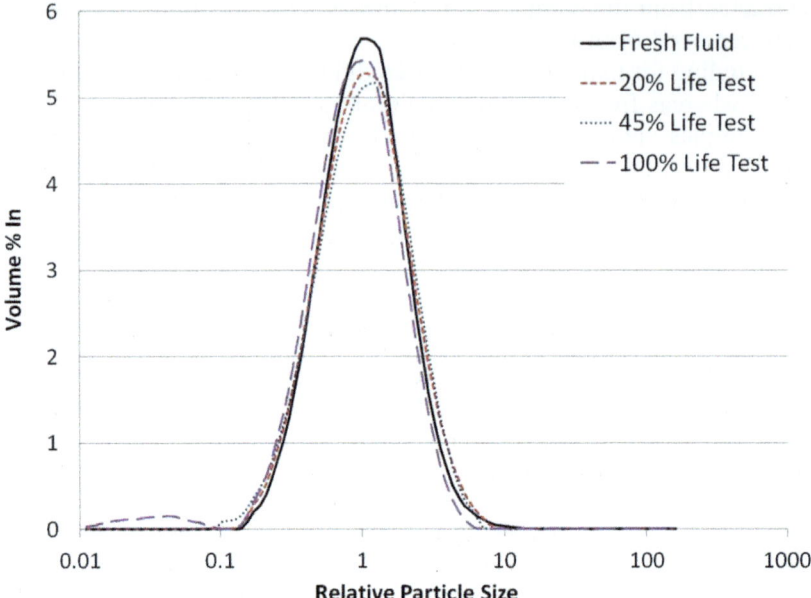

Figure 4.1 Relative particle size distributions (laser light scattering data) from fresh and used LORD® MR fluids after automotive durability life test.

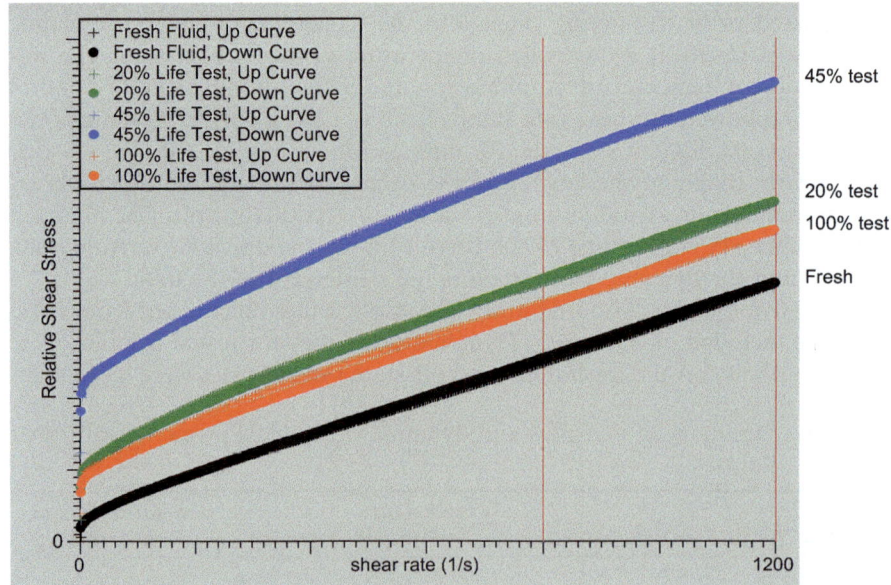

Figure 4.2 Dynamic rheological measurements (relative shear stress *vs.* shear rate) for fresh and used LORD® MR fluids. The range for calculation of viscosity and yield stress (800–1200 s^{-1}) is indicated by the vertical red lines.

times higher. The maximum values for the dynamic yield stress and viscosity were observed at 45% of life. Indeed, the fully-tested fluid (100% life test) had dynamic properties most similar to those of the fresh fluid.

4.2.2.2 Static Rheology

To investigate the observed thickening upon standing of used fluids, a test was developed to measure the time-dependent static yield stress. The vendor-recommended test for yield stress was an oscillatory test starting from very low displacement and required approximately one hour to complete. Because the MR fluids were found to have a changing yield stress value during this time, a modified test was developed that could measure the approximate yield stress within about two minutes. The modified test involved a sequence of three steps, repeated at each desired static hold time: first a conditioning step to return the fluid to its dynamic sheared state (100 s^{-1}, one minute), then a rest period (no shear) for the desired hold time, then a shear rate ramp up (0–250 s^{-1}, one minute) and back down (250–0 s^{-1}, one minute). The yield stress was chosen as the maximum shear stress observed in the up curve at very low shear rate (generally less than 5 s^{-1}). This shorter test was found to give a good approximation of the yield stress obtained from the vendor-recommended test (data not shown).

Figure 4.3 shows the shear stress *versus* shear rate curves used to determine the static yield stress of each fluid after hold times of 1 minute, 1 hour, 4 hours, and 12 hours. It is immediately apparent that the fresh fluid had no time-dependent change in yield stress, whereas the used fluids had substantial time-dependent increases in yield stress. The down curves for each fluid sample were nearly superimposed regardless of hold time, and the 0 s^{-1} shear stress from the down curve was almost the same as the dynamic yield stress for each fluid (compare to Figure 4.2). Thus, the yield stress developed during the hold time was overcome with relatively low shear to return the fluid to its dynamic state.

The relative yield stresses for each fluid are plotted *versus* hold time in Figure 4.4. The trend in static yield stress at each hold time was similar to that observed for the dynamic yield stress, with the fluids at intermediate life test having the highest yield stress values. The time dependence of the increase in yield stress appeared to be approximately the same for all life-tested fluids. Eqn (4.1) was used to fit the time-dependent change in yield stress, where YS(0) is

$$\text{YS}(t) = \text{YS}(0) + 10^{[A-(\tau/t)]} \tag{4.1}$$

the initial yield stress (hold time $= 1$ min), τ is the time constant and A is an empirical scaling parameter. Values for these parameters for each fluid sample are summarized in Table 4.2, and best-fit lines calculated using these parameters are shown with the data in Figure 2.4. Note that the YS(0) values follow the same trend but are not exactly the same as the dynamic yield stress values, likely because of the different shear histories of the samples in the different tests.

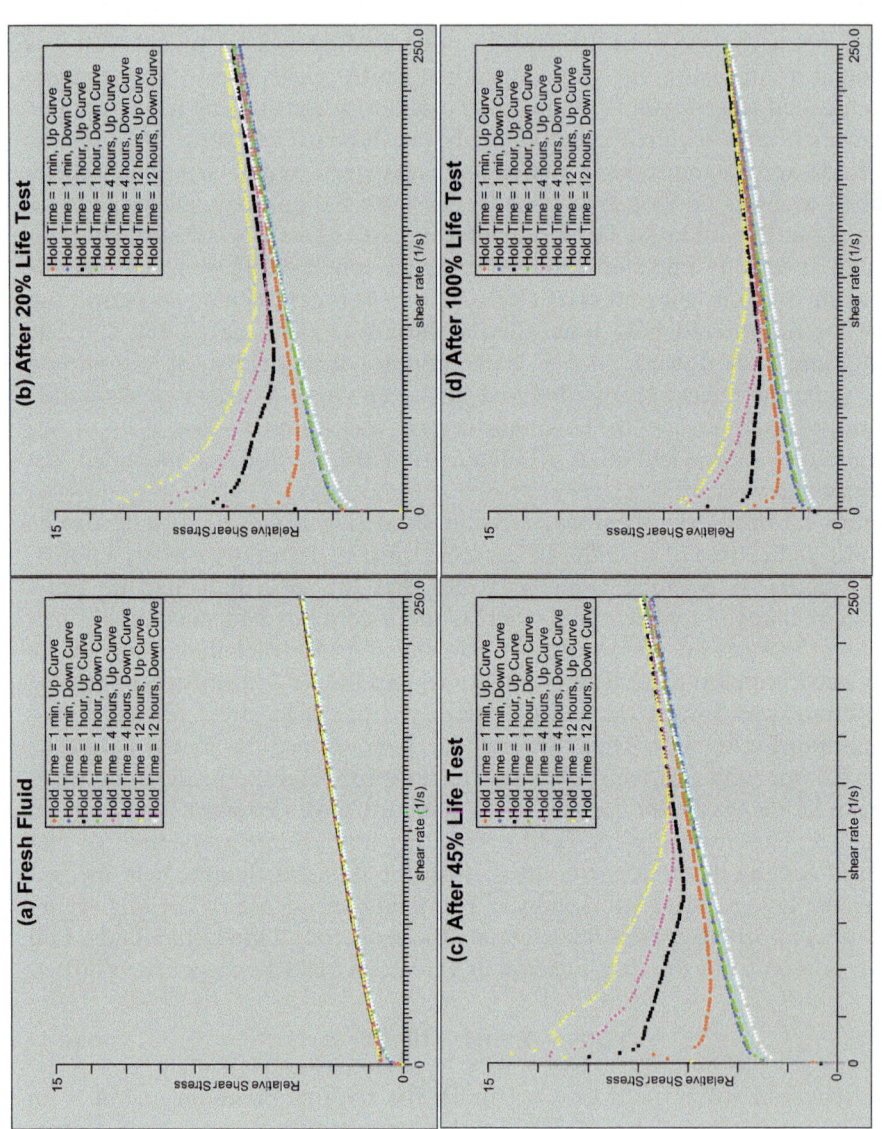

Figure 4.3   Relative shear stress *vs.* shear rate after various static hold times for (a) fresh and (b)–(d) durability-tested MR fluids.

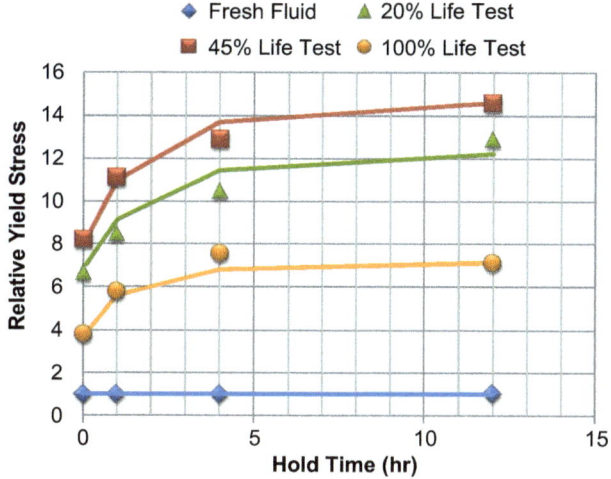

Figure 4.4   Static yield stress versus hold time for fresh and durability-tested MR fluids. Points are measured data, and lines are fits calculated from eqn (4.1) using the parameters listed in Table 4.2.

Table 4.2   Best-fit parameters for relative static yield stress *versus* hold time data shown in Figure 4.4.

Test Duration (% of Life Test)	YS(0)	τ	A	YS(∞)
0%	1	na	0	1
20%	7.0	0.42	0.75	13
45%	8.2	0.40	0.84	15
100%	3.8	0.30	0.55	7.4

Also shown in Table 4.2 are the ultimate yield stress values, YS(∞), calculated from eqn (4.1) at infinite time, where the term (τ/t) approaches zero and eqn (4.1) reduces to

$$YS(\infty) = YS(0) + 10^A \tag{4.2}$$

These values are plotted *versus* the percentage of life test complete in Figure 4.5(a), and in Figure 4.5(b) the results of chemical analyses for certain proprietary additives in the LORD MR fluid samples are shown. The large increase in static yield stress correlates with the disappearance of Additive 2 up to 45% life test, and then the static yield stress decreases more gradually with life, similarly to the decrease in concentration of Additive 1.

4.2.2.3   Summary and Conclusions

LORD MR fluids routinely survive multiple lifetimes in stringent automotive durability tests, allowing the commercial success of MR dampers. Chemical changes are observed that are a natural consequence of the activity of the various anti-wear and friction-reducing additives used to formulate these

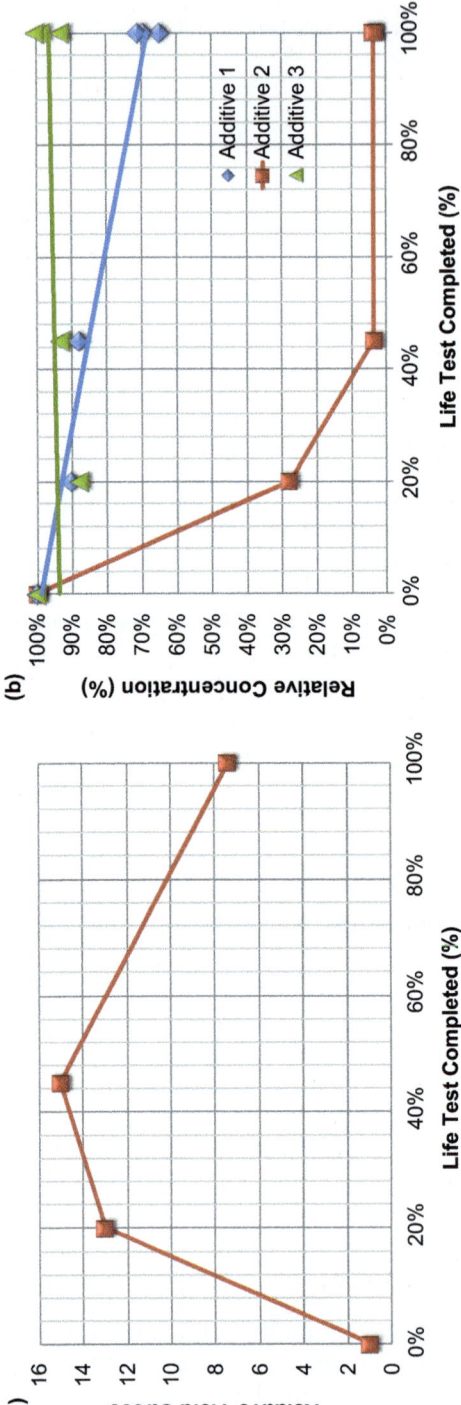

Figure 4.5 (a) Ultimate MR fluid yield stress and (b) relative change in proprietary additive concentrations in LORD® MR fluid *versus* percentage of life test completed.

durable fluids. The chemical changes do not adversely affect the dynamic fluid rheological properties, so the performance characteristics of the dampers are not degraded. However, changes in additive concentrations are correlated with substantial increases in the static yield stress of the used MR fluids, resulting in a reversible gelation of the MR fluid that will markedly enhance the settling stability of the fluid.

4.2.3 MR Fluids from High-Energy Damper Tests

LORD Corporation has been testing MR fluids and dampers in high energy applications for military and other off-road vehicles for many years,[7,8,11,13] most recently on the armor-reinforced HMMWV (High Mobility Multipurpose Wheeled Vehicle) and the LMTV (Light Medium Tactical Vehicle).[14,15] MR dampers have consistently provided substantial improvements in the comfort and safety of driver and crew over punishing terrain, and vehicle handling stability and allowable maximum speed have both been increased relative to vehicles equipped with conventional dampers. The vehicles require large dampers with piston speeds up to several meters per second and resulting shear rates of the MR fluids on the order of 10^4 s^{-1}.

Similarly strenuous laboratory tests have been developed at LORD Corporation for these types of dampers. Such tests assess not only the MR fluid durability but also the durability of seals, rods, piston materials, and other components of the device. While the focus of the current chapter is on fluid durability, the integrity of seals and other components can have a major influence on the lifetime performance of the MR fluid. In the tests described below, a number of LORD dampers (model numbers RD-8016-13 and RD-8016-16), intended for use in the armored HMMWV, were tested until damper failure, and the MR fluids were then removed and analyzed. The proprietary test protocol was a sinusoidal test with constant stroke amplitude, in which both high- and low-energy test segments were repeated continuously for the required number of cycles. The damper skin temperature was maintained at 120–130 °C with a temperature-controlled fan. It is likely that internal damper temperatures experienced by the fluid were significantly higher for some time period.

4.2.3.1 Physical Properties of Heavily-Used MR Fluids

Density and dynamic viscosity of used MR fluids, as well as the damper failure mode, are reported in Table 4.3 for several tested fluids. *Seal failure* refers to leakage of MR fluid (with iron particles) past device seals. A small amount of clear fluid leakage (containing no iron particles) is not unusual and does not necessarily lead to failure, although excessive clear fluid loss will result in a higher volume percentage of particles in the fluid and may lead to premature fluid wear. *High force* refers to failure due to excessive off-state force that is usually associated with IUT of the MR fluid, although there are other device-related issues (*e.g.*, high friction) that can also cause a higher off-state force. It is noteworthy that all dampers went substantially beyond the required test limit (100% test) before failure was observed.

As shown in Table 4.3, dampers that failed due to high force had increases in relative viscosity of 56% and 587%. The extremely high viscosity fluid, used fluid 2, also had a density 41% higher than the fresh fluid, equivalent to an approximately 60% increase in the volume fraction of iron particles in the used fluid. An increase in density cannot entirely explain the 56% increase in viscosity for used fluid 4, because used fluid 1 had the same 18% increase in density but almost the same viscosity as the fresh fluid. The substantially longer run time for used fluid 4 likely caused other changes in the fluid that resulted in higher viscosity. It was not possible to test reliably the density and viscosity of used fluid 3 due to the nature of the seal failure, although sufficient fluid was obtained to measure particle size distribution and chemical changes in this fluid.

The relative particle size distributions of the fresh and used fluids are shown in Figure 4.6. All used fluids show some decrease in the maximum particle size

Table 4.3   Summary of physical properties of heavily-used MR fluids.

Sample	Life Test Complete (%)	Failure Mode	Relative Density	Relative Viscosity
Control Fluid	0	–	1	1
Used Fluid 1	170%	Seal Failure	1.18	0.96
Used Fluid 2	203%	High Force	1.41	5.87
Used Fluid 3	256%	Seal Failure	na	na
Used Fluid 4	288%	High Force	1.18	1.56

na = data not available.

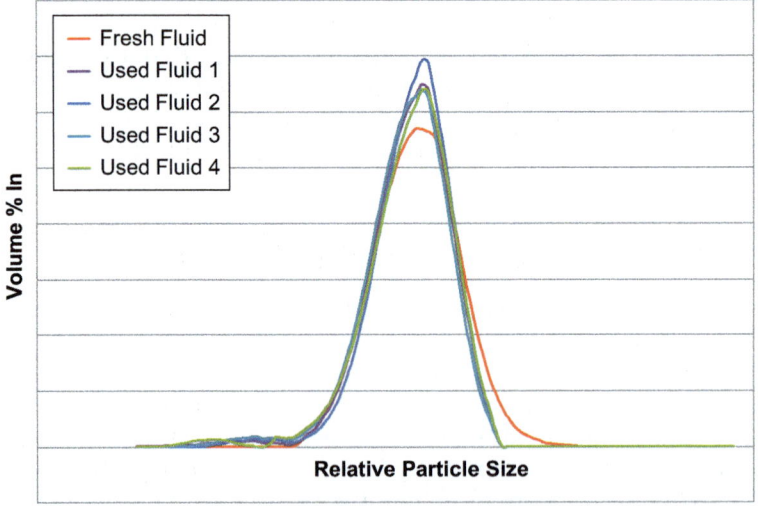

Figure 4.6   Relative particle size distributions (laser light scattering data) of heavily-used LORD® MR fluids.

as compared to an unused MR fluid. This is likely due to more complete dispersion of soft agglomerates typically found in fresh, unused MR fluid. All used fluid distributions are quite similar, although there may be subtle differences between the different samples in their fine particle distributions. Carlson[1] has described the effect of a small amount of fine iron oxide particles on fluid viscosity and attributed IUT in early MR fluid formulations to the formation of such fine materials during use, so it is conceivable that a similar effect is occurring in the current heavily used systems once they are driven to failure. If this is the cause for the differences between the fluids, the effect is quite subtle because two of the four used fluids did not cause damper failure by IUT.

4.2.3.2   Chemical Changes in Heavily-Used MR Fluids

Chemical changes in the proprietary additives of the LORD MR fluids were also assessed. Aside from increase in density and creation of fine particles, polymerization of the base oils in MR fluids during use has also be cited as a potential cause of IUT. Figure 4.7 shows the molecular weight (MW) distribution curves of fresh and used MR fluids as measured by gel permeation chromatography (GPC). Included in the data is an experimental MR fluid made with slightly higher MW base oil for comparison purposes. The main base oil peaks for all samples are essentially identical, indicating no substantial change in the base oil molecular weight distribution. The small vertical arrow to

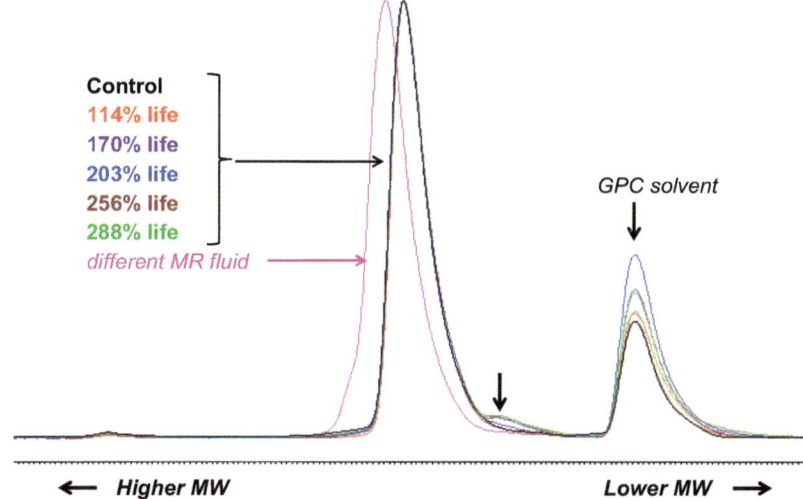

Figure 4.7   Changes in relative molecular weight of the base oils from heavily-used LORD® MR fluids (number average molecular weight, M_n, using gel permeation chromatography *versus* poly(styrene) standards). The curve labelled "different MR fluid" is an experimental fluid made with slightly higher molecular weight base oil for comparison purposes. The vertical arrow indicates a low-molecular weight breakdown component of Additive 1.

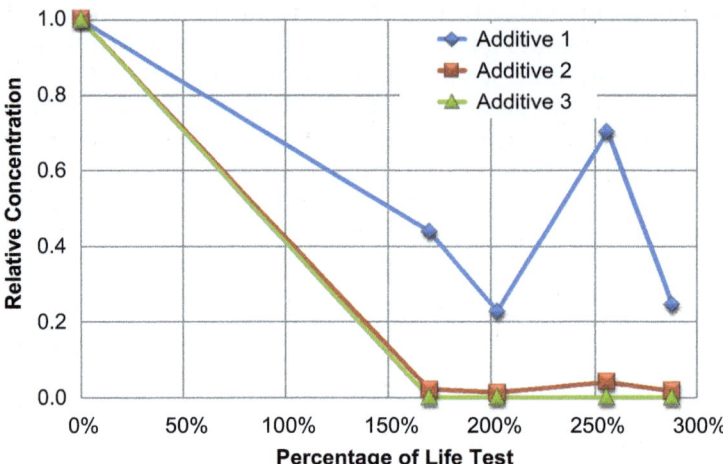

Figure 4.8   Changes in the relative concentrations of proprietary additives in heavily-used LORD® MR fluids.

the right of the main peak indicates a small amount of a lower molecular weight material that was detected in samples that had undergone 200% or more of life test. This material has been identified as a breakdown product of Additive 1.

The relative concentrations of proprietary additives for the various samples are shown in Figure 4.8. Since all fluids were run to failure, it is not surprising that additive 2 was nearly consumed considering that it was depleted relatively early during the less demanding automotive durability test; see Figure 4.5(b). Additive 3, which was barely depleted during the automotive test, was un-detectable in any of the heavily used fluid samples. In two of the four samples, fluid IUT was not the cause of failure, and in one of the two IUT failures (used fluid sample 2), loss of base oil by leakage was a primary contributor as shown by the high density of this sample. Therefore, it cannot be concluded defini-tively that loss of additive 3 is the cause of fluid failure. Additive 1 decreased in the highly used samples by 60–80%, with the exception of the fluid from 256% life test, which had a 30% decrease (similar to that of a 100% automotive test fluid). Additionally, the concentration of Additive 2 was slightly higher for this sample than for the other used samples. The fluid sample from the 256% test therefore appears to have experienced less wear than would typically be ex-pected. This emphasizes the importance of the interplay between the fluid and the device design in determining the overall reliability of an MR device.

4.2.3.3   Summary and Conclusions

LORD MR fluids have demonstrated long-term durability in severe damper applications involving high temperatures and shear rates on the order of 10^4 s^{-1}. MR dampers that are run to end of life in such tests may fail due to fluid IUT or mechanical failure, and it is often difficult to assign a single root cause due to the complex interplay between fluid and device performance. No evidence of polymerization of the base oils was observed even under the extreme test conditions. Iron particle wear under these high energy conditions was surprisingly low and can likely be attributed to a combination of optimal device design and the action of proprietary additives in formulations that were developed over many years of dedicated research and experience. Decreasing concentrations of these additives are inevitable because their mode of action is to react with and stabilize various fluid and device components, so their disappearance in chemical tests does not necessarily imply that an MR fluid is nearing its breakdown point. Overall, the LORD MR fluid displayed remarkable stability under extreme use conditions.

4.2.4 Oxidation Stability of MR Fluids

4.2.4.1 Modification of Commercial LORD® MR Fluids

In a recent study, Lee *et al.* reported on tribological studies, including friction, wear, and oxidation resistance, using a commercial MR fluid, LORD MRF-132DG, modified with the additives zinc-dithiophosphate (Zn-DTP), molybdenum-dithiocarbamate (Mo-DTC), and an aromatic amine antioxidant to improve tribological performance.[16] The additives were not tested individually but as an additive package containing 0.5% Zn-DTP, 0.5% Mo-DTC, and 1% amine antioxidant (percentages by weight based on total formulation). Both modified and unmodified MRF-132DG were tested for off-state tribological performance by the SRV test, a high-frequency linear oscillation test (ASTM D5707), and by the four-ball wear test (ASTM D2266). Oxidation stability of the fluids was assessed using a pressure differential scanning calorimeter (PDSC) to measure the oxidation induction time (OIT) according to ASTM D5483. On-state tribological properties were measured using a pin-on-disk tester that was modified to allow application of magnetic field.

In off-state tests, the coefficient of friction for the commercial MRF-132DG, as measured by SRV, was significantly improved by the addition of the additive package, increasing the amount of time the friction remained low and stable by almost sevenfold (from about 300 s for MRF-132DG to about 2000 s for the modified MR fluid). The four-ball wear test showed a smaller scar for the modified fluid as compared to the commercial fluid (0.54 mm *versus* 1.03 mm diameter, respectively). In the on-state pin-on-disk tests, the modified MR fluid had a slightly lower coefficient of friction at high sliding speeds but little difference at lower speeds, and the material removal rate from the pin was about the same for both fluids. The worn surfaces after the pin-on-disk test were generally smoother for MRF-132DG as compared to the modified fluid. Overall, there was much less difference in the fluids' tribological performance in the on-state as compared to the off-state.

Oxidation stabilities of the commercial and modified fluids were significantly different in the OIT test. The measured OIT of the commercial fluid was about 2.5 minutes, whereas the modified fluid did not undergo oxidation during the entire two hour duration of the test. Thus, the additive package significantly enhanced the oxidation stability of the commercial fluid, although which additive(s) in the package were responsible for the improvement could not be determined from this study.

4.2.4.2 Designed Experiments to Improve Oxidation Resistance of MR Fluids

Several patents from LORD Corporation[17] describe the use of various additives to improve the durability of LORD MR fluids, including antioxidants, anti-wear agents, and friction reducers similar to those used in the lubricant industry. We previously described a set of designed experiments that were conducted to determine the impact of some of these lubricant additives on the oxidative stability of MR fluids.[18] Oxidative stability was assessed by measuring the OIT using PDSC according to a procedure modified from ASTM D6186-98, in which a sample is heated under a pressurized oxygen atmosphere and the time required for oxidation to occur is recorded. The ASTM method was modified by increasing the oxygen pressure to 525 ± 25 psi and increasing the peak hold temperature to $180\,°C$.

A variety of antioxidants, anti-friction agents, and metal passivators, summarized in Table 4.4, were screened using design-of-experiments (DOE) methodology. Samples were made by first preparing a stock MR fluid from iron powder mixed with proprietary base oils and suspension aids, then adding the desired additives in the required amounts. Metal passivators were found to have little effect on OIT, indicating that the iron particles with their high surface area are not catalyzing oxidation of the base oils to an appreciable extent. Similarly, anti-friction agents had no significant effect on OIT.

Table 4.4 Additive types used in the design-of-experiments study to improve MR fluid oxidation stability.

Commercial Name	Additive Type	Function
Vanlube PCX	2,6-Di-*t*-butyl-*p*-cresol	Primary antioxidant (radical scavenger)
Vanlube NA, Vanlube 961	Alkylated diphenylamine	Primary antioxidant (radical scavenger)
Vanlube 7723	Methylene bis(dibutyl-dithiocarbamate)	Secondary antioxidant (peroxide decomposer), extreme pressure agent
Vanlube 7611M	Ashless phosphorodithioate	Antiwear agent
Vanlube 887	Tolutriazole compound in oil	Antioxidant, synergist with phenols and/or dithiocarbamates
Vanlube 601	Heterocyclic sulfur-nitrogen compound	Non-ferrous metal passivator, corrosion inhibitor
Vanlube 871	Dimercapto-thiadiazole	Film-forming metal passivator, antioxidant/antiwear
Cuvan 303	Amino-alkylated tolutriazole	Non-ferrous metal passivator, corrosion inhibitor

All additives provided courtesy of R.T. Vanderbilt.

Using a combination of particular primary and secondary antioxidants, alkylated diphenylamine and methylene bis(dibutyldithiocarbamate), respectively, it was possible to prepare MR fluids with an OIT of almost three hours, as compared to an OIT of seven minutes for the stock MR fluid containing no additives. The approach of using both primary and secondary antioxidants to improve oil stability is well known in the lubrication industry,[19] so it is likely that oxidation of MR fluids under the conditions of the OIT test proceeds by a similar mechanism to conventional hydrocarbon oxidation. It was also found that one type of diphenylamine was more effective than the other, and this was attributed to a solubility effect. No durability test results were reported.

4.2.4.3 Summary and Conclusions

Experiments have shown that the oxidation resistance of the commercial LORD MR fluid MRF-132DG can be significantly improved by the addition of appropriate additives, and designed experiments have shown that a combination of primary and secondary antioxidants can substantially improve the oxidation stability of MR fluids. For applications in which base oil oxidation is a significant contributor to MR fluid failure, addition of antioxidants would be expected to improve MR fluid lifetime.

It is noteworthy that no polymerization of base oils was observed in the LORD MR fluids tested to failure in the high energy damper tests. Because the additives used in these fluids are proprietary, it is not possible to determine whether these fluids survive because the additives were effective in preventing such oxidation or because base oil oxidation is not an important failure mechanism in this application.

4.3 Low-Temperature Requirements

4.3.1 General Fluid and Device Considerations

Low-temperature fluid and device considerations were considered in some detail previously by Black and Carlson.[20] They make the following points about low-temperature fluid and device performance:

- The magnetically-induced (on-state) force is nearly independent of temperature. Only the off-state force is dependent on temperature and varies with the off-state viscosity.
- The off-state viscosity of MR fluids is largely dependent on the type of carrier oil used. Silicone oils have the least dependence of viscosity on temperature, but elsewhere Black and Carlson noted that these oils have poor stability in high-temperature and high-shear environments.
- As long as the MR fluid remains above its freezing point or pour point, any movement in the device will lead to viscous heating of the fluid, and the off-state viscosity will decrease as the temperature increases.

   – Design engineers should choose designs and geometries that minimize the off-state viscous contribution to the device force in order to minimize the effects of low-temperature viscosity increase.

   By successfully applying the above principles, LORD MR fluid formulators and BWI MagneRide™ damper design engineers have together created a commercial MR damper that has met the low-temperature requirements of the automotive industry for a decade.[11] Furthermore, BWI has recently introduced a next-generation MagneRide damper for the Range Rover Evoque, and the dampers were required to meet "the full range of performance down to − 40 °C."[12] The work described below documents for the first time a direct comparison of low-temperature performance between dampers containing either conventional hydraulic oil or LORD MR fluid.

4.3.2   Low-Temperature Soak Test of MR Dampers

Two dampers for the HMMWV vehicle, one a conventional damper and the other a LORD-designed MR prototype damper, were cold-soaked to −40 °C. Both dampers were removed from the low temperature chamber and their off-state forces and skin temperatures were measured over a two-hour period as the dampers were constantly operated with a stroke of ± 25 mm at a frequency of 0.2 Hz. Figure 4.9 shows the damper force and skin temperature measurements *versus* time. The MR damper force is symmetric about the zero force axis, whereas the hydraulic damper force is higher in compression than in extension.

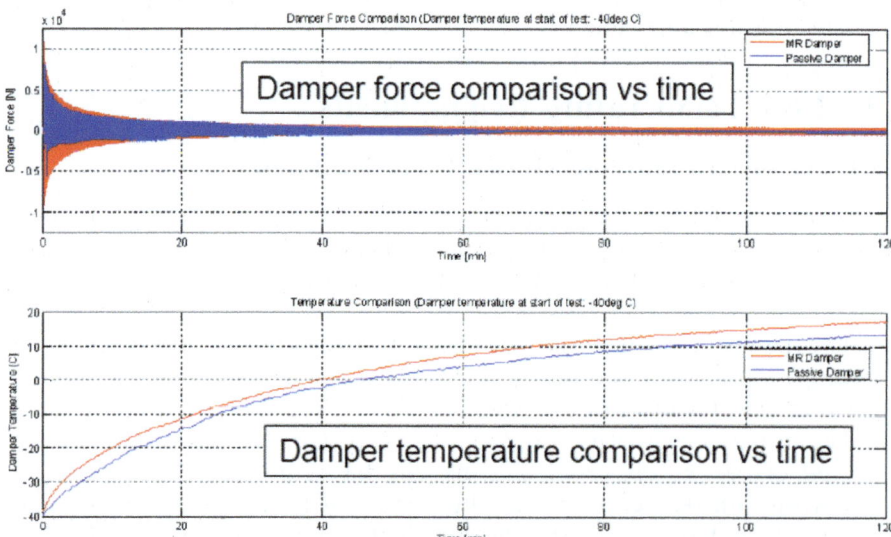

Figure 4.9   Low-temperature test results for hydraulic and MR dampers. The upper graph shows force and the lower curve shows skin temperature over the two-hour test period.

During the first 10–15 minutes of test, the hydraulic damper force is a little lower than the MR damper force in compression, with a larger difference in extension due to the asymmetric force of the hydraulic damper. During this time the damper temperatures increase from –40 °C to about –20 °C for the MR damper and about –25 °C for the hydraulic damper; the approximately 5 °C temperature difference is maintained for the remainder of the test. The damper forces are almost the same from about 15 minutes until about 55 minutes (hydraulic and MR damper temperatures about 2 °C and 7 °C, respectively), after which time the hydraulic damper force continues to decrease until it stabilizes at about 65 minutes, or about 5 °C. The MR damper force stabilized at about 45 minutes, or about 2 °C. The final damper force for the hydraulic damper is a little lower than that for the MR damper, but in general the force profiles are similar. The hydraulic damper force change *versus* time (temperature) is somewhat more erratic than that for the MR damper, which gives a smoother profile. The stable and predictable damper force *versus* temperature is an aid to damper control engineers who can take the temperature dependence of the force into account in the damper control algorithms.

Figure 4.10 shows the damper forces relative to their respective 15 °C values. The relative force for the MR damper increased by about 30 times, much less than the 100-fold increase in force at low temperature for the hydraulic damper. Hydraulic dampers have fixed orifice sizes tailored to the desired force curve for that particular damper application, so their force *versus* temperature performance is constrained by the device design and fluid property. Because the on-state force of the MR damper is adjustable, damper designs have been

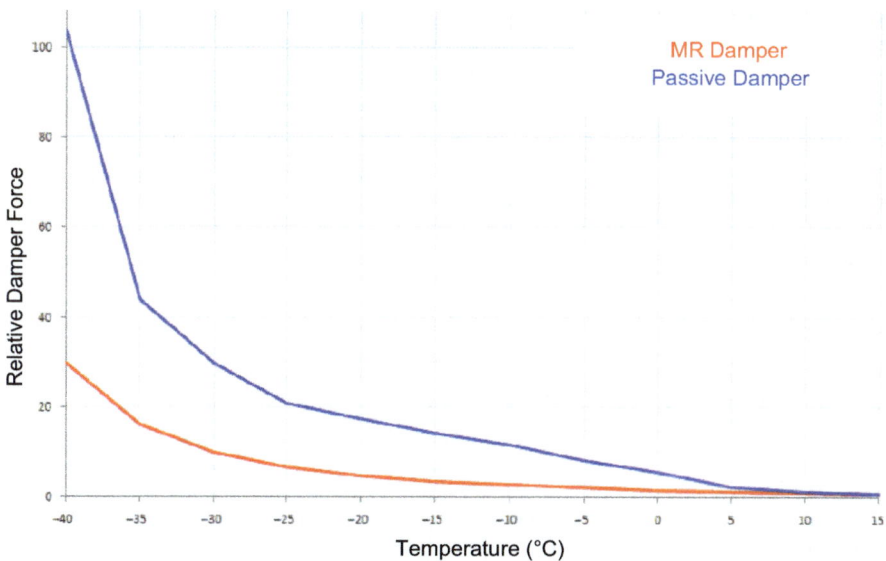

Figure 4.10   Relative damper force (relative to 15 °C) for standard and MR dampers as a function of temperature during the low-temperature test.

developed to minimize the off-state force, and the advantage is apparent in the substantially lower relative force at low temperatures. Thus, while the low-temperature viscosity may be higher for MR fluids than for hydraulic fluids, an optimized damper/fluid combination allows for a much narrower variation in force between low and high temperatures.

4.3.3   Trade-offs in Developing Low-Temperature MR Fluids

Fluid formulation changes that are made to decrease the low-temperature viscosity of MR fluids will inevitably involve compromises with other fluid properties. The main influences on LORD MR fluid viscosity are the base oil (as noted previously), the iron concentration, and the concentration of proprietary thickeners used to enable good settling performance. Table 4.5 summarizes the results of a designed experiment in which the iron content was held constant and the type of base oil and the relative amount of thickener were varied. Fluid dynamic viscosity, overnight settling, and low-temperature viscosity properties were measured and reported relative to a fluid with standard base oil and thickener. The low-temperature viscosity was measured using a rheometer with parallel plate geometry. The sample was loaded into the gap and allowed to temperature equilibrate at $-30\,^{\circ}$C with no shear stress applied, and then the shear rate was immediately increased to 200 s^{-1} and the viscosity was monitored over the first 30 seconds of test using the high-speed measurement mode of the rheometer. Initial viscosity could be measured within the first 0.5 second of test using this method. This test is referred to as a *breakaway* test.

Using a lower-viscosity base oil, or decreasing the thickener content by more than 50%, had the same effect on the dynamic viscosity at 40 °C, but the change in base oil had a more substantial effect than decreasing the thickener in the low temperature breakaway test. There appears to be almost a synergistic effect of the base oil/thickener combination at low temperature. With the standard base oil, decreasing thickener content improved the low-temperature viscosity by only about 25–30%, but with the lower-viscosity base oil the low-temperature viscosity was decreased by about 50%. The combined effects of lower thickener

Table 4.5 Summary of designed experiment to reduce low-temperature viscosity.^{a}

Base Oil	Boiling Point	Volatility at 150 °C	Thickener Amount	Viscosity at 40 °C	Settling Clear Layer	Initial Viscosity at −30 °C	15-sec Viscosity at −30 °C
Low-Viscosity Base Oil	0.85	2.78	0.48	0.52	1.90	0.18	0.28
			1	0.69	1.25	0.33	0.58
Standard Base Oil	1	1	0.48	0.69	1.60	0.67	0.75
			1	1	1	1	1

^{a}All values are relative to the standard fluid with standard thickener. Viscosity at 40 °C is the dynamic viscosity measured as described in section 4.2.2.1. Settling clear layer is the relative amount of clear layer above the iron sediment after 24 hours.

and lower viscosity base oil resulted in a 70–80% reduction in low-temperature breakaway viscosity and a reduction of about 50% in the 40 °C dynamic viscosity.

The compromises in MR fluid performance were found in the fluid volatility and settling performance. The lower viscosity base oil was about 2.8 times more volatile than the standard oil, which could become an issue for applications in which good performance is required at both low and high temperatures. Decreasing the thickener content by 52% resulted in a 60–65% increase in the settling clear layer, regardless of the base oil type; the lower-viscosity base oil increased the settling clear layer by about 25%. The combination of lower-viscosity oil and low thickener resulted in a settling clear layer nearly double that of a standard fluid. While the amount of settling is usually not as important as the dispersibility of the fluid, a high and/or fast settling will create additional problems in fluid and device production processes, and decreasing the thickener below a minimum value will eventually result in poor dispersibility.

4.4   Summary and Conclusions

This chapter has shown that LORD MR fluids have excellent performance in high-energy, high-temperature applications and that MR fluid devices have low-temperature performance characteristics better than those of a conventional hydraulic damper designed for the same application. MR dampers have met the performance needs of the automotive industry for over a decade, and they continue to perform well in even more demanding off-road and military applications. MR fluid often exceeds the performance of seals and other elastomeric components in the damper. Improvements in oxidation resistance and tribological performance have been demonstrated with various additives. However, it is important to note that the relevance of standard tribological tests to MR damper durability has not been demonstrated, and modification of existing commercial MR fluid formulations may upset the balance of additives already present in the fluid and thereby degrade actual performance in a damper.

Superior low-temperature performance of MR dampers has been demonstrated in this work and by successful implementation in numerous commercial automotive applications. Both fluid properties and device design are responsible for the lower variation in force with temperature as compared to conventional hydraulic dampers. Further improvements in low-temperature MR fluid performance are possible, but such improvement involve compromises with other properties, particularly high-temperature performance and dispersibility, and should therefore be undertaken with caution.

Acknowledgements

The author gratefully acknowledges BWI Group for commercial MR damper information, and the following personnel at LORD Corporation: Joey Faulk

(Chemical Research & Development), Brian Perry (Analytical Services), Steve Hildebrand, Rick Stys, and other members of the MR Suspensions team.

References

1. J. D. Carlson, *J. Intell. Mater. Syst. Struct.*, 2002, **13**, 431.
2. J. D. Carlson, *Int. J. Vehicle Des.*, 2003, **33**, 207.
3. B. Munoz, US Patent 5 683 615, 1997; B. Munoz, A. J. Margida and T. J. Karol, US Patent 5 705 805, 1998; T. J. Karol, B. C. Munoz and A. J. Margida, US Patent 5 906 767, 1999; K. A. Kintz and T. L. Forehand, US Patent 6 395 193, 2002.
4. J. D. Carlson and J. L. Sproston, *Proceedings of the 7th International Conference on New Actuators*, ed. H. Borgmann, Messe Bremen GmbH, 2000, 126–130.
5. J. D. Carlson, *Proceedings of the 3rd World Conference on Structural Control*, ed. F. Casciati, John Wiley, Chichester, 2003, 227–236.
6. R. Marjoram, K. St. Clair, B. McMahon, A. Goelz and A. Achen, *Proceedings of the 11th International Conference on New Actuators*, ed. H. Borgman, HVG Hanseatische Veranstaltungs-GmbH, Bremen, 2008, 503–506.
7. J. D. Carlson and F. D. Goncalves, *Proceedings of the 11th International Conference on New Actuators*, ed. H. Borgman, HVG Hanseatische Veranstaltungs-GmbH, Bremen, 2008, 477–480.
8. D. E. Barber and P. Sheng, *Actuator 10: International Conference and Exhibition on New Actuators and Drive Systems, Conference Proceedings*, ed. H. Borgmann, WFB GmbH- Messe Bremen, Bremen, 2010, 533–537.
9. V. R. Iyengar and R. T. Foister, US Patent 6 599 439, 2003.
10. J. D. Carlson, K. A. St. Clair, M. J. Chrzan and D. R. Prindle, US Patent 5 878 851, 1999.
11. D. E. Barber and S.-B. Choi, *Actuator 12: International Conference and Exhibition on New Actuators and Drive Systems, Conference Proceedings*, ed. H. Borgmann, WFB GmbH-Messe Bremen, Bremen, 2012, 76–80.
12. O. Raynauld and S. Fath, *ATZ*, 2012, **114**, 16.
13. D. E. Barber and J. L. Faulk, *ERMR 2012: 13th International Conference on Electrorheological Fluids and Magnetorheological Suspensions*, University of Gazi, Ankara, Turkey, July 3, 2012.
14. D. Ivers, S. Hildebrand and O. Molins, RTO-MP-AVT-170, *Active Suspension Technologies for Military Vehicles and Platforms*, Sophia, Bulgaria, May 2011.
15. D. Ivers and D. LeRoy, *ASME 2011 Conference on Smart Materials, Adaptive Structures and Intelligent Systems*, SMASIS2011-5058, September 18–21, 2011, Phoenix, Arizona, USA.
16. C. H. Lee, D. W. Lee, J. Y. Choi, S. B. Choi, W. O. Cho and H. C. Yun, *J. Tribol.*, 2011, **133**, 031801-1.
17. B. Munoz, US Patent 5 683 615, 1997; B. Munoz, A. J. Margida and T. J. Karol, US Patent 5 705 805, 1998; T. J. Karol, B. C. Munoz and

A. J. Margida, US Patent 5 906 767, 1999; K. A. Kintz and T. L. Forehand, US Patent 6 395 193, 2002; K. A. Kintz and T. L. Forehand, US Patent 7 087 184, 2006.

18. D. E. Barber, *Electrorheological Fluids and Magnetorheological Suspensions: Proceedings of the 12th International Conference*, ed. R. Tao, World Scientific Publishing, New Jersey, 2011, 318–324.

19. J. Braun, in *Lubricants and Lubrication*, ed. T. Mang and W. Dresel, Wiley-VCH GmbH, Weinheim, 2007, 2nd edn, Chapter 6.

20. T. Black and J. D. Carlson, *Synthetics, Mineral Oils, and Bio-Based Lubricants*, ed. L. R. Rudnik, CRC Press, Taylor and Francis Group, Boca Raton, FL, 2006, 565–583.

CHAPTER 5

Surface Effect on Flow of Magnetorheological Fluids: Featuring Modified Mason Number

BARKAN KAVLICOGLU,[a] FARAMARZ GORDANINEJAD*[b] AND XIAOJIE WANG[b]

[a] Advanced Materials and Devices, Inc., 4750 Longley Lane, Suite 104, Reno, Nevada, 89502, USA; [b] Department of Mechanical Engineering, University of Nevada, Reno, Reno, Nevada 89557, USA
*Email: faramarz@unr.edu

5.1 Introduction

For engineering analysis and design of magnetorheological (MR) fluid devices, it is customary to assume a constitutive model for MR fluids, such as Bingham Plastic or Herschel–Bulkley models.[1,2] Both of these models use shear yield stress, which is obtained by an extrapolation of shear stress–shear strain rate experimental data (flow curve) to zero shear strain rate value for a specific MR fluid. The definition of shear yield stress of MR fluids has generated discussions between engineers and scientists, because if the material and geometric characteristics of the wall surface change, different material properties can be obtained.[3–14] The dependency of MR fluid behavior on the surface properties makes it impossible to define a unique shear yield stress. In such a case

RSC Smart Materials No. 6
Magnetorheology: Advances and Applications
Edited by Norman Wereley
© The Royal Society of Chemistry 2014
Published by the Royal Society of Chemistry, www.rsc.org

engineers and scientists need to find a method to understand and analyze the MR fluid flow behavior in devices.

This work presents a new methodology to define MR fluid properties and a new approach for flow analysis of MR fluids through channels with various surface topologies. The goal is to introduce a phenomenological relation for the friction factor of a MR fluid in terms of applied magnetic field, surface morphology and flow rate without using the concept of yield stress. In order to achieve this, the first objective is to conduct an experimental study, where a piston driven flow type rheometer with a rectangular channel is built to examine the surface effects on the MR fluid channel flow. The second objective is to introduce methods to obtain the flow behavior as a function of surface morphology. Based on these methods, non-dimensional friction factor equations are developed for various surface morphologies, without using the concept of yield stress. This attempt is analogous to the Moody diagram for flow analysis of Newtonian fluids. It is well known that in laminar flow the friction factor of a Newtonian fluid is only a function of the Reynolds number which is defined in terms of the fluid properties of density and viscosity, a velocity and a characteristic length or characteristic dimension. However, in turbulent flow, the friction factor of fluids not only depends on the Reynolds number, but also on the surface roughness. In this study, it will be demonstrated that, due to the magnetic field, surface effects also play an important role on the flow behavior of MR fluids, even in the laminar flow region. Therefore, rather than utilizing the Reynolds number another dimensionless parameter, called the Mason number, is utilized in modeling and characterizing MR fluid flow. The Mason number is defined as a ratio of viscous force to magnetic force that can represent the flow conditions of MR fluids subjected to various magnetic fields.

5.2   Literature Review

Bingham Plastic or Herschel–Bulkley models[1,2] are commonly used to model the behavior of MR fluids. The Bingham plastic model for MR fluids is defined, as follows:[1]

$$\tau(\dot{\gamma}, B) = \tau_y(B) + \mu_p \dot{\gamma} \quad \text{for} \quad \tau > \tau_y$$

$$\dot{\gamma} = 0 \quad \text{for} \quad \tau \le \tau_y \tag{5.1}$$

where, $\tau(\dot{\gamma}, B)$ is the shear stress, $\tau_y(B)$ is the dynamic yield stress induced by the magnetic flux density B, $\dot{\gamma}$ is the shear rate, and μ_p is the plastic viscosity independent of magnetic field strength. In order to incorporate the shear-thinning behavior of MR fluids, the Herschel–Bulkley model can be utilized, as follows:

$$\begin{cases} \tau(\dot{\gamma}, B) = \tau_y(B) + k|\dot{\gamma}|^{n-1}\dot{\gamma} & |\tau| > \tau_y \\ \dot{\gamma} = 0 & |\tau| \le \tau_y \end{cases} \tag{5.2}$$

where, k and n are fluid index parameters. The Herschel–Bulkley model assumes a nonlinear post-yield behavior for non-Newtonian fluids.[2]

At the macroscopic level, the shear yield stress is experimentally obtained by using shear rheometers to measure steady flow over a steel surface. However, if the material and geometric characteristics of the wall surface change, different values of shear yield stresses can be obtained.[3–5] Therefore, if the shear yield stress is surface dependent, it would be difficult to define the shear yield stress solely as a material property. Consequently, one might ask if it is possible to predict the MR fluid pressure drop in various design environments, where the surface type of the device might change. Previous studies have focused on the effects of surfaces on the behavior of MR fluids and how the yield stress is affected by surface modifications.[3–5]

At the microscopic level, a shear yield stress relation of MR fluids was developed by Ginder *et al.*,[6] which relates the shear yield stress to the applied magnetic field flux density by examining the interparticle forces. They developed an analytical model for the static shear yield stress of MR fluids under various field strengths and showed that the shear yield stress is directly proportional to $\varphi \mu_0 M_s^2$, where ϕ is the volume fraction of particles, μ_0 is the vacuum permeability and M_s is the magnetic saturation. Models for the shear yield stress with similar definitions have also been reported by other researchers.[5,15–18] A review by Wang and Gordaniejad[1] provides detailed information about these models.

None of the current constitutive models for MR fluids consider the effect of interaction of the surface with the particles. Studies performed on surface effects include macroscopic analysis of MR fluid performance by altering the surface properties that the MR fluid is flowing through.[12,13] To examine different surface effects on the static yield stress of different magnetic colloidal suspensions, the plate roughness and plate material of a parallel-plate rheometer are varied.[3,4] Two paramagnetic plate surfaces, stainless steel and glass and one ferromagnetic plate surface, iron, are used in these experiments. It is shown that due to the smooth surface of the glass plates, there is nearly zero shear yield stress at the wall. The stainless steel plates resulted in higher shear yield stress, due to the roughness of their surface. The roughness of the stainless steel and iron plates are identical; however, due to the wall interaction in iron plates, the highest shear yield stress is observed in measurements using iron plates.

In efforts to increase the performance of electrorheological (ER) fluids (the electrical analogs of MR fluids), the effects of electrode pattern on channel flow of ER fluids have been analyzed by Abu-Jdayil *et al.*,[19,20] Otsubo[21] and Lee *et al.*[22]

Abu-Jdayil *et al.*[19,20] presented an experimental study on ER fluid flow through oblique and corrugated electrodes *via* laser Doppler anemometry. For different types of electrodes, a relative increase in the ER effect is observed compared to smooth electrodes under an alternating current field. The greatest increase is observed in symmetric corrugated electrodes. Another study on electrode pattern effects is performed by Otsubo.[21] The striped electrodes provided a non-uniform electric field. They found that the non-uniform electric field results in local high particle concentration, which contributes to an improved ER fluid shear yield stress performance. It was stated that the particle concentration in striped electrode sections were about 50% higher than the particle concentration in flat parts.

Lee *et al.*[22] performed an experimental study on the grooved electrode surfaces encapsulating ER fluids within a sandwich beam. It is shown that the shear modulus of the ER fluid could be increased by 30% by modifying the electrode surfaces.

For surface effects on MR fluid behavior, Gorodkin *et al.*[23] conducted an experimental study, where the plates of a rotational magneto-rheometer were modified with circumferential and radial grooves to study the surface effects on the shear stress of MR fluids under an applied magnetic field. An increase in the static shear yield stress of MR fluids by a factor of 2.8 compared with non-grooved surfaces is reported. It is suggested that the grooves might create a magnetic entrapment to prevent the slip of particles on the plate surfaces; thus, resulting in an enhancement of yield stress.

Kavlicoglu *et al.*[8,13] and Gordaninejad *et al.*[9] were the first to develop a methodology to eliminate the shear yield stress from defining the performance of a MR fluid. Based on a series of channel flow tests performed with iron particle impregnated non-ferrous channel surfaces[8] and grooved channel surfaces, it was observed that the resistance of MR fluid to flow greatly depends on the surface topology. Friction factor equations and plots (analogous to Moody diagram for Newtonian fluids) were generated. This methodology estimated the friction factor associated with MR fluid flow, using only the knowledge of surface geometry or topology, as well as the flow rate, magnetic field strength, and other actual material properties. Details of the work[8,9,13] are also provided in this chapter.

Nishiyama *et al.*[10,11] experimentally investigated the effect of surface roughness on rheological properties of MR fluids, under MR fluid plugging performance. Under a rotating shear mode, an increase in the yield stress with roughness was observed.[10] The effect of acrylic and steel smooth and grooved surfaces was examined,[11] and the increase in plug endurance pressure with the introduction of grooves was demonstrated.

Laun *et al.*[12] performed magnetorheometry tests for nonmagnetic and ferromagnetic plates, and concluded that, *via* the surface roughness and grooves on the surface, the shear stress of nonmagnetic plates can be increased up to the level of ferromagnetic plates. It is also demonstrated that if a layer of carbonyl iron particles is coated on a brass surface, the shear stress can also be increased.

Although, the effects of MR fluid–wall interactions are examined by many researchers, the authors[13] carried out a comprehensive study to experimentally investigate the behavior of MR fluid flow through channels with surface effects for various types of surface topologies, and surface materials. In addition, we also developed a model based on experimental data to approach and to understand the effects of wall surface on the pressure loss across a channel, under an applied magnetic field. This chapter is a summary of our work on flow analysis of MR fluids under surface effects. Experimental analyses are conducted to measure the pressure loss of a MR fluid for various applied magnetic field strengths, channel surface properties and flow rates. A flow type rheometer that allows installation of different surface topographies in the activating channel area is built. Different surface topographies, such as, various surface roughnesses, as well as, geometrically designed grooves, are utilized in the experimental study. Models for pressure loss of MR fluids under the effects of the wall topographies are developed. Theoretical analyses are conducted to define the shear strain rate and friction factor of MR fluid channel flow in terms of applied magnetic field, channel surface properties and flow rate without using the concept of shear yield stress. A theoretical analysis is performed to obtain shear stress–shear strain rate data from the experimental results. A non-dimensional pressure loss equation is developed which features a modified Mason number. It is demonstrated that by using this modified Mason number, one can obtain flow properties of a MR fluid on a single master curve.

5.3 Experimental Study

The main objective of the experimental study was to analyze the flow of MR fluids over various surface geometries and topologies, and layout a foundation for the development of a non-dimensional unified flow analysis model, without utilizing the concept of shear yield stress. The first steps in the experimental study included the rheological measurements of LORD MRF-132AD fluid using a Paar Physica MCR-300 shear rheometer (Figure 5.1) which has an

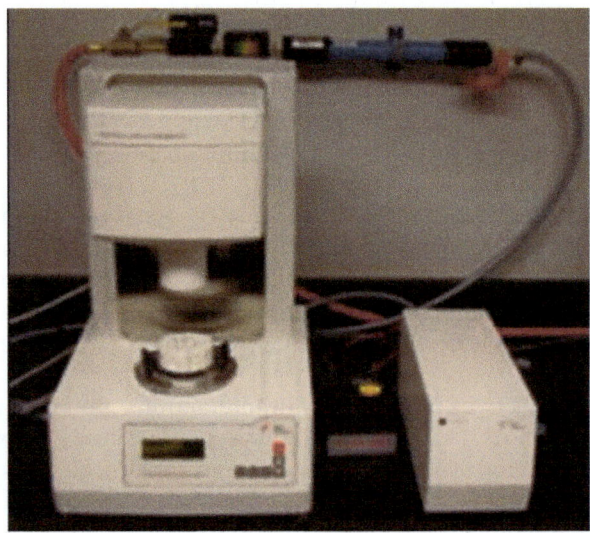

Figure 5.1   Paar Physica MCR 300 Shear Rheometer.

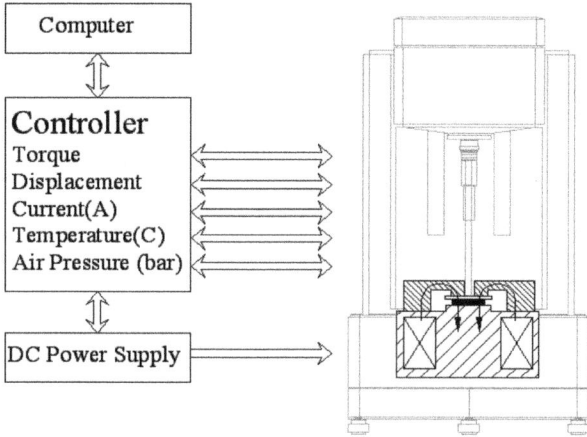

Physica MCR300

Figure 5.2   Schematic of the MCR300 Shear rheometer.

operating shear rate range from 0.1 s^{-1} to 900 s^{-1}. This shear rheometer consists of four main parts, *i.e.* the measuring drive system for testing specimen (Physica MCR300), electronic control unit for data acquisition and processing, interface software, and the DC power supply to provide and control the coil current. The rheometer system is equipped with a magnetic cell having a parallel plate configuration. Approximately 0.315 mL of sample is filled in a constant gap of 1.0 mm between two parallel plates during the experiment. In order to obtain the flow curves of the MR fluids, a linearly increasing shear rate from 1 s^{-1} to 400 s^{-1} is applied and shear stress, τ, is obtained. The schematic of the rheometer is given in Figure 5.2.

5.3.1   Experimental Setup

In order to further examine the rheological behavior of MR fluids at relatively higher shear strain rates (up to 9000 s^{-1}), a piston driven flow type rheometer with a rectangular cross-section channel is built.

The MR fluid is pressurized to flow through the channel between two parallel-arranged magnetic poles by means of a hydraulic actuator. The schematic of the experimental setup is presented in Figure 5.3. An electromagnetic coil is built and installed at the middle of the channel to allow the application of magnetic flux density normal to the flow direction. With this flow rheometer it would be possible to examine the surface roughness and grooved channel wall effects on the MR fluid channel flow. The channel is composed of two pieces and each piece has a recess that allows the installation of different surface property samples, as shown in Figure 5.4. A gauss meter measures the magnetic field strength inside the channel induced by the coil. The pressure drop, ΔP, is measured using two pressure transducers across the channel. The flow

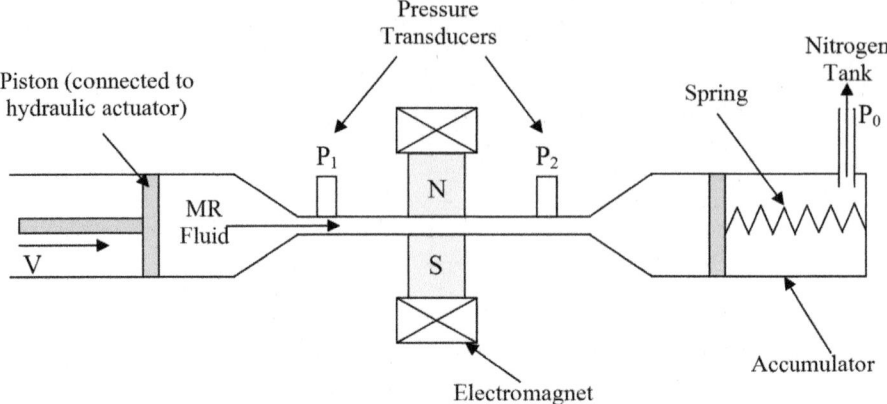

Figure 5.3   Schematic of the experimental setup.

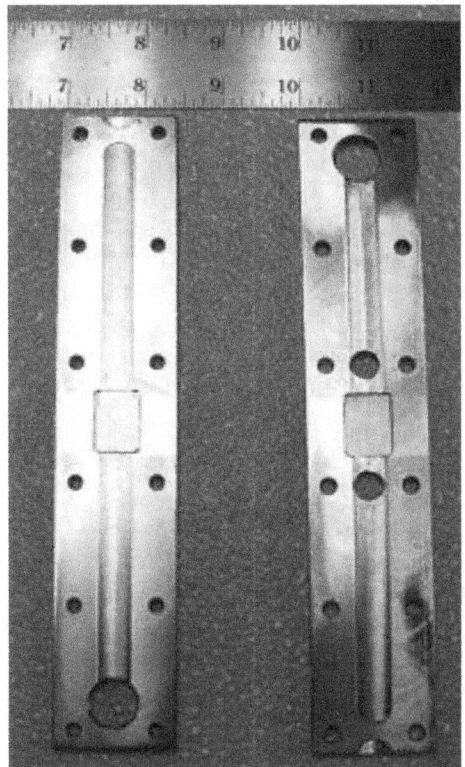

Figure 5.4   Two-piece test channel with interchangeable wall sections.

rheometer is connected to an Instron Model 8821S servo hydraulic actuator system. The input profile is a double ramp displacement profile, which generates constant velocities. The actuator head travels 19 mm upwards and returns to its original position. Each set of tests includes different velocities varying from 0.13 mm s^{-1} to 12.7 mm s^{-1} to examine various shear strain rates and various volumetric flow rates. The electromagnet activation input current for each velocity profile is varied as 0 A, 0.5 A, 1.0 A, 2.0 A and 3.0 A. The experimental setup is shown in Figure 5.5.

The experimental setup consists of an actuator, the test channel, two pressure transducers, a displacement transducer, a nitrogen pressurized accumulator, a power supply and the data acquisition system. The actuator is filled with MR fluid. The test channel is connected to the actuator, and the accumulator, which is pressurized to 0.35 MPa *via* a nitrogen tank. The dimensions of the test channel are given in Table 5.1.

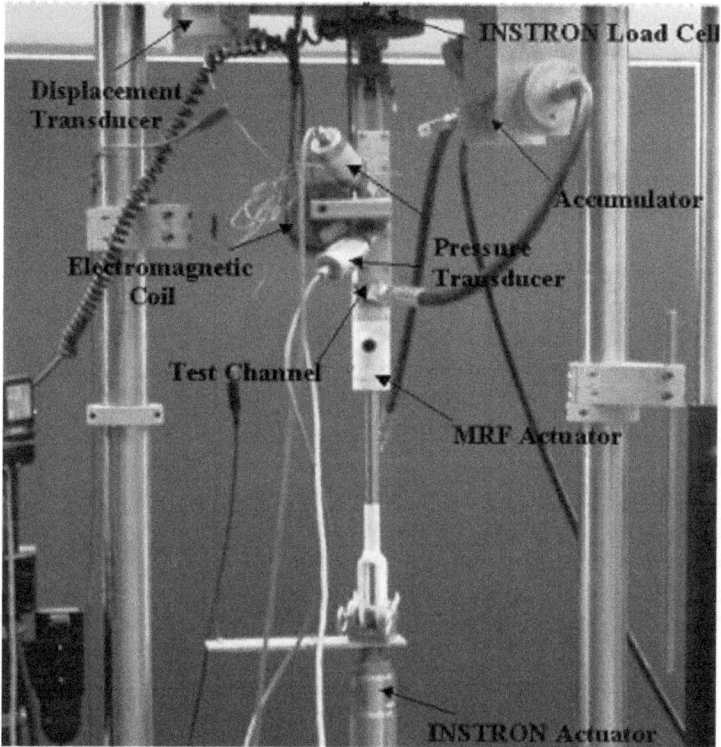

Figure 5.5 The experimental setup.

Table 5.1 Dimensions of the Test Channel.

Dimension	(mm)
Channel length, l_c	14
Channel width, w_c	10
Channel gap, g_c	1

5.3.2   Surface Topologies

Three different sets of surface topologies are examined. The first set of tests focus on non-ferrous porous surfaces impregnated with MR fluid with no flow through the porous media. Different porosity sizes are investigated. The pressure drop and flow rate data are obtained for different porosities. The second set of tests examines the effects of channel surface roughness on the MR fluid flow. For these tests, both the surface topology and magnetic properties of the surface are varied from non-ferrous to ferrous surfaces. The final set of tests examines the effects of grooved surfaces on the friction factor of MR fluid flow. Similarly, the surface material permeability is also changed for this set of tests. The following sections present more details on the selected surface topologies and surface materials.

5.3.2.1   *Non-Ferrous Porous Surface Specimens*

The experiments include 5 different surface specimens which are smooth surface, 5 μm, 20 μm, 40 μm and 100 μm porous surface. The size of the specimens indicates the average size (or diameter) of the cavities of the porous surfaces. All specimens are made of stainless steel and supplied by Mott Corporation. The specimens are cut to the geometric dimensions of 14 mm long, 10 mm wide and 1.5 mm thick to fit into the recesses in the two-piece channel. The porous samples are impregnated with MR fluid using a through flow impregnation technique, where MR fluid is pushed through the porous specimens, before being installed on the experimental setup channel. In order to better understand the mechanism behind the porous media interaction with a MR fluid, Scanning Electron Microscope (SEM) images of different samples are taken. This is done to visually examine the iron particle penetration mechanism, penetration depth and iron particle concentration in the porous media. For each sample 5 SEM images are taken at different locations.

Figures 5.6a and 5.6b schematically present the locations at which the SEM images are taken. The samples used in the experiments are broken in half at plane $Y = L/2$. The images are taken at the X–Z plane where $Y = L/2$ and $X = w/2$. The first image shows the entire cross-section of the sample. The second image focuses on the surface ($Z = 0$) (Section A). The other images scan the thickness of the sample to examine the change in the iron particle concentration. Three images are taken at $Z = 0.1$ mm (Section B), $Z = t/2$ (where $t = 1.6$ mm) (Section C) and $Z = 1.5$ mm (Section D).

Figure 5.7 presents the cross-section image (50×) of 20 μm porous media without impregnation. Figures 5.8 and 5.9 are the two SEM images for impregnated specimens. Figure 5.8 shows the overall cross-section. Figure 5.9 is a 550×magnified image showing the penetration of iron particles close to the surface at Section A. It is evident that, the iron particles are trapped inside the porosities and create a surface effect when the MR fluid flows over this surface.

(a)

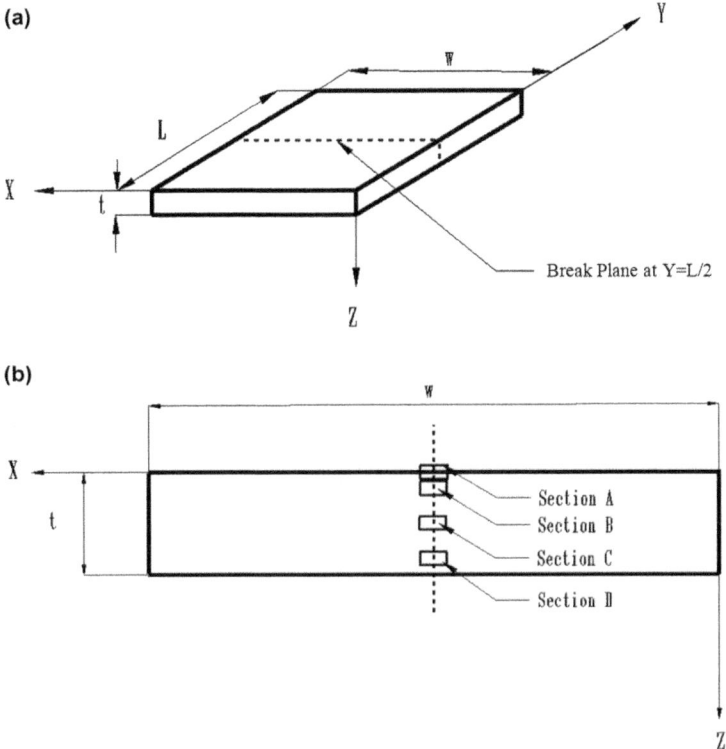

(b)

Figure 5.6 (a) Geometric dimensions of porous media samples. (b) Sections at which SEM images are taken ($Y = L/2$ plane).

Figure 5.7 SEM image of 20 μm porous media without MRF impregnation (50×).

Figure 5.8   SEM image of porous media impregnated by vacuum purge process (50×).

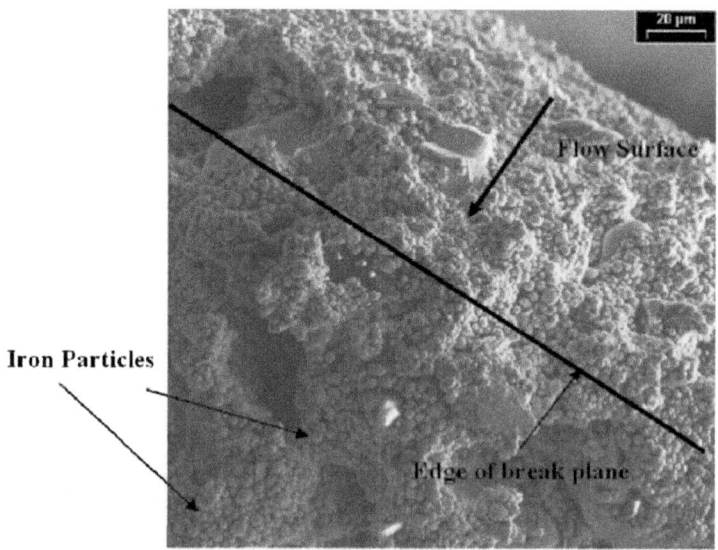

Figure 5.9   550× magnified SEM image of porous media impregnated by vacuum purge process (Section A).

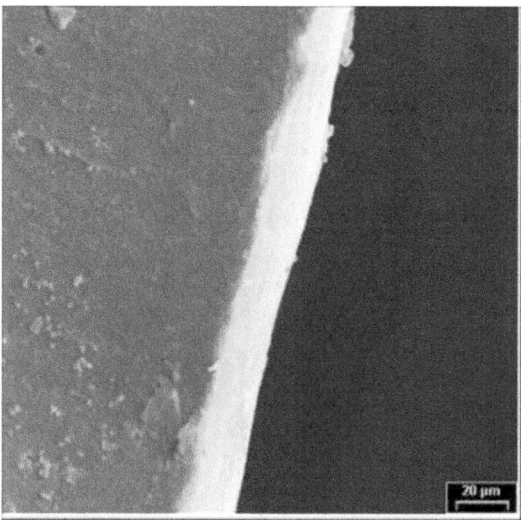

Figure 5.10   SEM image of 1.6 μm rough plastic specimen.

5.3.2.2   Rough Surface Specimens

In the second set of tests, various surface roughness and magnetic permeability materials are examined. For non-ferrous plastic surfaces and ferrous nickel surfaces, the surface roughness of the channel wall is varied as 0.4, 1.6, 3.2, 6.4, and 12.7 μm. Standard rough surface samples (surface comparators) are commercially available for plastic and nickel surfaces for these given surface roughness values. The plastic surface has a relative magnetic permeability of 1, whereas nickel has a relative permeability of 600. The iron particles suspended inside the MR fluid have a relative magnetic permeability of approximately 4000. Therefore a third surface type made out of steel with a relative magnetic permeability of 2000 is selected, so that three different magnetic permeable materials can be tested. SEM images of plastic specimens are also taken for rough surface specimens. For comparison of the surface topography, Figures 5.10 and 5.11 provide the SEM images of the plastic specimen of 0.4 μm and 12.7 μm rough surfaces.

5.3.2.3   Grooved Surface Specimens

In the last set of tests, it is decided to modify the channel walls with grooves to examine how the machined groove dimensions affect the pressure loss of MR fluid. Nine different groove configurations made out of aluminium (magnetic permeability of 1) are tested; additionally an aluminium un-grooved configuration is also tested to compare the groove width and depth effects. The schematic in Figure 5.12 presents the groove cross-section and geometric dimensions. The groove depth, d, and width, w, are varied to determine the effects of groove depth and width on the MR fluid pressure drop across the channel. The groove configurations are given in Table 5.2.

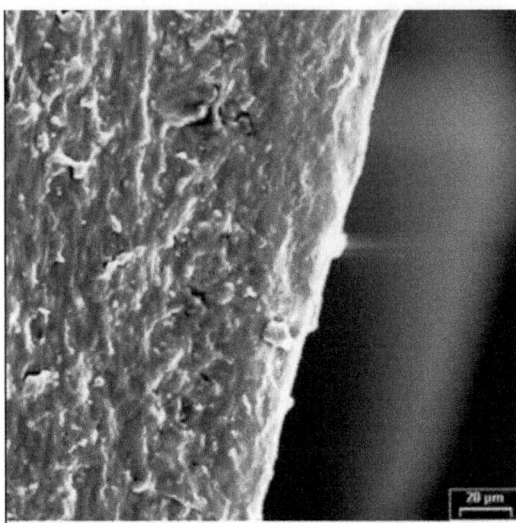

Figure 5.11   SEM image of 12.7 μm rough plastic specimen.

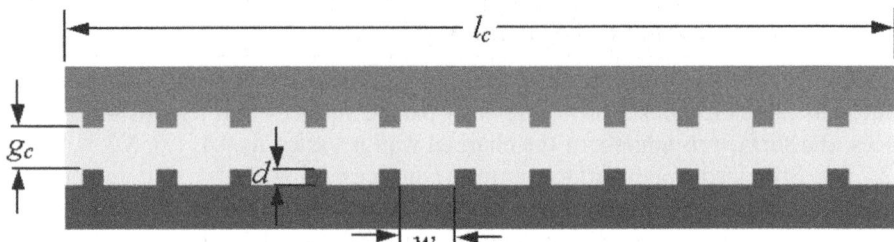

Figure 5.12   Grooved channel cross-section and dimensions.

Table 5.2   Dimensions of different groove configurations tested
in the flow type rheometer.

	(w/mm)	*(d/mm)*
G1	0.76	0.38
G2	0.76	0.64
G3	0.76	0.76
G4	1.02	0.38
G5	1.02	0.64
G6	1.02	0.76
G7	1.27	0.38
G8	1.27	0.64
G9	1.27	0.76
G0	ungrooved surface	

Once the results for aluminium are analyzed, it is observed that the width of
the grooves have minimal effect on the pressure drop for the range tested. The
groove depth effects are dominant. Therefore, as a second part of the grooved

surface specimens, ferrous cold-rolled steel surfaces are manufactured and tested. For cold-rolled steel surface tests the dimensions of configurations G1, G2 and G3 are selected.

5.4 Experimental Results

5.4.1 Mason Number

For Newtonian fluid flow, the friction factor is a function of Reynolds number. For MR fluid flow used in this work, (as will be explained in the following sections), the experimental results show that the Reynolds number has less effect on the friction factor of MR fluid under an applied magnetic field. The range for the Reynolds number observed in the flow rheometer is 5–200, which is in the laminar flow range.

The Mason number, which is a dimensionless measure of viscous forces to magnetic forces, is considered as a key parameter to develop the friction factor of MR fluid in channel flow. Since the friction factor of MR fluid flow is due to the interaction of the fluid with the wall surface, the related Mason number should be obtained at the wall, and therefore, is defined, as follows:

$$Mn_w = \frac{8\eta_f \dot{\gamma}_w}{\mu_0 \mu_f \beta^2 H_{MR}^2} \tag{5.3}$$

where, η_f is the MR fluid viscosity, $\dot{\gamma}_w$ is the shear rate of MR fluid flow at the wall, μ_0 is the vacuum permeability, μ_f is the MR fluid relative permeability, H_{MR} is the magnetic field strength inside the MR fluid and β is defined as:

$$\beta = \frac{\mu_p - \mu_f}{\mu_p + 2\mu_f} \tag{5.4}$$

where, μ_p is the iron particle relative permeability. The magnetic field strength, H_{MR}, can be approximated as $H_{MR} = B_{air}/\mu_0\mu_f$ with negligible error, where B_{air} is the magnetic field measured in the channel without the existence of MR fluid. Therefore, the Mason number can also be written as:

$$Mn_w = \frac{8\eta_f \dot{\gamma}_w \mu_0 \mu_f}{\beta^2 B_{air}^2} \tag{5.5}$$

The MR fluid relative permeability, μ_f, can be determined graphically from the B–H curve of a given MR fluid for different magnetic fields. The magnetic B–H relationship of a MR fluid can be defined as:[24]

$$B = 1.91\phi^{1.133}\left(1 - \left(\mu_0 e^{-10.97\mu_0 H_{MR}}\right)\right) + \mu_0 H_{MR} \tag{5.6}$$

Therefore, a MR fluid's relative permeability can be defined as:

$$\mu_f = \frac{dB}{dH_{MR}} = 1.91\phi^{1.133}\left(10.97\mu_0 e^{-10.97\mu_0 H_{MR}}\right) + \mu_0 \tag{5.7}$$

where, B is in Tesla, H_{MR} is in ampere m^{-1}, and:

$$\mu_0 = 4\pi \times 10^{-7}\, Wb/_{ampere.m} \qquad (5.8)$$

ϕ is the iron particle volume percentage in the MR fluid. The MR fluid relative magnetic permeability in eqn (5.7) includes the iron particle volume effects. Therefore, the Mason number defined in eqn (5.3) includes the magnetic field related parameters when different particle percentage MR fluids are used. In other words the Mason number includes a rheological property (viscosity of the fluid) and magnetic properties. The rheological properties of the MR fluid used in this study are first presented in the following sections. These properties are obtained using a commercial rheometer and then the results obtained from the high shear rate flow-type rheometer are presented.

5.4.2   Parallel Plate Shear Rheometer Tests

A Paar Physica MCR300 is used to obtain the flow results for the MR fluids for shear strain rates of 0.1–400 s^{-1}. The experimental data is processed based on the method of the Halsey et al.,[25] where the normalized apparent viscosity of the MR fluid is plotted as a function of Mason number. Halsey et al.[25] performed rheological measurements on an electrorheological fluid that behaves as a power law fluid. The measurements are performed using a parallel annulus cell geometry stress rheometer. The apparent viscosity of the ER fluid is normalized to the infinite-shear-rate viscosity of the fluid and plotted as a function of the Mason number, where the Mason number for an ER fluid is the ratio of hydrodynamic forces to electric forces. Halsey et al.[25] defined the slope of normalized apparent viscosity vs. Mason number plots as the shear-thinning exponent, Δ.

In this study, the measurements using a MCR-300 rheometer are performed for input electric currents of 0–2 A, and shear rates of up to 400 s^{-1}. The processed data according to method presented in ref. 28 is given in Figure 5.13. In Figure 5.13, the apparent viscosity is defined as:

$$\mu_{app} = \frac{\tau}{\dot{\gamma}} \qquad (5.9)$$

For normalization the μ_∞ for the MR fluid is determined as the viscosity at 400 s^{-1} shear rate from the shear rheometer tests and $\mu_\infty = 0.29$ Pa s for Lord MRF 132-AD. The shear thinning exponent, Δ, is defined as:

$$\Delta = 1 - n \qquad (5.10)$$

where, n is the fluid flow behavior index, and is defined as:

$$n = \frac{d(\ln \tau_w)}{d(\ln Q)} = \frac{d(\ln \Delta P)}{d(\ln Q)} \qquad (5.11)$$

In these experiments the shear thinning exponent is not a strong function of the applied magnetic field, as presented in Table 5.3. The magnetic properties of the MR fluid used in Mason number calculation are listed in Tables 5.4 and 5.5.

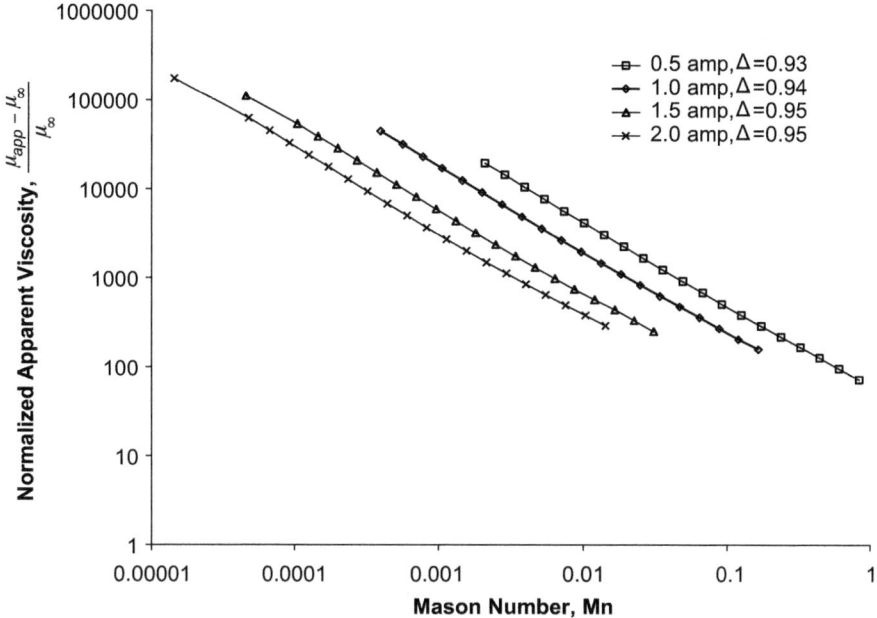

Figure 5.13 Normalized apparent viscosity of Lord MRF-132AD as a function of Mason number.

Table 5.3 Shear thinning exponents for MRF 132-AD.

MCR-300 Electric Input Current/A	Shear Thinning Exponent, Δ 132-AD
0.5	0.93
1.0	0.94
1.5	0.94
2.0	0.95

Table 5.4 Magnetic properties of MRF 132-AD for Mason number calculation.

MCR-300 Input Current/A	132-AD	
	μ_f	β
0.5	5.86	0.996
1.0	4.84	0.996
1.5	3.43	0.997
2.0	2.40	0.998

Table 5.5 H_{MR} for MCR-300 shear rheometer.

MCR-300 Electric Input Current/A	H_{MR}/Am^{-1}
0.5	13 400
1.0	30 500
1.5	63 800
2.0	104 000

5.4.3   Flow Rheometer Tests

5.4.3.1   Non-Ferrous Porous Surfaces

Figure 5.14 presents the pressure drop for 5 μm, 20 μm and 100 μm porous surfaces impregnated with through flow technique for B_{air} of 0 T, 0.191 T, 0.308 T and 0.405 T which correspond to 0 A, 0.5 A, 1.0 A and 3.0 A of input electric current, respectively. For the off-state case (0 A) the pressure losses are almost identical in the tested range of the volumetric flow rate for all samples. This would suggest that the impregnated porous surface does not affect the pressure loss of MR fluid without the applied magnetic field. In other words, the viscous pressure drop of MR fluid flow is not sensibly affected by the surface topography. However, when the magnetic field is applied, there is an apparent increase in the pressure loss when compared to the smooth surface. An increase in the pressure loss is also observed with increased porosity size. This is due to the fact that with larger porosities it is easier to impregnate the sample and trap more MR fluid, and thus have increased amount of iron particles inside the porosities. With an increased entrapment of iron particles a more effective magnetic field can be obtained on the surface of the porous samples, due to increased permeability of the magnetic field path across the porous sample.

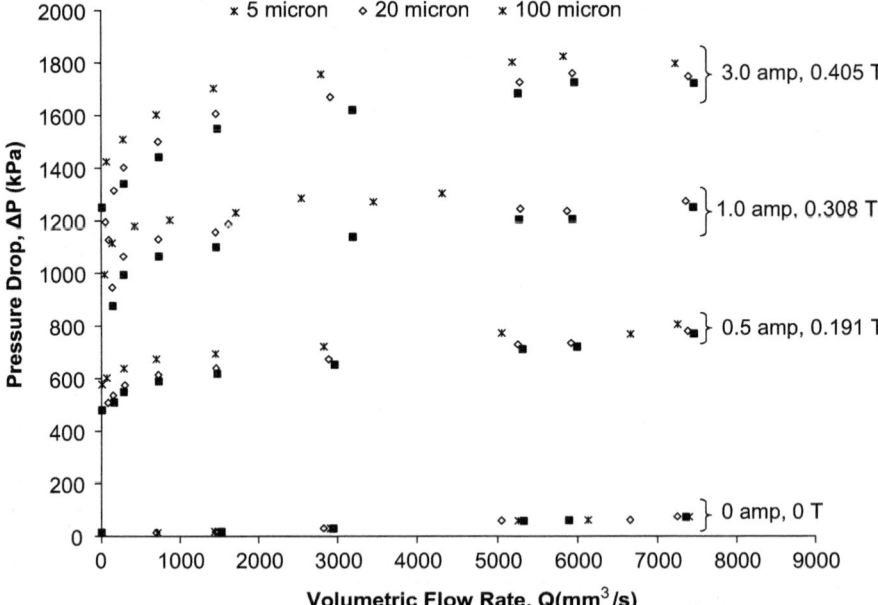

Figure 5.14   Comparison of pressure drop as a function of volumetric flow rate for 5, 20 and 100 μm porous surfaces impregnated using the through flow technique.

5.4.3.2 Rough Surfaces

The second part of the experimental study is the examination of rough surface effects on the pressure drop of MR fluids. As mentioned previously, in analysis of MR fluid devices it is customary to assume a shear yield stress to determine the pressure drop across an MR valve. However, the first part of the experimental study has shown that the surface treatment also affects the pressure drop. Therefore, it is decided to investigate the surface effects in depth by varying the surface roughness as well as the surface permeability.

5.4.3.2.1 Plastic Surfaces. Figure 5.15 presents a typical experimental result for various magnetic fields, B_{air}, that are measured using a gauss meter probe inside the air channel (without MR fluid). These raw data for pressure drop and piston moving displacement history are obtained for all surface morphologies and different applied magnetic fields. The results presented in Figure 5.15 are for a channel with plastic walls and a 3.2 μm surface roughness. The pressure loss and displacement profile are presented for a piston velocity of 2.54 mm s^{-1}. As can be seen in Figure 5.15, for a given surface roughness value, the pressure drop increases as the magnetic field increases. Similar results are obtained for all configurations and different velocities. The pressure drop for different plastic surface roughness values as a function of the volumetric flow rate are presented in Figure 5.16 for magnetic field strengths of 0 T, 0.191 T and 0.405 T corresponding to 0, 0.5 and 3.0 A input currents, respectively. It can be concluded that for the off-state case (0 T) the viscous pressure losses are almost identical in the tested range of the volumetric flow rate for all the samples, which suggests that the surface

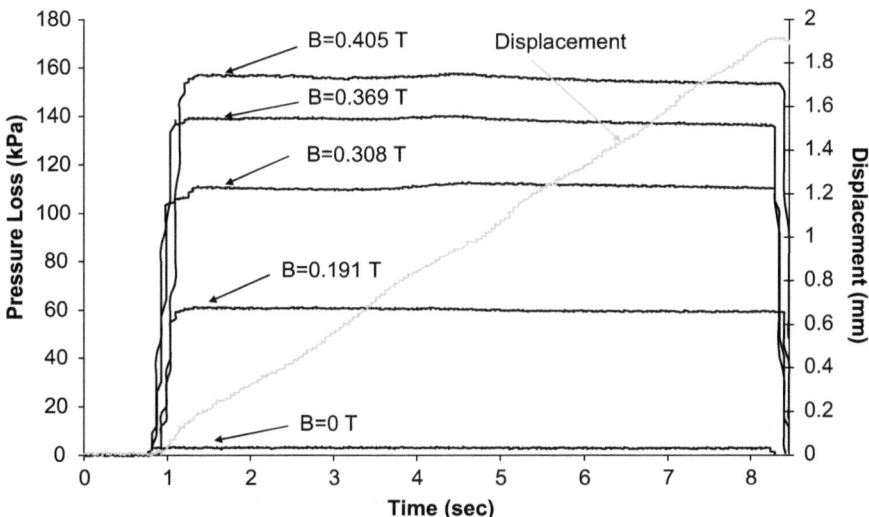

Figure 5.15 Pressure drop and displacement profile for 3.2 μm rough plastic surface for various magnetic fields for 2.54 mm s^{-1} actuator piston velocity.

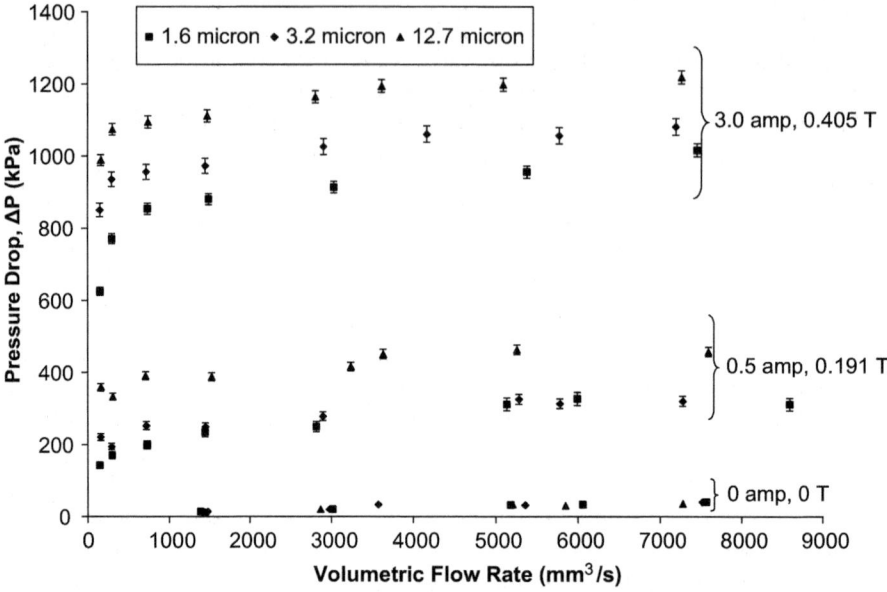

Figure 5.16   Comparison of pressure drop as a function of volumetric flow rate for 1.6 μm, 3.2 μm and 12.7 μm rough plastic surfaces.

roughness does not affect the viscous pressure loss. However, when the magnetic field is applied, the effect of surface roughness is apparent. For a given magnetic field as the surface roughness increases, an increase in the pressure loss is observed. The increase in the pressure loss is more effective at high magnitude magnetic fields, such as 0.4 T.

Figures 5.17 to 5.19 present the MR fluid pressure loss for all plastic surface roughness values and magnetic fields, for actuator piston velocities of 0.25 mm s^{-1}, 2.54 mm s^{-1} and 12.7 mm s^{-1}, respectively. As can be seen from Figures 5.17–5.19, the viscous pressure drop is not affected by the surface roughness for the given volumetric flow rate range. However, once an external magnetic field is applied, it is apparent that for a non-ferrous plastic surface the pressure drop is greatly affected by the surface roughness. One of the important conclusions that can be reached is that, by increasing the surface roughness it would be possible to obtain a higher pressure drop without increasing the magnetic field. This conclusion is essential in the design of highly efficient MR fluid devices.

5.4.3.2.2   Nickel Surfaces.   The pressure drop for different nickel surface roughness values as a function of the volumetric flow rate are presented in Figure 5.20 for 0, 0.5 and 3.0 A input electric currents, corresponding to magnetic field flux densities of 0, 0.267 and 0.469 T, respectively. Similarly, in the off-state case (0 T) the viscous pressure losses are almost identical in the tested range of the volumetric flow rate for all the samples, which

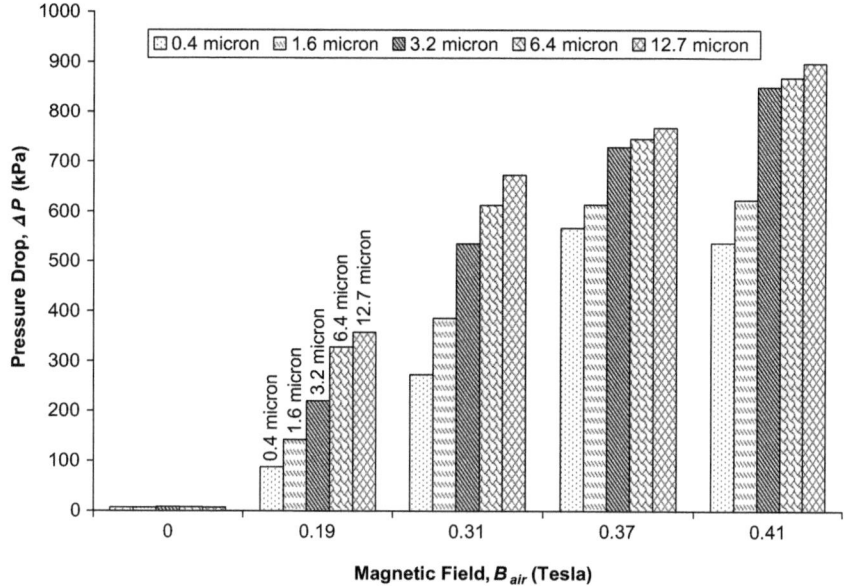

Figure 5.17 Pressure loss for various magnetic fields for a piston velocity of 0.25 mm s^{-1}.

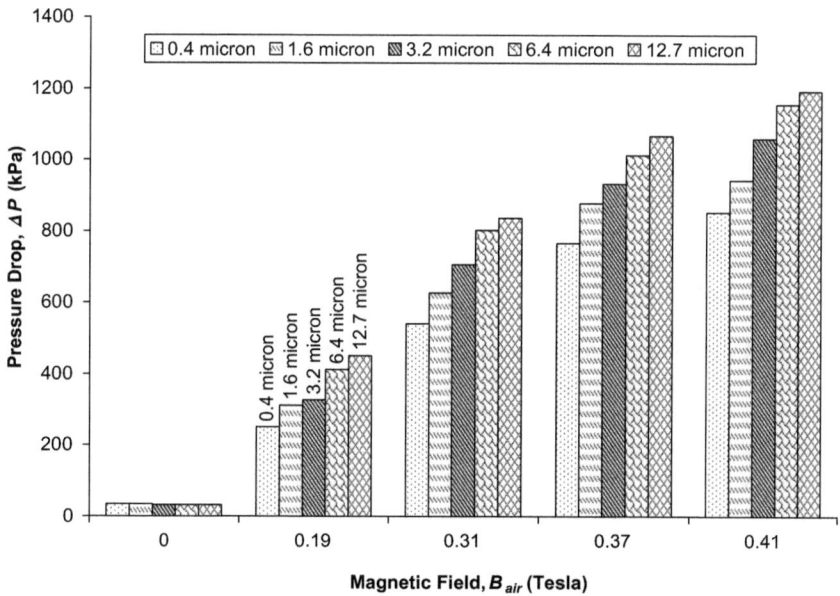

Figure 5.18 Pressure drop for various magnetic fields for a piston velocity of 2.54 mm s^{-1}.

suggests that the surface roughness does not affect the viscous pressure loss, regardless of the surface material being ferrous or not. However, when the magnetic field is applied, the effect of surface roughness is apparent. For a

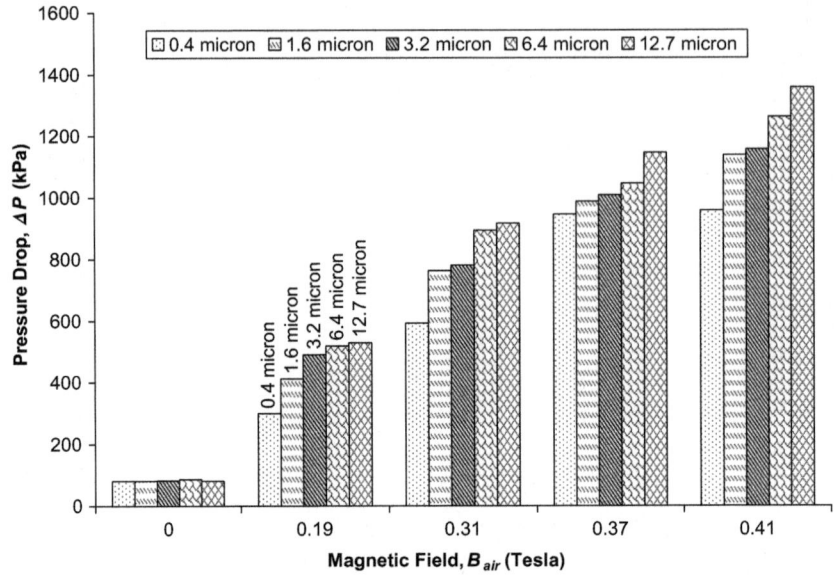

Figure 5.19   Pressure drop for various magnetic fields for a piston velocity of 12.7 mm s^{-1}.

given magnetic field, as the nickel surface roughness increases, an increase in the pressure loss is observed.

Although nickel is a ferromagnetic surface, due to its lower permeability than iron particles (600 compared to 4000), the surface effects are still playing an important role. This can be explained as the entrapment of iron particles in the peaks and valleys of the rough surface and creating localized high magnetic field areas. As the surface roughness increases, the amount of particles trapped inside the rough surface causes a higher pressure drop.

5.4.3.2.3   Cold Rolled Steel Surfaces.   The pressure drops for different steel rough surfaces as a function of the volumetric flow rate are presented in Figure 5.21 for 0, 0.5 and 3.0 A input electric currents, corresponding to magnetic field flux density of 0, 0.291 and 0.577 T, respectively. For steel surfaces the surface effects diminishes due to increased surface relative magnetic permeability of 4000. The interactions between the wall and the adjacent particles are strong even with a very smooth steel surface; therefore, the entrapment of particles inside the peaks and valleys of the rough surface does not have an effect on the increase of the pressure drop.

5.4.3.3   Grooved Surfaces

5.4.3.3.1   Aluminium Grooved Surfaces.   Figure 5.22 presents a typical experimental result for configuration G1 and for various magnetic fields,

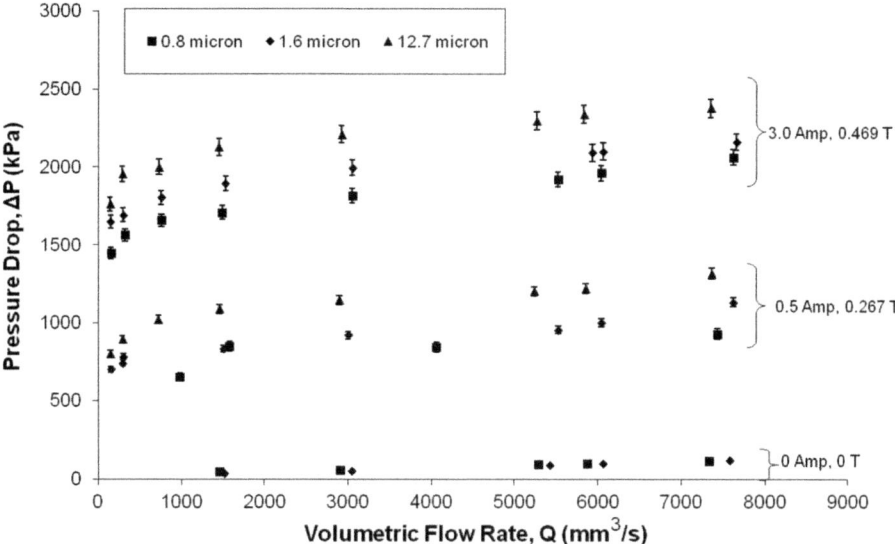

Figure 5.20 Comparison of pressure drop as a function of volumetric flow rate for 0.8 μm, 1.6 μm and 12.7 μm rough nickel surfaces.

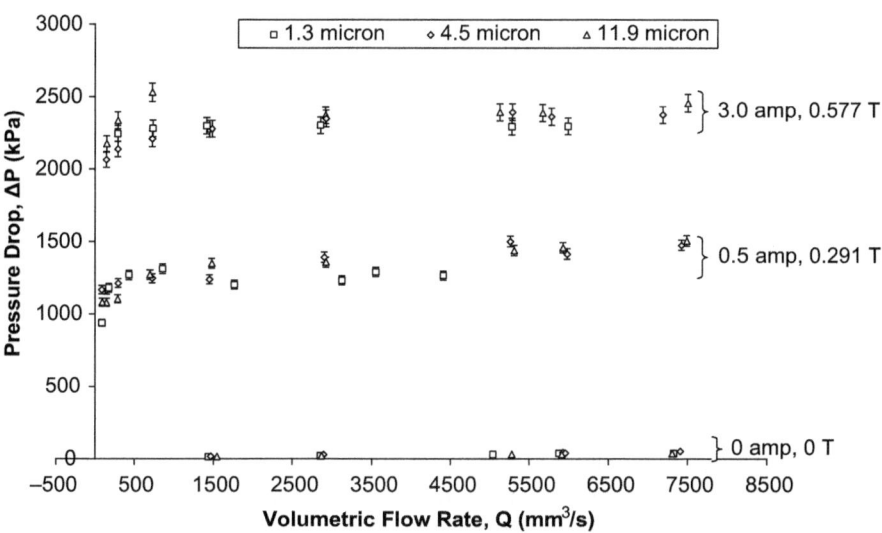

Figure 5.21 Comparison of pressure drop as a function of volumetric flow rate for 1.3 μm, 4.5 μm and 11.9 μm rough steel surfaces.

B_{air}. The pressure loss and displacement profile are presented for a piston velocity of 1.27 mm s^{-1}. As can be seen, for a given profile, the pressure drop increases as the magnetic field increases. Similar results are obtained for all configurations and different volumetric flow rates. To compare the

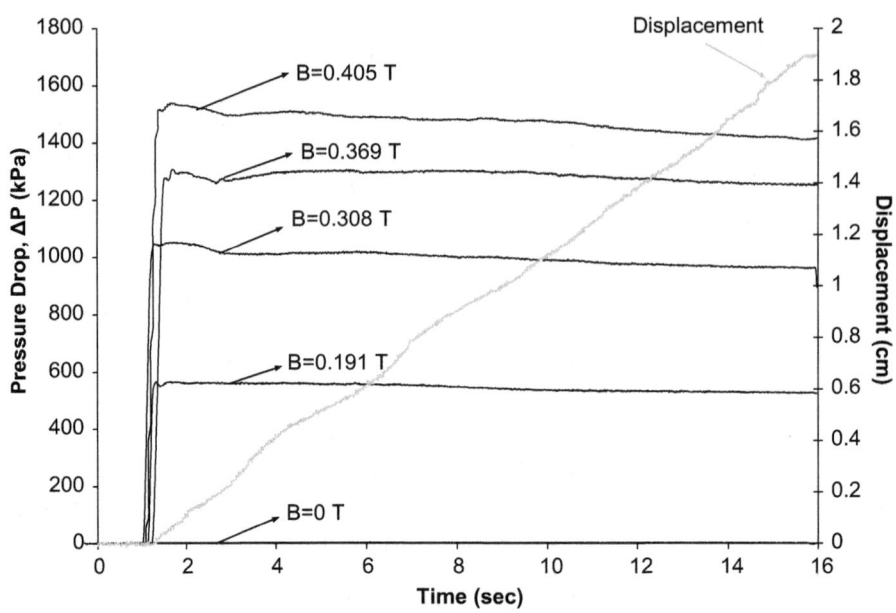

Figure 5.22   Pressure drop and displacement profile for aluminium grooved surface (G1) for various magnetic fields for 127mm s^{-1} actuator piston velocity.

performances of different configurations the pressure drop is plotted as a function of the volumetric flow rate for various magnetic fields.

Figure 5.23 presents the pressure drop for configurations G0, G1, G2 and G3 for B_{air} of 0 T, 0.19 T and 0.40 T, corresponding to 0 A, 0.5 A and 3.0 A of input electric current, respectively. For the off-state case (0 A) the pressure losses are almost identical in the tested range of the volumetric flow rate for all samples. This would suggest that the grooved surface configurations have less effect on pressure loss of MR fluid without the applied magnetic field. However, when the magnetic field is applied, there is an apparent increase in the pressure loss when compared with the smooth surface. The pressure drop also increases with increasing depth of the grooves while keeping the width of the grooves constant (when G1, G2 and G3 are compared). The increase in the pressure loss with changing width is relatively small and negligible when compared to the groove depth effect (when G3 and G6 are compared). The increase in the pressure loss with changing width is relatively small and negligible when compared to the groove depth effect when G3, G6 and G9 are compared as shown in Figure 5.24.

5.4.3.3.2   Cold Rolled Steel Grooved Surfaces.   As a next step in grooved channel wall experiments, it is decided to conduct flow rheometer experiments on steel grooved walls. G1, G2 and G3 configurations are tested and compared with an "un-grooved" steel surface, G0. Figure 5.25 presents the

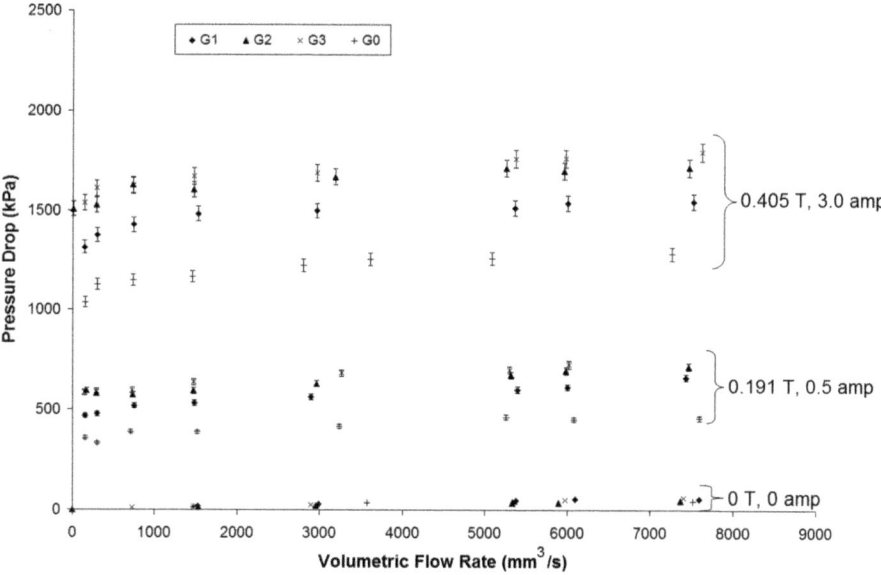

Figure 5.23 Comparison of pressure drop for aluminium grooved channel wall configurations G0, G1, G2, G3 and G6 for various magnetic fields.

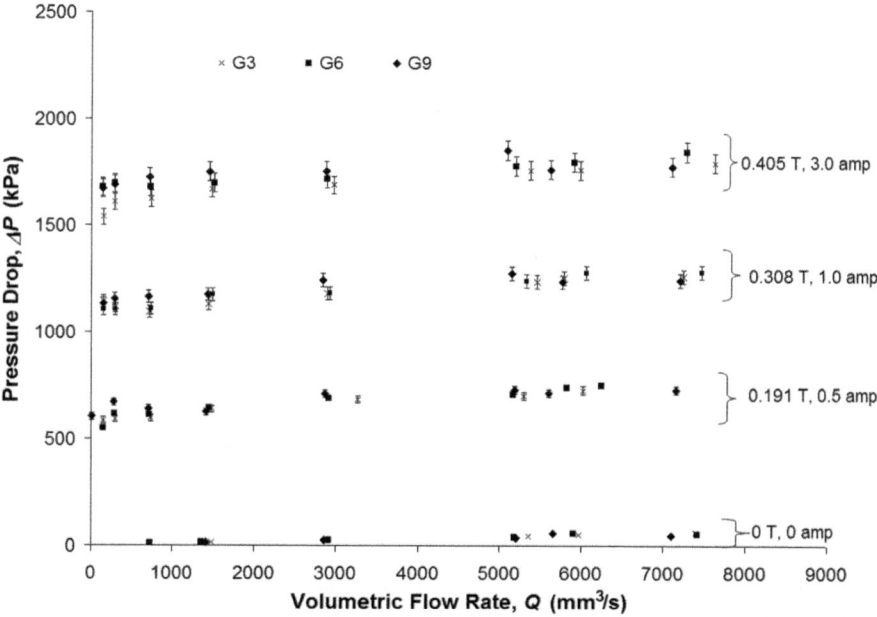

Figure 5.24 Comparison of pressure drop for aluminium groove configurations G3, G6 and G9 (same depth, different width) for various magnetic fields.

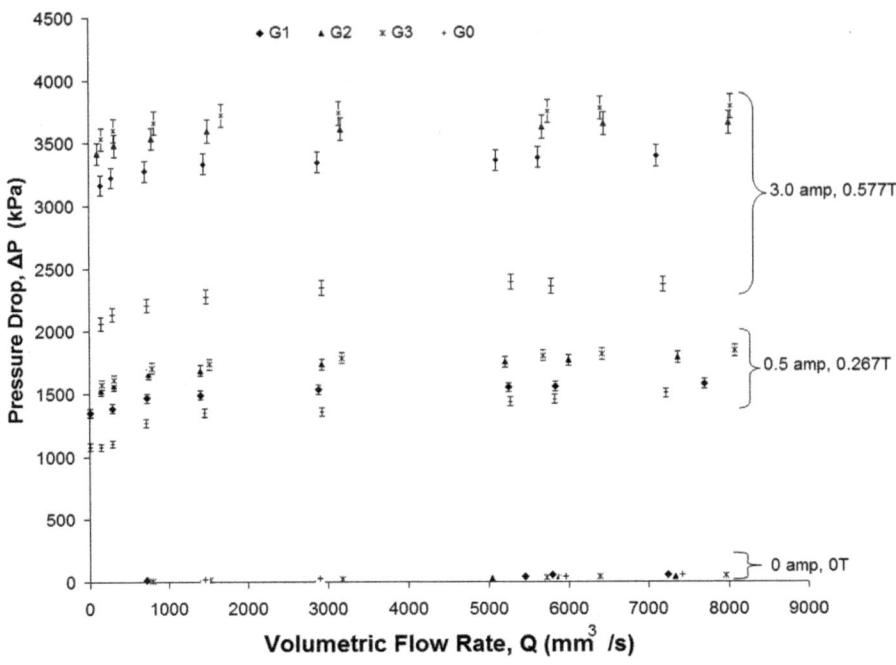

Figure 5.25   Comparison of pressure drop for steel grooved channel wall configurations G0, G1, G2 and G3 for various magnetic fields.

pressure drops for configurations G0, G1, G2, G3 and G6 for B_{air} of 0 T,
0.19 T and 0.40 T, corresponding to 0 A, 0.5 A and 3.0 A of input electric
current, respectively. For the off-state case (0 A) the pressure losses are almost identical in the tested range of the volumetric flow rate for all samples.
However, when the magnetic field is applied, there is an increase in the pressure loss, when compared with the smooth surface.

Previously it was concluded that the surface roughness has minimal effect
when a steel channel wall is utilized. However, when grooves are machined on a
steel channel wall, the pressure drop is greatly affected. This might be due to the
fact that, when a sharp corner is present on the surface, the magnetic field
strength is highly elevated at that corner, which might lead to an elevated
pressure drop. In addition, as the groove depth increases, more iron particles
are present in the groove, which also elevates the magnetic field and results in
further increases in the pressure drop.

To further examine this claim, a three-dimensional electromagnetic finite
element model of the channel and the electromagnet is developed. The objective
of the finite element analyses (FEA) is to evaluate the magnetic field distribution inside the grooved section and compare it with the "un-grooved" configuration, when an external magnetic field is applied by means of an
electromagnet. Figure 5.26 presents the magnetic field distribution along the
mid-plane of the channel. The higher magnetic field areas right above the

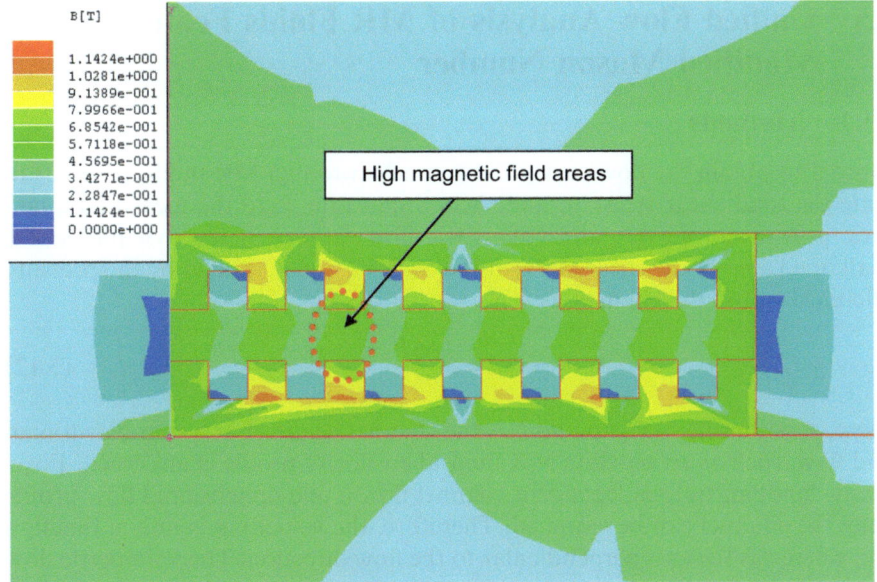

Figure 5.26 Magnetic field distribution along the channel for a grooved configuration.

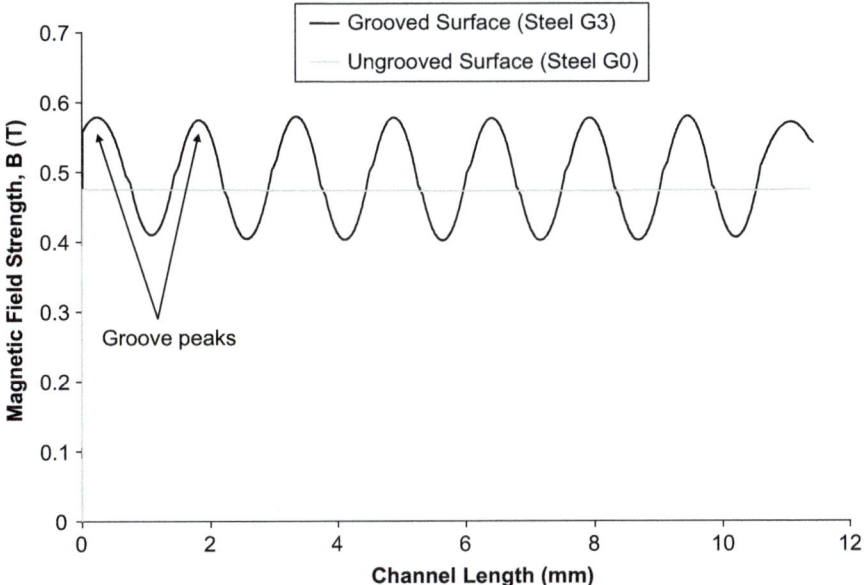

Figure 5.27 Magnetic field distribution along the channel for a grooved configuration.

groove peaks can be observed in Figure 5.26. The magnetic field distributions along the centerline of the grooved channel and un-grooved channel are compared in Figure 5.27.

5.5 Unified Flow Analysis of MR Fluids Featuring Modified Mason Number

5.5.1 Methods

In order to obtain the flow curves for a non-Newtonian MR fluid one needs to relate the total shear stress, τ, and shear strain rate, $\dot{\gamma}$, to the measured quantities: pressure drop, ΔP, and volumetric flow rate, Q. If the dynamic loss is small, the value of the shear stress at the wall, τ_w, is determined by the pressure difference across the channel as:[26]

$$\tau_w = \frac{g_c \Delta P}{2 l_c} \tag{5.12}$$

where, g_c is the channel gap and l_c is the channel length. For a non-Newtonian fluid flow, such as an on-state MR fluid, the velocity profile is unknown. For a steady, laminar, time-independent channel flow a one-dimensional flow profile along the channel can be assumed. Therefore, the velocity u is only a function of z, where z is the axis perpendicular to the flow direction. The volumetric flow rate can be defined as:

$$Q = w_c \int_{-g_c/2}^{g_c/2} u \, dz \tag{5.13}$$

where, w_c is the channel width. Equation (5.13) can be rewritten by performing integration by parts and using symmetry as:

$$Q = 2 w_c \left([uz]_0^{g_c/2} - \int_0^{g_c/2} z \frac{du}{dz} dz \right) \tag{5.14}$$

Assuming no-slip condition at the wall of the channel, the first term in the parentheses on the right hand side of eqn (5.14) is zero, therefore:

$$Q = -2 w_c \int_0^{g_c/2} z \frac{du}{dz} dz \tag{5.15}$$

Since $\dfrac{du}{dz} = -\dot{\gamma}$ and $\tau = \dfrac{2 \tau_w z}{g_c}$, by changing the variable in the integral, one has:

$$Q = \frac{g_c^2 w_c}{2 \tau_w^2} \int_0^{\tau_w} \tau \dot{\gamma}(\tau) d\tau \tag{5.16}$$

To establish a relationship between pressure drop, ΔP, and volumetric flow rate, Q, a constitutive model is required, for the fluids with yield stress, to integrate eqn (5.16). For MR fluids it is customary to employ models such as the

Bingham Plastic model or Herschel–Bulkley model.[1,2] Since a model that does not utilize the concept of shear yield stress to eliminate the confusions in shear yield stress definitions for MR fluids is to be developed, a different method has been followed. Equation (5.16) can be differentiated with respect to τ_w to give:

$$\frac{dQ}{d\tau_w} = -\frac{2Q}{\tau_w} + \frac{\dot{\gamma}(\tau_w)}{\tau_w} \cdot \frac{g_c^{\,2} w_c}{2} \tag{5.17}$$

or:

$$\dot{\gamma}(\tau_w) = \frac{2Q}{g_c^{\,2} w_c}\left(2 + \frac{d(\ln Q)}{d(\ln \tau_w)}\right) \tag{5.18}$$

If n is defined as:

$$n = \frac{d(\ln \tau_w)}{d(\ln Q)} = \frac{d(\ln \Delta P)}{d(\ln Q)} \tag{5.19}$$

then, the shear strain rate at the wall can be defined as:

$$\dot{\gamma}_w = \frac{6Q}{g_c^{\,2} w_c}\left(\frac{2n+1}{3n}\right) \tag{5.20}$$

The term in parentheses in eqn (5.20) can be treated as the correction factor for flow in uniform rectangular channels for non-Newtonian fluids, which is similar to the Rabinowitch correction used for conventional capillary viscometer.[27]

Alternatively, one can model the post-yield behavior of MR fluids with the power law model, which does not require the shear yield stress term, *i.e.*:

$$\tau = C\dot{\gamma}^{n'} \tag{5.21}$$

where, C is the consistency index and n' is the flow behavior index. If eqn (5.21) is substituted into eqn (5.16), one can obtain the same relation as eqn (5.20). Therefore, n in eqn (5.21) is identical to the flow behavior index, n'. Eqn (5.12) and (5.20) provide the relationship for the wall shear stress, τ_w, and wall shear strain rate, $\dot{\gamma}_w$ as functions of the measurable properties, pressure loss and volumetric flow rate. For a Newtonian fluid, $n=1$; for an MR fluid, n can be determined as a function of magnetic field, B, from the flow curve plots of $\ln \tau_w$ (or $\ln \Delta P$) against $\ln Q$ for various magnetic fields, as given in eqn (5.19).

5.5.2 Non-Dimensional Friction Factor Equation

Once the relationship between the wall shear stress and shear strain rate is established, a non-dimensional equation can be introduced for MR fluid flow analysis that includes the applied magnetic field intensity, channel geometry and surface topography effects, based on the experimental results. When developing these equations, it is assumed that the MR fluid flow across the channels is isothermal and is a uniform laminar flow with unchanging cross-section. The fluid is also assumed to have time-independent properties such as density and viscosity. The friction factor, c_f, is:[28]

$$c_{f,MR} = \frac{\tau_w}{\frac{1}{2}\rho_f V_m^2} \tag{5.22}$$

where, τ_w is the wall shear stress given in eqn (5.12), ρ_f is the MR fluid density and V_m is the mean flow velocity in the channel and is defined as $V_m = \frac{Q}{A_{ch}}$. A_{ch} is the cross-sectional area of the channel. As will be explained in the following sections, based on the experimental study, it is concluded that the friction factor associated with the MR fluid flow can be defined as:

$$c_{f,MR} = A(\text{surface topology and geometry, } B) \, \mathrm{f}(Mn) \tag{5.23}$$

where, A is a function of channel surface topology, geometry and magnetic field. The Mason number, Mn is defined in eqn (5.3). From eqn (5.21) and the definition of the Mason number, it is evident that one needs to calculate the shear rate of MR fluid flow at the wall, $\dot{\gamma}_w$, and the shear stress at the wall, τ_w. The definition of $\dot{\gamma}_w$ also includes the flow behavior index, n, defined in eqn (5.19).

The flow behavior index is determined for all surface topologies and input electric currents, based on the flow curves obtained in the experimental study. Some sample plots of flow behavior index for various surface topologies are presented in Figures 5.28–5.32. As seen in Figures 5.28–5.32, the flow behavior

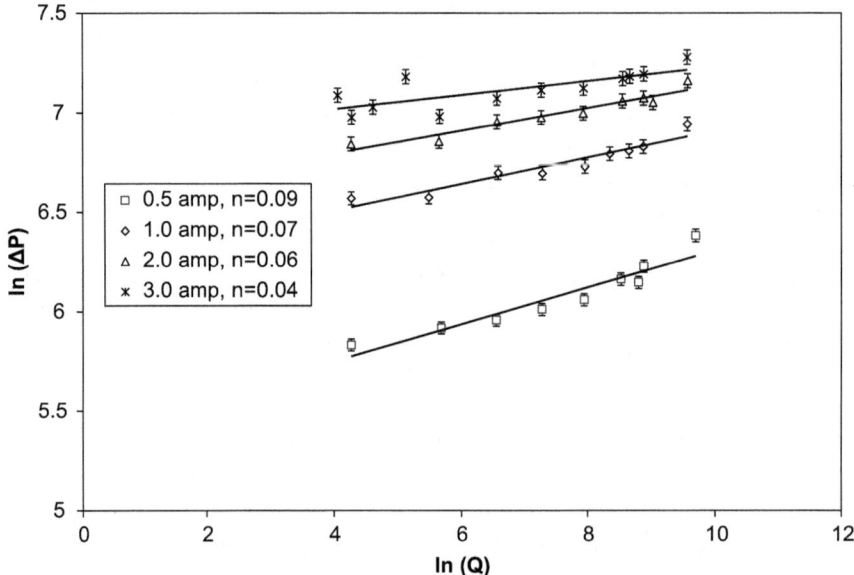

Figure 5.28   ln (ΔP) *vs.* ln (Q) curves to determine n for 5 μm porous surface impregnated with through-flow technique.

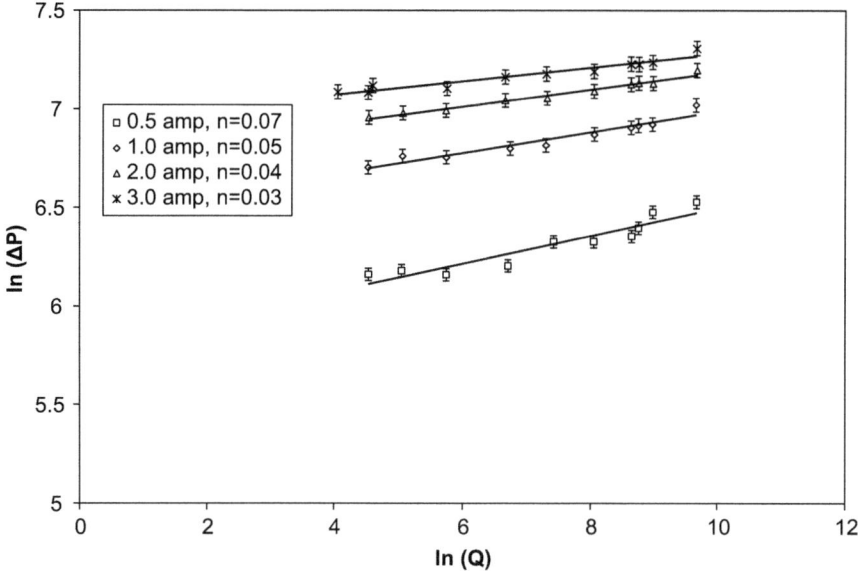

Figure 5.29 ln (ΔP) *vs.* ln (Q) curves to determine *n* for 100 μm porous surface impregnated with through-flow technique.

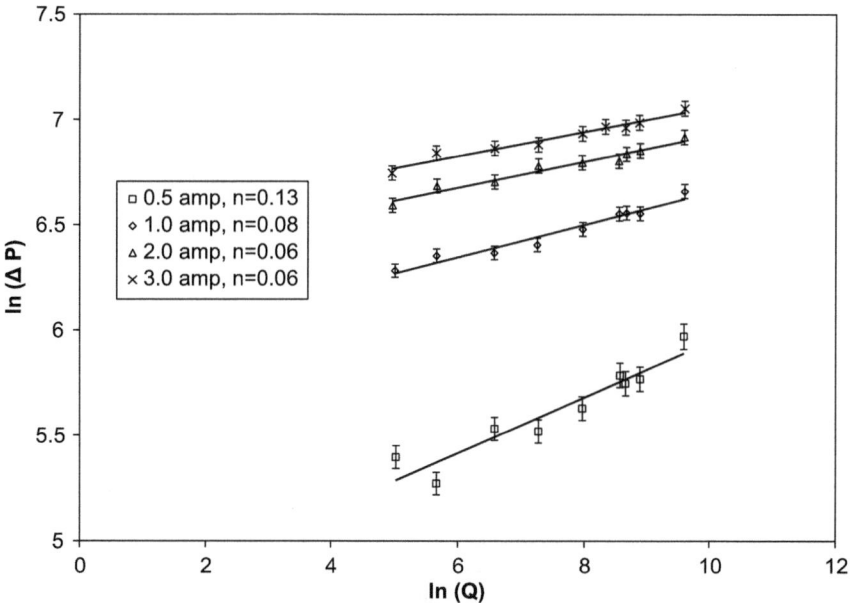

Figure 5.30 ln (ΔP) *vs.* ln (Q) curves to determine *n* for 3.2 μm rough plastic surface.

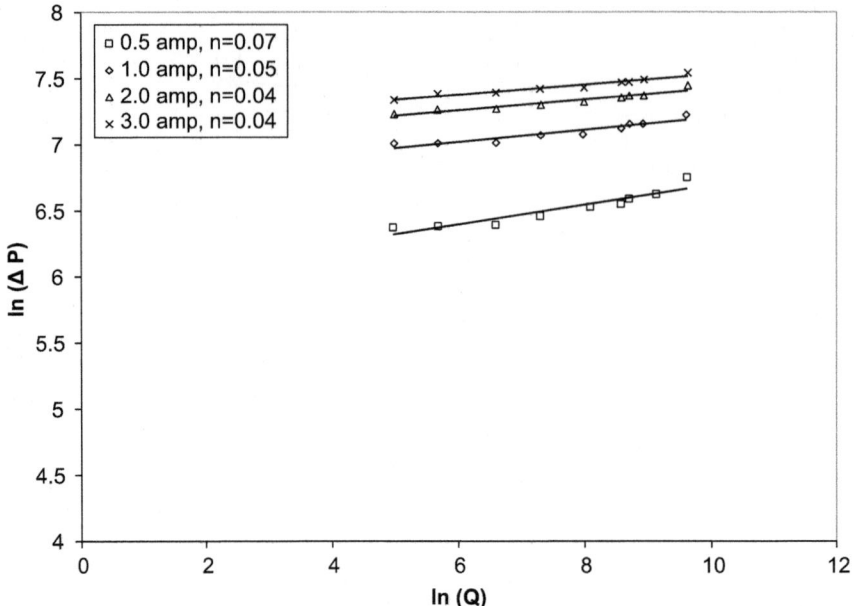

Figure 5.31 In (Δ*P*) *vs.* In (*Q*) curves to determine n for grooved aluminium G3 configuration.

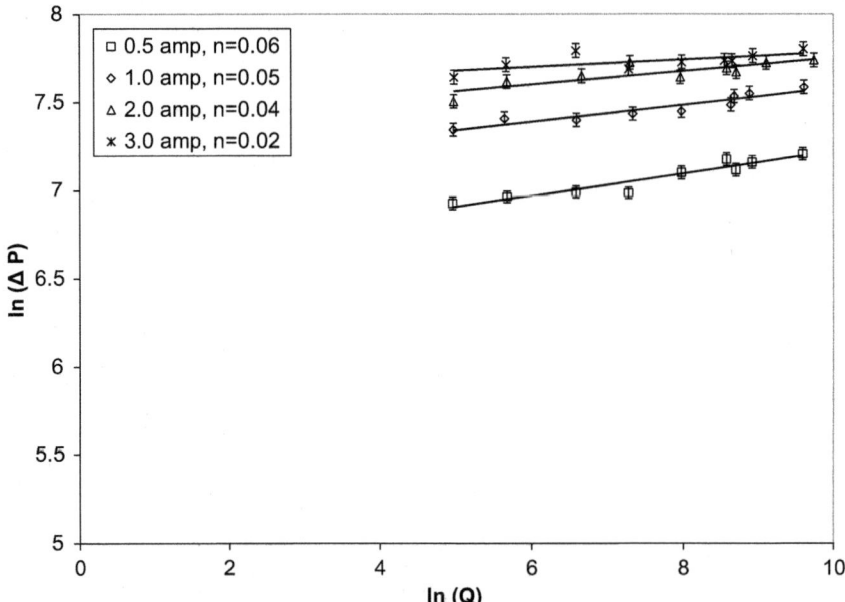

Figure 5.32 In (Δ*P*) *vs.* In (*Q*) curves to determine *n* for smooth cold rolled steel surface.

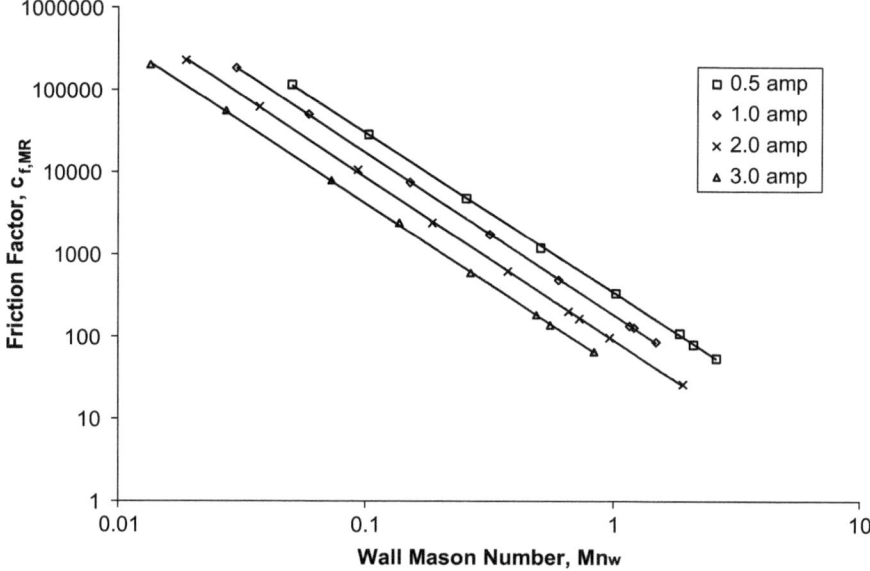

Figure 5.33   Friction factor of MR fluid as a function of Mason number at the wall for cold rolled steel surface.

index (thus the apparent shear thinning exponent, Δ) changes with the magnetic field strength.

The value of n decreases with increasing magnetic field for all surface types. In general, the n values are in the range of 0.03–0.09. No obvious tendency of n is observed as a function of surface morphology. This suggests that the surface type does not affect the post yield rheological behavior of Lord MRF 132AD.

Once, the shear rate and shear stress at the channel wall are determined it is possible to plot the friction factor as a function of the Mason number, *Mn*. Figure 5.33 shows one such plot for cold rolled steel surface, in log–log scale. Figures 5.34–5.36 present similar results for various surface topologies. As can be seen, the friction factor curves shift as the input current changes. It is determined that the slope of these curves in Figures 5.34–5.36 is equal to $(2 - n)$ for a given input current and surface type. Therefore, it is possible to redefine eqn (5.23) as:

$$c_{f,MR} = A(\text{surface topology and geometry, } B)Mn^{-\lambda} \qquad (5.24)$$

where λ is defined as the power index for the Mason number at the wall. From definitions of the friction factor and Mason number, proportionality equations can be written as:

$$c_f \propto \frac{\tau_w}{\dot{\gamma}^2} \quad \text{and} \quad Mn \propto \dot{\gamma} \qquad (5.25)$$

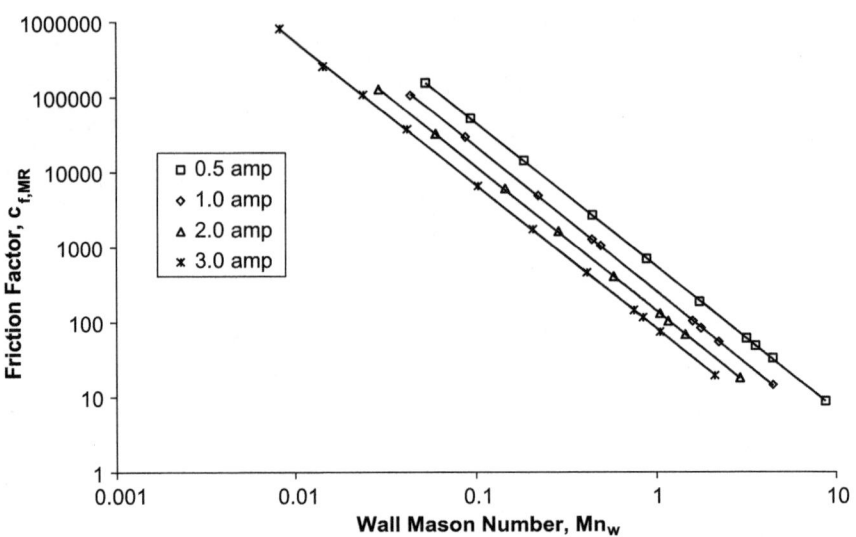

Figure 5.34   Friction factor of MR fluid as a function of Mason number at the wall for 20 μm porous surface impregnated by through-flow technique.

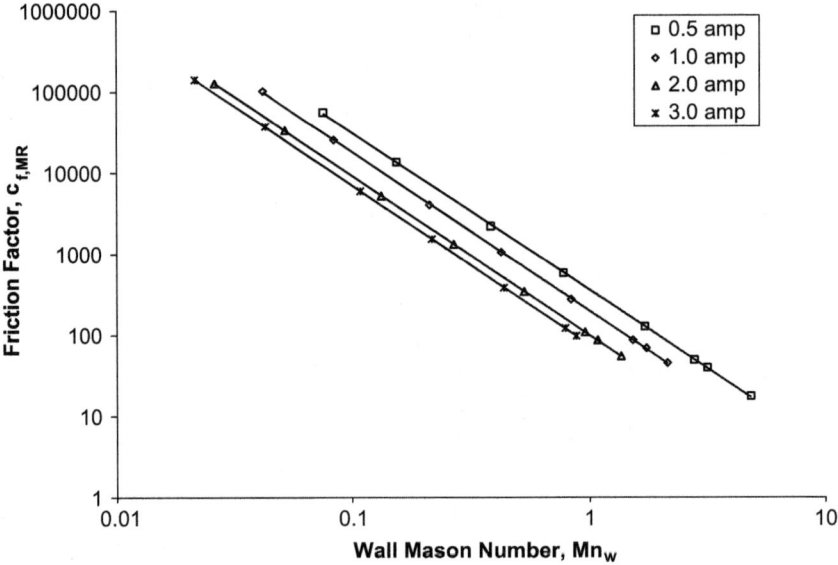

Figure 5.35   Friction factor of MR fluid as a function of Mason number at the wall for aluminium grooved surface (G3 configuration).

Therefore, it is possible to show that:

$$c_f \propto \mathrm{Mn}^{-\lambda} \Rightarrow \tau_\mathrm{w} \propto \dot{\gamma}^{(2-\lambda)} \Rightarrow \ln(\tau_\mathrm{w}) = (2 - \lambda)\ln(\dot{\gamma}) \qquad (5.26)$$

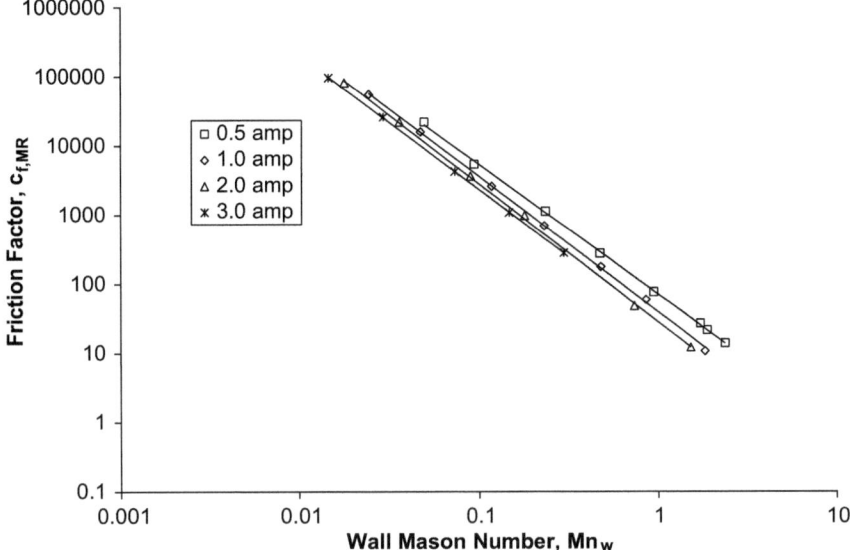

Figure 5.36 Friction factor of MR fluid as a function of Mason number at the wall for 3.2 μm plastic rough surface.

In eqn (5.19) the flow behavior index is defined as the slope of $\ln(\tau_w)$ *vs.* $\ln(\dot{\gamma})$ curve. Equation (5.26) shows that the same slope is equal to $(2 - \lambda)$. Therefore:

$$n = 2 - \lambda \quad \text{or} \quad \lambda = 2 - n \qquad (5.27)$$

Since it is shown that $\lambda = 2 - n$, this index is only a function of the MR fluid itself. The experimental results have shown no dependence of n to the surface topology.

Both the function, A, in eqn (5.24) and the Mason number involve magnetic field terms. Therefore, it is required to collect all magnetic field terms under one function. In order to achieve this, A can be defined as follows in order to divide the magnetic terms and surface topology terms in two different functions:

$$A = g(\text{magnetic terms}) \times h(\text{surface topology terms}) \qquad (5.28)$$

After careful examination of the experimental data it is possible to define the function A as:

$$A = k \left(\frac{B_{air}}{B_{sat}} \right)^{-m} = k \left(\frac{\mu_0 \mu_f H_{MR}}{B_{sat}} \right)^{-m} \times g(\text{surface topology terms}) \qquad (5.29)$$

where, B_{sat} is the saturation magnetic field of the MR fluid and is considered to be 0.8 T, k and m are constants to be determined by the least square regression method applied to function A for each surface type. Now, it would be possible to re-arrange the friction factor equation, so that all magnetic field terms are included in the Mason number definition, such that:

$$C_{f,MR} = k \times g(\text{surface topology terms}) \times \left(\frac{8\eta_f \dot{\gamma}_w B_{sat}^{m/\lambda}}{(\mu_0 \mu_f) \beta^2 H_{MR}^2 B_{sat}^{m/\lambda}} \right)^{-\lambda}$$

or   (5.30)

$$C_{f,MR} = k \times g(\text{surface topology terms}) \times \left(\frac{8\eta_f \dot{\gamma}_w}{(\mu_0 \mu_f)^{1-m/\lambda} \beta^2 H_{MR}^{(2-m/\lambda)} B_{sat}^{m/\lambda}} \right)^{-\lambda}$$

The term in the last parenthesis of eqn (5.30) is defined as the modified Mason number at the wall, $Mn_{w,\text{mod}}$, *i.e.*:

$$Mn_{w,\text{mod}} = \left(\frac{8\eta_f \dot{\gamma}_w}{(\mu_0 \mu_f)^{1-m/\lambda} \beta^2 H_{MR}^{(2-m/\lambda)} B_{sat}^{m/\lambda}} \right) \qquad (5.31)$$

By defining the modified Mason number at the wall, all magnetic field terms and surface topology terms are now presented in two separate functions.

In the following sections, eqn (5.31) will be developed for each of the surface topologies tested in this study and the final form of the friction factor equations will be presented. It will be also demonstrated that by using an average value of n (instead of using n as a function of the magnetic field) the friction factor of a MR fluid can be estimated using the friction factor equations with minimal error. The mean value of n is calculated as 0.063 which makes $\lambda = 1.937$.

5.5.2.1   Non-Ferrous MR Impregnated Porous Surface Friction Factor

From the experimental results, it is proposed to model the friction factor for non-ferrous porous surfaces impregnated with MR fluids as:

$$c_{f,MR} = k \times g(p_s) Mn_{w,\text{mod}}^{-\lambda} \qquad (5.32)$$

where, p_s is the porosity size. After a trial and error process to determine dimensionless $g(p_s)$, eqn (5.32) can be written as:

$$c_{f,MR} = k \exp\left(n \frac{p_s}{d_p} \right) \left(\frac{8\eta_f \dot{\gamma}}{(\mu_0 \mu_f)^{1-m/\lambda} \beta^2 H_{MR}^{(2-m/\lambda)} B_{sat}^{m/\lambda}} \right)^{-\lambda} \qquad (5.33)$$

where, d_p is the average iron particle size of the MR fluid and taken as 5 µm. The parameters k, m and n in eqn (5.33) are determined by least squares regression method applied to the experimental data.

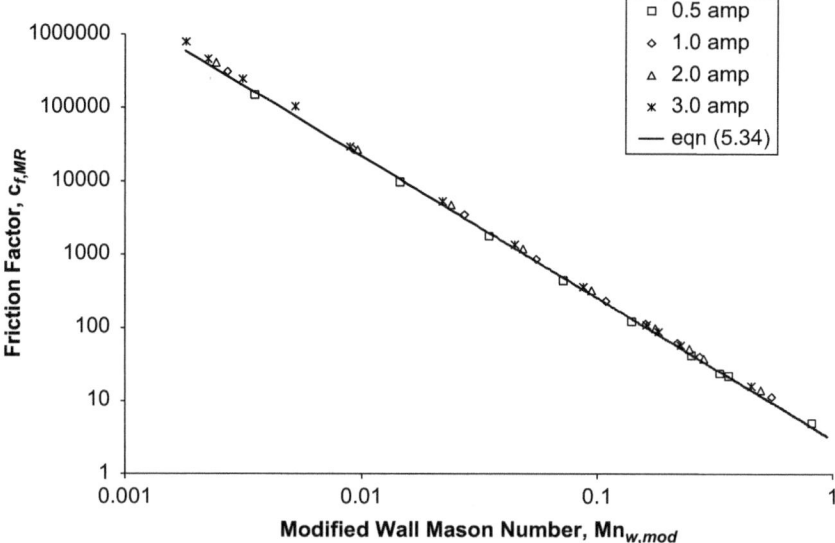

Figure 5.37 Comparison of experimental data and eqn (5.34) for 5 μm porous surface impregnated by through flow technique.

After all experimental data are processed and the parameters in eqn (5.33) are determined, the friction factor for MR fluid flow for through flow impregnation technique the friction factor can be written as:

$$c_{f,MR,porous,throughflow} = 7.57\exp\left(-9.382x10^{-4}\frac{p_s}{d_p}\right)\left(\frac{8\eta_f\mu_0\mu_f\dot{\gamma}_w}{\beta^2\left(\mu_0\mu_f\right)^{-0.394}H_{MR}^{0.606}B_{sat}^{1.394}}\right)^{-1.937}$$

(5.34)

where m is equal to 2.701 for through flow impregnated porous surfaces. Figures 5.37 and 5.38 present how eqn (5.34) represents the experimental data with a single curve for 5 μm and 40 μm porous surfaces.

5.5.2.2 Rough Surface Friction Factor

From the experimental results, it is proposed to model the friction factor for rough surfaces as:

$$c_{f,MR} = k \times g(\varepsilon)Mn_{w,mod}^{-\lambda}$$

(5.35)

where, ε is the surface roughness. The following sections will explain the development of eqn (5.35) for plastic, nickel and cold rolled steel surfaces. It is seen that the parameter, k and $g(\varepsilon)$ vary for surfaces with different magnetic properties. Finally, it is attempted to define a unified equation by defining k and

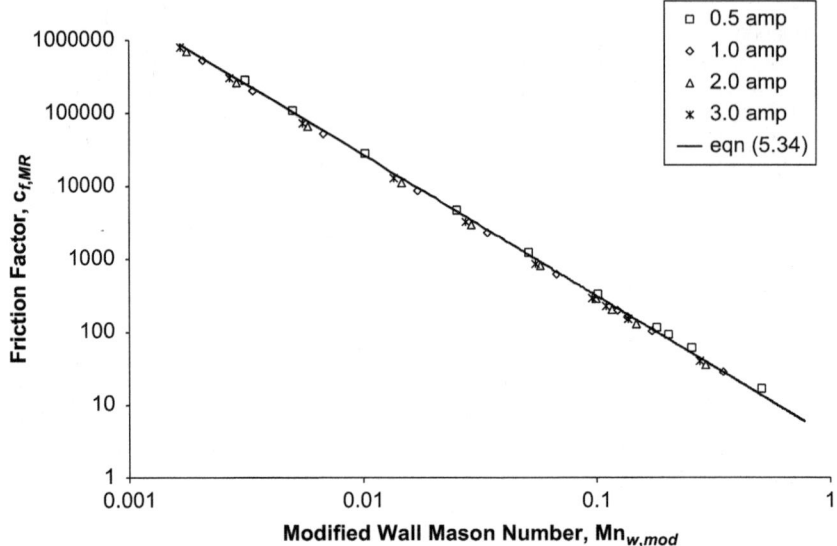

Figure 5.38   Comparison of experimental data and eqn (5.34) for 40 μm porous surface impregnated by through-flow technique.

$g(\varepsilon)$ in terms of surface material magnetic permeability. After a trial and error process to determine $g(\varepsilon)$, eqn (5.35) can be written as:

$$c_{f,MR} = k \left(\frac{\varepsilon}{g_c}\right)^{\mathrm{p}} \left(\frac{8\eta_f \dot{\gamma}_w}{(\mu_0\mu_f)^{1-m/\lambda} \beta^2 H_{MR}^{(2-m/\lambda)} B_{sat}^{m/\lambda}}\right)^{-\lambda} \tag{5.36}$$

5.5.2.2.1   Plastic Rough Surfaces.   Performing least squares regression of all experimental data for plastic rough surfaces, eqn (5.36) can be written as:

$$c_{f,MR,rough,plastic} = 341.1 \left(\frac{\varepsilon}{g_c}\right)^{0.719} \left(\frac{8\eta_f \dot{\gamma}_w}{\beta^2 (\mu_0\mu_f)^{0.119} H_{MR}^{1.119} B_{sat}^{0.881}}\right)^{-1.937} \tag{5.37}$$

Figure 5.39 compares the experimental data with eqn (5.37) for plastic rough surfaces. It can be seen that for a given surface type it is possible to collapse all corresponding data to a single curve using eqn (5.37).

5.5.2.2.2   Nickel Rough Surfaces.   Performing least squares regression of all experimental data for nickel rough surfaces, eqn (5.36) can be written as:

$$c_{f,MR,rough,nickel} = 8.84 \left(\frac{\varepsilon}{g_c}\right)^{0.263} \left(\frac{8\eta_f \dot{\gamma}_w}{\beta^2 (\mu_0\mu_f)^{-0.422} H_{MR}^{0.578} B_{sat}^{1.422}}\right)^{-1.937} \tag{5.38}$$

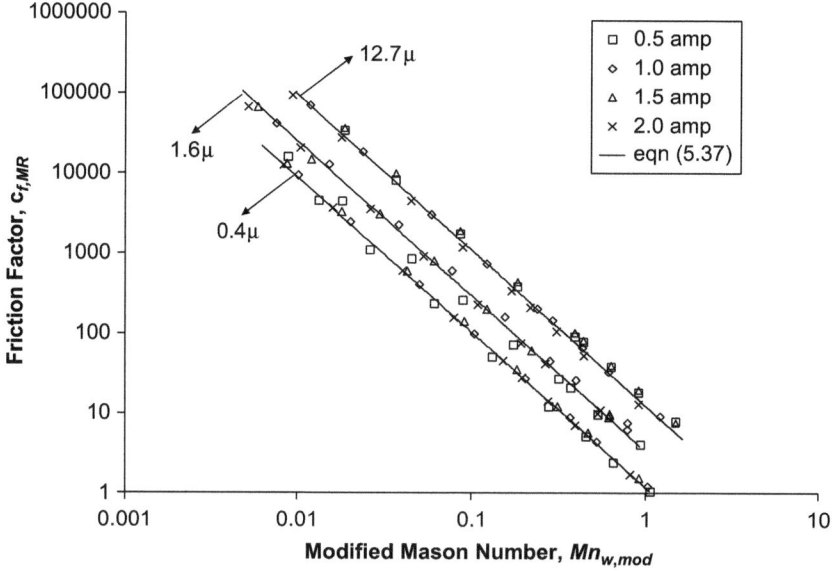

Figure 5.39 Comparison of the friction factor from the experimental results and eqn (5.37) for plastic rough surfaces.

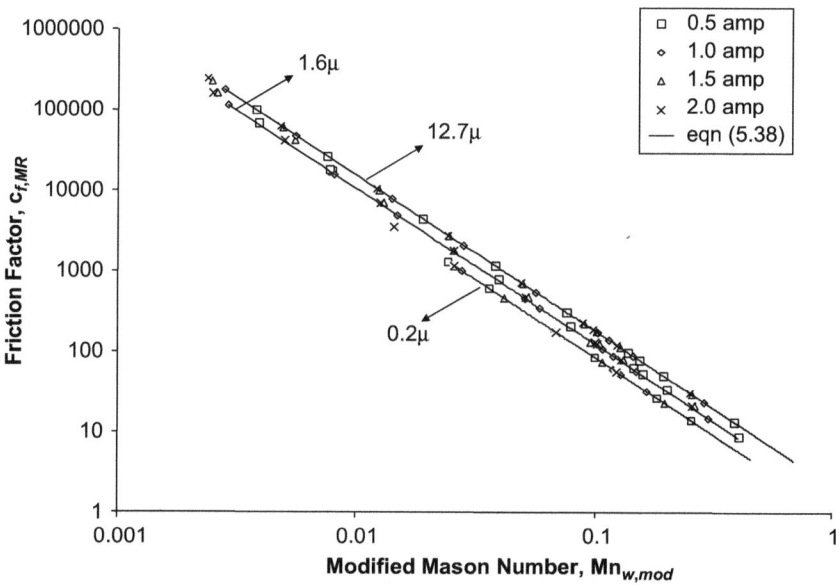

Figure 5.40 Comparison of the friction factor from the experimental results and eqn (5.38) for nickel rough surfaces.

Figure 5.40 compares the experimental data for friction factor of MR fluid with eqn (5.38) for nickel rough surfaces. Similarly, it is possible to represent a given surface type by a single curve using eqn (5.38).

5.5.2.2.3 Steel Rough Surfaces. Performing least squares regression of all experimental data for steel rough surfaces, eqn (5.36) can be written as:

$$c_{f,MR,rough,steel} = 5.54 \left(\frac{\varepsilon}{g_c} \right)^{0.038} \left(\frac{8\eta_f \dot{\gamma}_w}{\beta^2 \left(\mu_0 \mu_f \right)^{-0.851} H_{MR}^{0.149} B_{sat}^{1.851}} \right)^{-1.937} \tag{5.39}$$

Figure 5.41 presents the friction factor of MR fluid for steel rough surface with a roughness of 4.5 μm.

Table 5.6 summarizes the parameters that are used in eqn (5.39) for various rough surface types. Table 5.6 is a useful tool to determine these parameters for surfaces that have relative magnetic permeability between 1 and 2000.

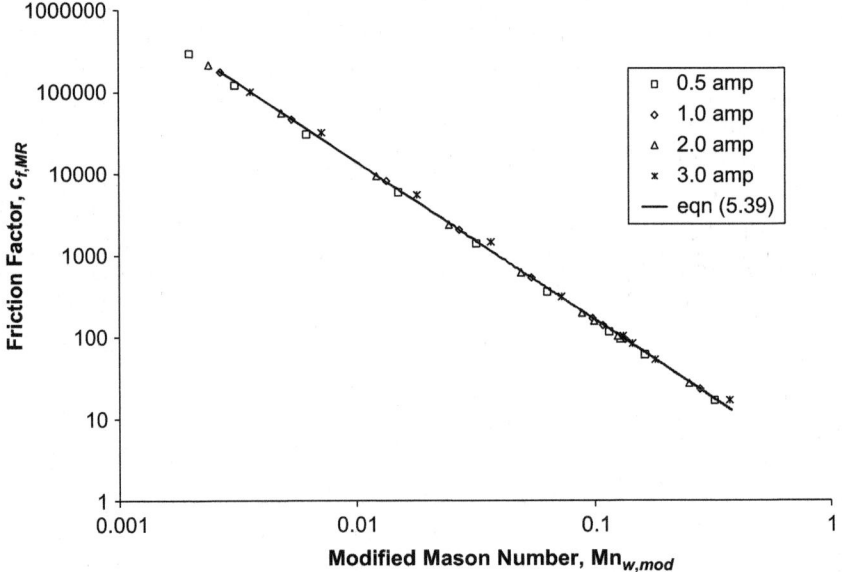

Figure 5.41 Comparison of the friction factor from experimental results and eqn (5.39) for 4.5 μm rough steel surface.

Table 5.6 Parameters of eqn (5.39) for various surfaces.

Surface Material	Relative magnetic permeability	Parameter		
		k	p	m
Plastic	1	341.1	0.719	1.706
Nickel	600	8.84	0.263	2.754
Steel	2000	5.54	0.038	3.585

5.5.2.3 Grooved Surface Friction Factor Equation

From the experimental results, it is proposed to model the friction factor for grooved surfaces as:

$$c_{f,MR} = k \times g(d) Mn_{w,\text{mod}}^{-\lambda} \qquad (5.40)$$

where, d is the depth of the grooves. The following section discusses the development of eqn (5.40) for aluminium and cold rolled steel grooved surfaces. It is seen that the parameters, k and $g(d)$ vary for surfaces with different magnetic properties. Finally, it is attempted to define a unified equation by defining k and $g(d)$ in terms of surface material magnetic permeability. For the non-dimensional surface function, $g(d)$ is determined as:

$$g(d) = \left(\frac{d}{g_c}\right)^r \qquad (5.41)$$

Where, d is the groove depth and r is a parameter to be determined by least squares regression to the experimental data.

5.5.2.3.1 Aluminium Grooved Surfaces. Performing the least squares regression to the experimental data, eqn (5.40) for aluminium grooved surfaces can be written as:

$$c_{f,MR} = 28.7 \left(\frac{d}{g_c}\right)^{0.756} \left(\frac{8\eta_f \dot{\gamma}_w}{(\mu_0 \mu_f)^{0.011} \beta^2 H_{MR}^{1.011} B_{sat}^{0.989}}\right)^{-1.937} \qquad (5.42)$$

Figure 5.42 compares the friction factor determined from the experiments and from eqn (5.42) for G1, G2 and G3. It can be seen that eqn (5.42) represents all experimental data for a given groove configuration with only a single curve. The friction factor for the smooth surface, G0, is also presented in Figure 5.42 to compare the increase in the friction factor for different groove configurations.

5.5.2.3.2 Cold-Rolled Steel Grooved Surfaces. Performing the least squares regression to the experimental data, eqn (5.40) for steel grooved surfaces can be written as:

$$c_{f,MR} = 23.8 \left(\frac{d}{g_c}\right)^{0.224} \left(\frac{8\eta_f \dot{\gamma}_w}{(\mu_0 \mu_f)^{0.010} \beta^2 H_{MR}^{1.010} B_{sat}^{0.990}}\right)^{-1.937} \qquad (5.43)$$

Figure 5.43 compares the friction factor determined from the experiments and from eqn (5.43) for G1 and G2. It can be seen that eqn (5.43) represents all experimental data with only a single curve for a given surface configuration.

5.5.3 Verification

To verify the approach presented in this study, the normalized apparent viscosity of MRF-132AD, as obtained from the rheological measurements performed using the Paar Physica MCR-300 shear rheometer, is plotted as a

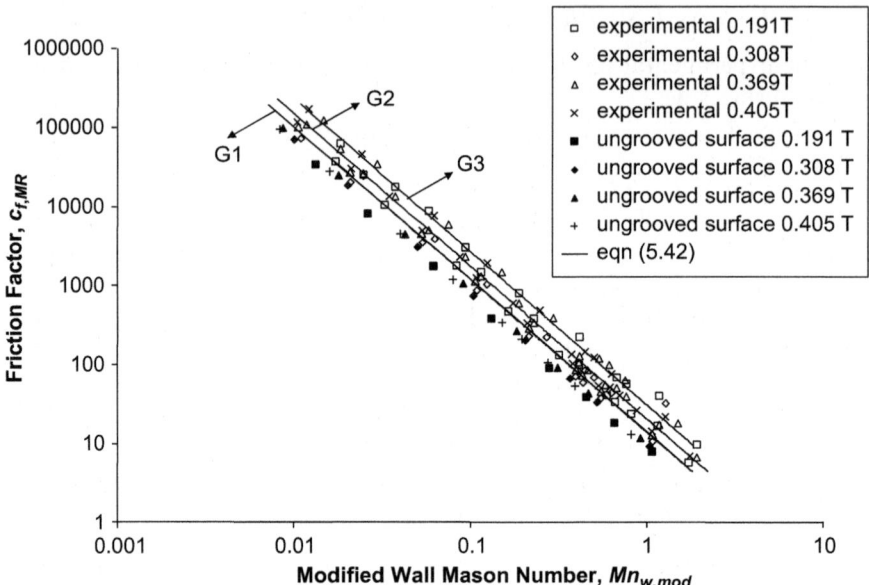

Figure 5.42   Comparison of the friction factor from the experimental results and eqn (5.42) for aluminium grooved surfaces.

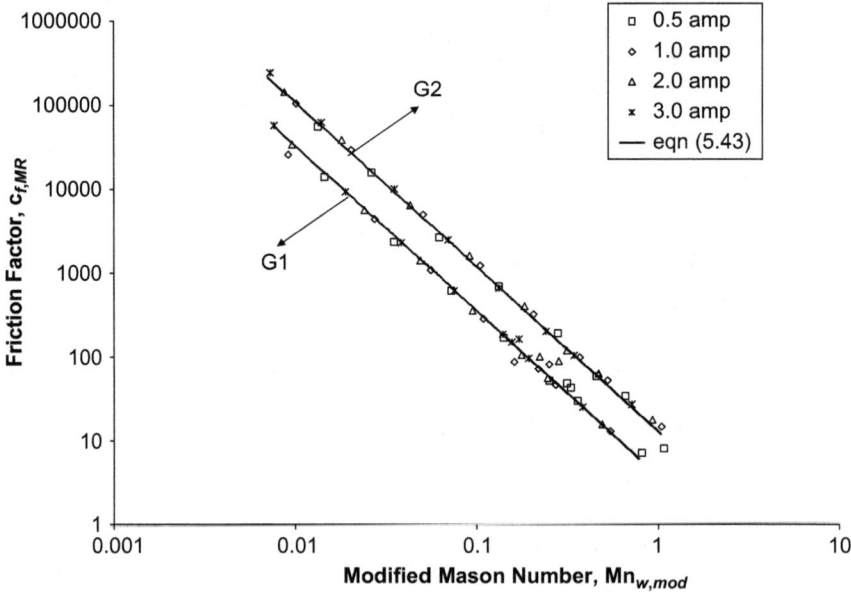

Figure 5.43   Comparison of the friction factor from the experimental results and eqn (5.43) for steel grooved surfaces.

function of regular Mason number and the modified Mason number (eqn (5.39) using parameters obtained for steel rough surfaces), in Figures 5.44 and 5.45, respectively. The infinite shear-rate viscosity is selected as, $\mu_\infty = 0.2$ Pa s and

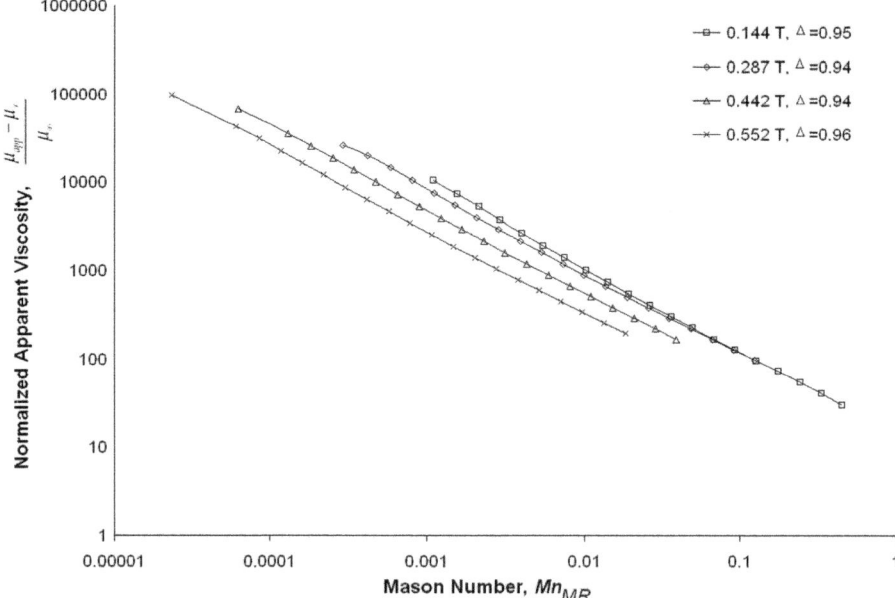

Figure 5.44 Normalized apparent viscosity as a function of Mason number obtained from the MCR-300 shear rheometer.

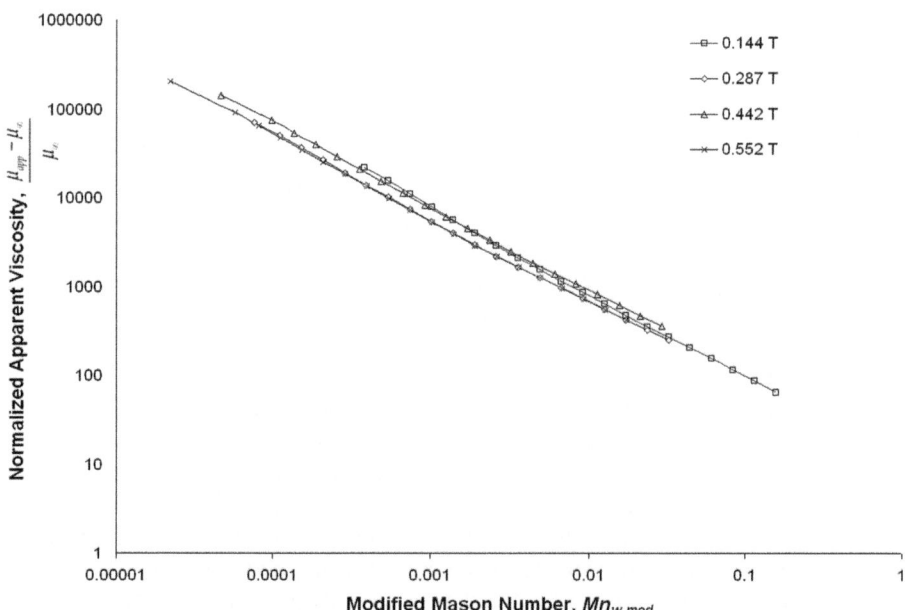

Figure 5.45 Normalized apparent viscosity obtained from MCR-300 shear rheometer tests as a function of modified Mason number.

the surface roughness is taken as 100 microns based on manufacturer's data. It can be seen that the normalized apparent viscosity can be represented as a single master curve when the modified Mason number is used, whereas when the regular Mason number is used, the curves for different magnetic fields are distinctively separated. This validates that the non-dimensional modified Mason number can be utilized for the proposed unified flow analysis in both shear and channel flow of MR fluids, without the need for the concept of shear yield stress.

5.6   Summary and Conclusions

This study presents a new approach for flow analysis of magnetorheological (MR) fluids through channels with various surface topologies. Based on an experimental study, an analytical method is developed to predict the pressure loss of a MR fluid as a function of the applied magnetic field strength, volumetric flow rate and surface topology, *without* utilizing the concept of shear yield stress. A channel flow rheometer with interchangeable channel walls is built to demonstrate that the pressure loss across the MR fluid flow channel is significantly affected by the channel surface properties. Different surface morphologies with various magnetic properties are used in these experimental studies. Rheology experiments are also conducted using a commercial shear rheometer at low shear rates. Based on the experimental results, non-dimensional friction factor equations are introduced.

In the initial stages of the experimental study, it is shown that when a non-magnetic porous channel surface is impregnated with MR fluid, the pressure drop increases significantly. The iron particles are impregnated in the porosities and create magnetic traps. Due to these traps, the shear stress at the channel wall is increased. The impregnation technique also plays an important role. As the amount of particles trapped inside the porosities are increased, it is possible to further increase the pressure drop.

The study on the surface roughness effects demonstrated that the pressure drop for a MR fluid across a channel with different surface properties is greatly dependent on the surface roughness for surface materials with low relative magnetic permeability. The pressure drop can be increased more than 100% through only surface treatment techniques. It is shown that the roughness effects are more apparent at low flow rates and low magnetic field strengths. As the magnetic permeability of the surface material increases, the surface effects tend to diminish.

For grooved surfaces, it is observed that the pressure drop across the channel depends strongly on the groove depth. As the depth of grooves increases, the magnetic field along the groove is intensified resulting in a higher pressure loss. However, the effect of groove width is minimal and negligible, for the tested width ranges. Unlike rough surfaces, the magnetic permeability of the surface material is not a factor in grooved surfaces. Both magnetic and non-magnetic channel materials showed a significant pressure drop increase as the groove depth is increased. This is explained by the localized high magnetic field strength areas at the corners of the grooves. These enhanced magnetic field areas result in an increased pressure drop regardless of the surface material. Preliminary electromagnetic finite element analysis is performed to show how the magnetic field distribution changes when grooves are machined in the channel walls.

A non-dimensional model is developed for each surface morphology, which relates the friction factor of MR fluid to the surface properties, volumetric flow rate and magnetic field. The experimental studies demonstrated that the pressure drop for a MR fluid across a channel with different surface properties is directly proportional to $Mn^{-\lambda}$, where Mn is the Mason number and λ is the power index that is determined from experimental results. It is proven that the power index, λ, is only a function of MR fluid properties, and is related to the flow behavior index, n, as $\lambda = 2 - n$ for a given input current. For the specific MR fluid studied in this chapter $\lambda = 1.937$. The Mason number is modified to incorporate surface effect terms. The modified Mason number is introduced as:

$$Mn_{w,\mathrm{mod}} = \left(\frac{8\eta_f \dot{\gamma}_w}{(\mu_0 \mu_f)^{1-m/\lambda} \beta^2 H_{MR}^{(2-m/\lambda)} B_{sat}^{m/\lambda}} \right)$$

The parameter m is affected by the surface morphology. Therefore, the surface topology affects the H_{MR} term and consequently affects the MR fluid friction factor.

The non-dimensional friction factor model defined the flow behavior of MR fluid as a single curve for all flow rates and magnetic field intensities, for a given surface type. Therefore, it is demonstrated that the MR fluid behavior can be defined by unified curves that depend on the surface topology. Using the relation developed for the friction factor of MR fluid channel flow, the pressure drop of an MR fluid in channel flow, or the torque or shear stress generated by an MR fluid in shear flow can be estimated without referring to a constitutive model or using the concept of shear yield stress.

References

1. X. Wang and F. Gordaninejad, Magnetorheological Materials and their Applications: A Review, in *Intelligent Materials*, ed. M. Shahinpoor and H.-J. Schneider, Royal Society of Chemistry Publishing, Cambridge, United Kingdom, 2007, ch. 14, pp. 339–385.
2. X. Wang and F. Gordaninejad, Flow Analysis of Field-Controllable, Electro- and Magneto-Rheological Fluids Using Herschel-Bulkley Model, *J. Intell. Mater. Syst. Struct.*, 1999, **10**(8), 601–608.
3. E. Lemaire and G. Bossis, Yield Stress and Wall Effects in Magnetic Colloidal Suspensions, *J. Phys. D, Appl. Phys*, 1991, **24**, 1473–1477.

4. G. Bossis, P. Khuzir, S. Lacis and O. Volkova, Yield Behavior of Magnetorheological Suspensions, *J. Magn. Magn. Mater.*, 2003, **258–259**, 456–458.
5. X. Tang and H. Conrad, Quasistatic Measurements on a Magnetorheological Fluid, *J. Rheol.*, 1995, **40**, 1167–1178.
6. J. M. Ginder and L. C. Davis, Shear Stresses in Magnetorheological Fluids: Role of Magnetic Saturation, *Appl. Phys. Lett.*, 1994, **65**(26), 3410–3412.
7. S. Gorodkin, N. Zhuravski and W. Kordonski, Surface Shear Stress Enhancement under MR Fluid Deformation, *Proceedings of 8th International Conference on Electrorheological (ER) Fluids and Magnetorheological (MR) Suspensions*, Nice, France, 2001, pp. 847–852.
8. B. Kavlicoglu, F. Gordaninejad, X. Wang and G. Hitchcock, Effects of Magneto-Rheological Fluid Impregnated Porous Surfaces on Channel Flow of Magneto-Rheological Fluids, *Proc. SPIE Smart Struct. Mater.*, 2005, **57460**, 434–445.
9. F. Gordaninejad, B. Kavlicoglu and X. Wang, Friction Factor of Magneto-Rheological Fluid Flow in Grooved Channels, *Int. J. Modern Phys. B*, 2005, **19**(7–9), 1297–1303.
10. H. Nishiyama, H. Takana, K. Mizuki, H. Weisbecker and S. Odenbach, Dynamic Response of Magneto-Rheological Fluid Channel Flow with Fluid-Wall Interactions, *J. Phys.: Conf. Ser.*, 2009, **149**, DOI: 10.1088/1742-6596/149/1/012077.
11. H. Nishiyama, H. Takana, K. Shinohara, K. Mizuki, K. Katagiri and M. Ohta, Experimental Analysis in MR Fluid Channel Flow Dynamics with Complex Fluid-Wall Interactions, *J. Magn. Magn. Mater.*, 2011, **323**, 1293–1297.
12. H. M. Laun, C. Gabriel and C. Kieburg, Wall Material and Roughness Effects on Transmittable Shear Stresses of Magnetorheological Fluids in Plate-Plate Magnetorheometry, *Rheol. Acta*, 2011, **50**(2), 141–157.
13. B. Kavlicoglu, F. Gordaninejad and X. Wang, A Unified Approach for Flow Analysis of Magnetorheological Fluids, *ASME J. Appl. Mech.*, 2011, **78**(4), 041008.
14. M. Ocalan and G. H. McKinley, Rheology and Microstructural Evolution in Pressure-Driven Flow of a Magnetorheological Fluid with Strong Particle–Wall Interactions, *J. Intell. Mater. Syst. Struct.*, 2012, **23**(9), 969–978.
15. Z. P. Shulman, V. I. Kordonsky, E. A. Zaltsgendler, I. V. Prokhorov, B. M. Khusid and S. A. Demchuk, Structure, Physical Properties and Dynamics of Magnctorhcological Suspensions, *Int. J. Multiphase Flow*, 1986, **12**, 935–955.
16. R. E. Rosensweig, On Magnetorheology and Electrorheology as States of Unsymmetric Stress, *J. Rheol.*, 1995, **39**(1), 179–192.
17. G. Bossis, E. Lemaire, O. Volkova and H. Clercx, Yield Stress in Magnetorheological and Electrorheological Fluids: A Comparison between Microscopic and Macroscopic Structural Models, *J. Rheol.*, 1986, **41**(3), 687–704.

18. J. Huang, J. Q. Zhang and J. N. Liu, Effect of Magnetic Field on Properties of MR Fluids, *Int. J. Modern Phys. B*, 2005, **19**, 597–601.

19. B. Abu-Jdayil and P. O. Brunn, Effects of Electrode Morphology on the Channel Flow of an Electrorheological Fluid, *J. Non-Newtonian Fluid Mech.*, 1996, **63**, 45–61.

20. B. Abu-Jdayil and P. O. Brunn, Effects of Nonuniform Electric Field on Channel Flow of an Electrorheological Fluid, *J. Rheol.*, 1995, **39**, 1327–1341.

21. Y. Otsubo, Effect of Electrode Pattern on the Column Structure and Yield Stress of Electrorheological Fluids, *J. Colloid Interface Sci.*, 1997, **190**, 466–471.

22. C.-Y. Lee and K.-L. Jwo, Experimental Study on Electro-Rheological Material with Grooved Electrode Surfaces, *J. Mater. Design*, 2001, **22**, 277–283.

23. S. Gorodkin, N. Zhuravski and W. Kordonski, Surface Shear Stress Enhancement under MR Fluid Deformation, *Proceedings of 8th International Conference on Electrorheological (ER) Fluids and Magnetorheological (MR) Suspensions*, Nice, France, 2001, pp. 847–852.

24. D. Carlson, Magnetorheological Fluids, in *Smart Materials*, ed. M. Schwartz, CRC Press, 2008, ch. 17.

25. T. C. Halsey, J. E. Martin and D. Adolf, Rheology of Electrorheological Fluids, *Phys. Rev. Lett.*, 1992, **68**(1), 1519–1523.

26. A. H. P. Skelland, *Non-Newtonian Flow and Heat Transfer*, John Wiley & Sons, Inc., 1967.

27. A. A. Zaman, Techniques in Rheological Measurements: Fundamentals and Applications, in *NSF Engineering Resource Center for Particle Science & Technology*, University of Florida, Gainesville, Florida, 1998.

28. B. R. Munson, D. F. Young and T. H. Okiishi, *Fundamentals of Fluid mechanics*, John Wiley & Sons, Inc., 3rd edn, 1998.

CHAPTER 6

Thin-film Rheology and Tribology of Magnetorheological Fluids

JUAN DE VICENTE[*a] AND ANTONIO J. F. BOMBARD[b]

[a] Department of Applied Physics, Faculty of Sciences, University of Granada, C/ Fuentenueva s/n, 18071 – Granada, Spain; [b] Universidade Federal de Itajubá, ICE/DFQ, Av BPS 1303, Itajubá/MG, Brazil
*Email: jvicente@ugr.es

6.1 Introduction

Ferrofluids and magnetorheological (MR) fluids are magnetic field-responsive colloidal suspensions. The main difference between them strikes in the size of the constituents. Strictly speaking, ferrofluids are based on superparamagnetic nanometre sized magnetic particles. On the other hand, MR fluids are constituted by micron-sized magnetizable particles. Ideally, ferrofluids remain in the "liquid" state even in the presence of large magnetic fields, while MR fluids exhibit a "liquid–solid" transition upon the application of moderate magnetic fields (approx. 10 kA m^{-1}). The reason for this is the formation of particle clusters aligned in the field direction.[1,2] By way of example, ferrofluids are currently employed to form liquid seals around the spinning drive shafts in hard disks, to remove heat from the voice coil in loudspeakers, as contrast agents for magnetic resonance imaging and experimental cancer treatment. On the other hand, the mechanical property of MR fluids has been exploited in numerous torque transfer automotive applications like the fabrication of

RSC Smart Materials No. 6
Magnetorheology: Advances and Applications
Edited by Norman Wereley
Published by the Royal Society of Chemistry, www.rsc.org

clutches, brakes and dampers. Currently, this technology has been incorporated, under the name Magneride™, in the primary suspension system of luxury vehicles since 2002, the most recent models involving the Cadillac DTS, Ferrari 599 GTB, Buick Lucerne and Audi R8.

Most of the past efforts have been devoted to understand the bulk rheological behavior of these materials in the presence of viscometric flows under external magnetic fields. However, from a practical point of view it is also interesting to study their thin-film rheological properties as most of the applications concern strongly confined geometries under exceedingly large shear rates 10^4–10^6 s^{-1}.[2] In this case, the interaction of MR fluids with the devices in which they are used is important. A well-known example is the so-called "in-use-thickening" that is presumably due to the wear between metallic parts and iron microparticles employed in commercial MR fluid formulations and that has important impact in a lower durability and reliability of MR fluids' applications.[3,4] Surprisingly, in spite of its importance, the literature on this topic is very scarce.

The term "tribology" describes the "science and technology of interacting surfaces in relative motion". It originates from the Greek "tribos" which means rubbing or attrition and includes the study and application of the principles of friction, wear and lubrication. The frictional behavior is typically quantified by the friction coefficient μ, which is basically the ratio between the tangential force and the applied normal force. In general, the friction coefficient strongly depends on the relative speed between rubbing surfaces, the high-shear lubricant viscosity in the contact and the sustained normal load (all three traditionally grouped in the so-called Gumbel number $Gu \approx speed \times viscosity/load$). Hence, lubrication properties are classically represented in the form of a Stribeck curve, where the coefficient of friction is given as a function of the Gumbel number, as shown in Figure 6.1. The Stribeck curve is usually divided

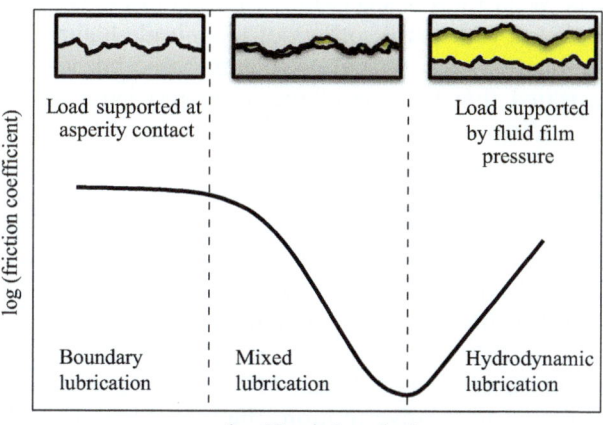

Figure 6.1   Typical Stribeck curve showing the dependence of the friction coefficient with the Gumbel number.
Adapted from Figure 1 in ref. 31.

into several regimes. In the boundary lubrication regime, which occurs at slow speeds when there is negligible fluid entrainment into the contact, the load is carried by the contacting asperities and is dependent on the surface and interfacial film properties at the molecular scale. In the hydrodynamic or elastohydrodynamic lubrication (EHL) regime, a film of lubricant, whose thickness depends on the viscosity and entrainment speed, is entrained to fully separate the solid surfaces (classical rheology). In the mixed lubrication regime, which lies between boundary lubrication and EHL, both the boundary film and bulk lubricant play a role in determining friction.[5]

The behavior and performance of lubricants containing dispersed particles has been a topic of research for many years.[6] A wide range of mechanisms of friction-reduction, enhanced load-carrying capacity and antiwear of nano-particles in lubricants have been described in the literature,[7,8] either forming transferred/deposited solid lubricant films on the rubbing surfaces that eventually serve as spacing, preventing solid–solid contact between asperities of the two mating surfaces ("mending"[9] and "polishing"[10] effects), or by shearing trapped nanoparticles at the interface without the formation of an adhered film. On the other hand, contrary to the case of dilute nanoparticle based lubricants, the action of large amounts of nanoparticles and microparticles is frequently detrimental as they typically sediment under gravity, and also act as grinding materials because the roughness of the friction surfaces are often several microns in size.[8,11–13] Previous work has shown that large particles penetrate and pass through low speed, rolling contacts, where they adhere to form a continuous film on the tracks.[14,15] This film slowly disappears with increasing speed.[15]

The tribological properties of magnetic field responsive suspensions have been vaguely investigated in the literature. A pioneering study by Bullough and coworkers[16] demonstrated that severe wear under boundary lubrication may preclude the use of electric field responsive (ER) fluids in applications involving metallic contacts. Wear rates of steels and brass sliding contacts lubricated with ER fluids were found to be three orders of magnitude higher than those produced under lubrication by a typical petroleum-based lubricant. Later, Wong et al.[17] demonstrated that because of the fewer restrictions on the choice of surfactants and additives, MR fluids can perform better than ER fluids in the unexcited mode of operation and under boundary lubrication conditions. A later review paper by the same authors demonstrated that the tribology of these fluids is a complex affair in the light of results from MR fluid, steel-on-steel, block-on-ring, sliding contacts tests.[18] The effect of particle concentration in a MR fluid on the performance of a boundary lubricated contact under sliding conditions was investigated by Leung et al.[19] Unexpectedly, they found that friction and wear were not monotonic and reached a maximum at a particle content of 10–100 g L^{-1}. When MR fluids are in contact with sealed devices and subjected to relative motion, the seal wears much faster than in a device that uses conventional lubricating oil due to the abrasive nature of the iron particles within the MR fluid. In this sense, a seal wear test method to simulate the wear of rod seals in MR fluid-based dampers was developed by Iyengar et al.[20] In 2011, Lee et al.[21] employed specific additives to modify a commercial MR fluid and improve friction and wear characteristics. The additives were: (a) zinc-dithiophosphate to prevent wear; (b) molybdenum-dithiocarbamate to reduce friction and (c) amine anti-oxidant, to slow down the oxidation of iron particles within the MR fluid.

The understanding of the tribological properties of magnetic fluids under the presence of magnetic fields is even more limited. Just recently, Li-jun *et al.*,[22] employing a MS-800 four-ball tester, modified to apply a controllable and variable magnetic field, studied the tribological properties of a magnetic fluid based on 6 wt% $Mn_{0.78}Zn_{0.22}Fe_2O_4$ nanoparticles. They found that under the effect of a magnetic field, the bearing capacity increased with increasing magnetic induction. More recently, Hu *et al.*[23] modified a four-ball tribological tester to apply magnetic fields. They demonstrated that in the presence of magnetic fields the friction coefficient in MR fluids was four times as large as that without magnetic fields.

Abrasive iron particles within MR fluids have been exploited for optical and jet finishing technology by Kordonski and coworkers[24–26] in what they called a field assisted finishing process. Also recently, a magnetic abrasive finishing (MAF) process has been reported that uses flexible brushes embedding abrasive SiC particles to adapt the contour of the workpiece surface to be finished.[27,28]

This chapter is organized as follows. In the first part we will review some recent advances on the tribological behavior of nanosized magnetic particles operating between interacting surfaces under a continuum approach. We will focus on the use of these particles to reduce friction and also to prevent lubricant starvation. In the second part we will move to a more complex scenario involving larger particles and report on the thin-film rheology and tribological properties of micron-sized carbonyl iron magnetic particles. We will mostly concentrate on isoviscous elastic contacts because friction in this case is not as severe as in conventional hard contacts.

6.2 Lubricating with Nanosized Magnetic Particles

The magnetic nanofluids used in this section were purchased from Ferrotech (APG series). For comparative purposes they have the same superparamagnetic Langevin type dependence (saturation magnetization 24.3 ± 0.7 kA m^{-1}) but different Newtonian viscosities ($\eta = 46$, 200 and 560 mPa s). The carrier fluids were synthetic ester and synthetic hydrocarbon, for the first and the last two ferrofluids, respectively. Also, density and surface tension were the same for all the fluids. Considering a particle size log-normal distribution and using the Langevin function, a value of 9 nm was obtained for the average particle diameter, with a polydispersity around 20%. Particle volume fraction was around 5.5 vol% for the three lubricants. Neither shear-thinning nor visco-elasticity were observed in the range of shear rates of interest.

Surface separations in this section will be small (around a few microns) but still large enough for the suspension to be treated as a continuous medium. Accordingly, its macroscopic behavior can be described by a modified incompressible Navier–Stokes equation including a magnetic body force.[29]

6.2.1 Thin-film Rheology in Face Seals

Friction reducing capabilities of ferrofluids have been extensively reported in the literature for a wide range of non-conformal counteracting geometries.[22,30] Andablo-Reyes *et al.*[31] conducted a careful investigation on the field-driven friction reduction for the case of two slightly tilted flat circular surfaces in relative torsional shearing motion. For this purpose, a commercial magnetorheometer (MCR501, Anton Paar) was adapted. In a typical experiment the (misaligned) upper plate was rotated from $\Omega = 0$ to 150 rad s^{-1} at a fixed gap distance under a non-uniform magnetic field. During the experiment, both the torque and normal force acting on the plate were measured. The concentricity error was measured to be less than 2 microns and misalignment was found to be 1.75 microns. In this apparatus, the friction coefficient μ was computed from the ratio between the shear stress at the rim of the plates and the average or nominal normal stress acting on the rotating plate. By solving a modified Reynolds equation it was possible to compute the pressure distribution and consequently determine both the shear stress and the normal force acting on the plates. A reduction in the friction coefficient by one order of magnitude was observed by increasing the magnetic field. Experiments were in good agreement with the predictions obtained by numerical solution of the Reynolds equation. The friction reduction was directly related to the fact that the normal force acting on the plates increased when increasing the magnetic field strength as a consequence of the appearance of magnetic body forces in the ferrofluid (see Figure 6.2).

6.2.2 Using Magnetic Fields to Control Starvation in Lubricated Contacts

Ferrofluids can also be used to control lubricant starvation in rubbing contacts. In a recent communication, Andablo-Reyes *et al.*[32] demonstrated the use of

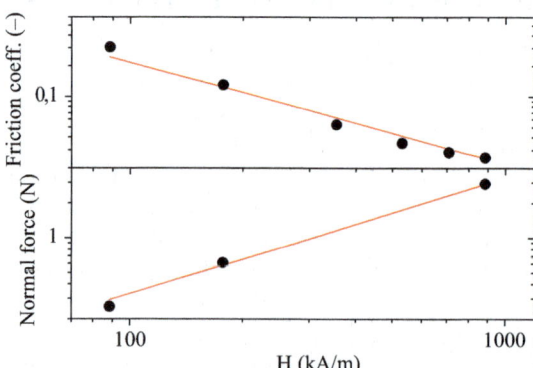

Figure 6.2   Friction coefficient and normal force acting on the upper plate as a function of the magnetic field strength for low angular speeds ($\eta\Omega = 0.88$ Pa): (Symbols) experimental data, (Lines) numerical solution of the Reynolds equation.
Adapted from ref. 31.

ferrofluids to improve the replenishment of lubricant to the contact in the absence of free bulk lubricant. An indirect method was employed that consisted of measuring the friction force *versus* the entrainment speed in an adapted Mini Traction Machine (MTM) under different running conditions. In this tribometer an elastomer ball (Sylgard™ 184, radius 9.5 mm) is loaded and rotated against a flat aluminum surface of a rotating disc. With this, the reduced elastic modulus is 8 MPa and the point contact safely operates in the isoviscous elastic lubrication regime. In the MTM both surfaces are independently driven to achieve any desired sliding/rolling speed combination. Experiments reported here were carried out at room temperature at a slide–roll ratio of 0.5. The frictional force was measured by a load cell attached to the ball motor. The abrupt change in the slope of the friction coefficient observed at large speeds was associated to the onset of the starved regime. The authors reported that the critical point between the fully flooded and starved lubrication regime was reached at lower speeds for the ferrofluid with the higher viscosity. On the contrary, the applied load had no influence in the range investigated (3–7 N). Upon the application of magnetic fields, the contacts were found to continue working in the fully lubricated regime up to larger speeds hence preventing starvation. A typical example is shown in Figure 6.3.

6.3   Lubricating with Micron-sized Magnetic Particles

Particles employed in the formulation of MR fluids analyzed in this section were kindly supplied by BASF SE. They were all hard grades. Table 6.1 contains size and chemical composition data for the particles. These particles

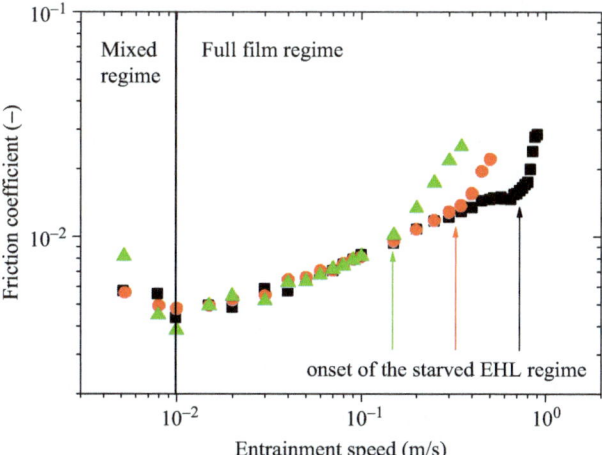

Figure 6.3   Stribeck curve for a suspension of nanosized magnetic particles for three different magnetic fields: (Green ▲) 19.9 kA m^{-1}, (Red ●) 39.8 kA m^{-1} and (■) 119.4 kA m^{-1}. The viscosity of the ferrofluid is 200 mPa s and the applied load is fixed at 7 N.
Adapted from Figure 7 in ref. 32, with kind permission from Springer Science and Business Media.

Table 6.1　Particle size and bulk composition of hard grade carbonyl iron powders employed in the preparation of MR fluids.

	OM	*OS*	*OX*	*HS*	*HSI*	*HQ*
Coating	None	SiO_2 (0.7%)	Fe_2O_3 (5%)	None	$FePO_4$ (1.0%)	None
$d_{50}/\mu m$	4.0	4.0	4.0	2.0	2.0	1.1
Polydispersity $(d_{90} - d_{10})/d_{50}$	1.75	1.5	1.75	1.0	1.0	1.5
Fe	>97.8%	>97%	>96%	>97.8%	>97.8%	>97.5%
C	0.7–0.9%	0.7–0.9%	0.8–0.9%	<1.0%	<1.0%	0.9%
N	0.6–0.9%	0.6–0.9%	0.7–0.9%	<1.0%	<1.0%	0.9%
O	0.2–0.4%	0.4–0.7%	≈0.3%	<0.5%	<0.5%	0.5%

were thoroughly dispersed in Newtonian fluids such as polyalphaolefin oil (PAO, Chevron Synfluid 2 cSt) and silicone oils (SO, Sigma-Aldrich). In some cases, an organoclay (Bentone SD3, Elementis) was also added to 0.3 wt% in MRF formulation. Of course, the formulation of a commercial MR fluid is known to be more complex. However, in general it is more convenient not to include further dispersing additives in the formulation to facilitate the interpretation of the results.

Compliant tribopairs were formed between stainless steel balls and PTFE surfaces. A non-conforming ball-on-three plates contact was used to ascertain the frictional properties of MR fluids. In this geometry, a ball (radius R) is pressed at a given normal force F_N against three plates that are mounted on a movable stage.[8] In a typical experiment, the ball is made to rotate at an increasing sliding speed V while the plates are held at rest. Geometrical arguments reveal that the friction coefficient can be obtained from the torque sensed by the ball (M) as follows:

$$\mu = \frac{F_F}{F_L} = \frac{M}{F_N R} \tag{6.1}$$

Here F_F and F_L are the total frictional force and the total normal load coming from the three point sliding contacts, respectively.

In this apparatus, the ball was made of stainless steel (AISI 316; radius $R = 0.25$ inch $= 6.35 \times 10^{-3}$ m) and the plates were equal sized parallelepipeds (3 mm × 6 mm × 16 mm) made of PTFE (Anton Paar). The root-mean-square roughness of the steel ball ($R_q \approx 10$ nm) was smaller than the PTFE roughness ($R_q \approx 700$ nm). The Young's modulus for the stainless steel and the PTFE were 193×10^9 Pa and 0.5×10^9 Pa, respectively while their Poisson's ratios were 0.3 and 0.46. With this, the reduced elastic modulus, E' was 1.265×10^9 Pa. The applied load used was $F_N = 10$ N which, according to Hertz theory, produced a large contact radius $a = 3.3 \times 10^{-4}$ m, an indentation depth $\delta = 1.7 \times 10^{-5}$ m, and maximum contact pressure $p_{max} = 2.1 \times 10^7$ Pa. All tests were carried out at a temperature of 25 °C and at a constant slide-to-roll ratio of SRR $= 2$ (*i.e.* pure sliding conditions). Friction measurements were taken over a wide sliding speed range while progressively increasing the sliding speed from $V = 5 \times 10^{-5}$ to 1 m s^{-1} so as to fully map the Stribeck friction curve.

The regime of lubrication was identified using the hydrodynamic lubrication regime map for circular contacts.[33] According to this, so long as a full fluid film is present, the contact should be operating in the isoviscous-elastic EHL lubrication regime. This was expected because of the low elastic modulus and low contact pressure in the contact.

6.3.1 Boundary Lubrication Regime: Particle Effects

When rubbing the surfaces at low speed and/or large load, friction is supported by asperity contacts in the boundary lubrication regime. This section will be devoted to the study of this regime of lubrication.

In the case of unlubricated contacts, the friction coefficient was found to be essentially constant with V up to speeds of the order of 0.01 m s^{-1}; severe wear of the PTFE surface occurred for larger speeds (see Figure 6.4). The low friction coefficient value achieved ($\mu = 0.02$–0.03) is explained because of the molecular structure of PTFE. At reasonably low speeds, a PTFE transfer film is deposited on the steel surface that reduces friction. In general, friction of dry contacts typically comes from deformation and interfacial adhesion contributions. In particular, in compliant contacts, the friction coefficient due to deformation energy losses is very small (one order of magnitude smaller than the measured friction coefficient) and, as a consequence, interfacial adhesion dominates.[8]

In the presence of Newtonian fluids, the contact might change regime of operation depending on the viscosity of the lubricant (see next section); however, as soon as the oil viscosity remains at a low value (*e.g.* using PAO), the contact will still operate in the boundary regime because the lubricant is not

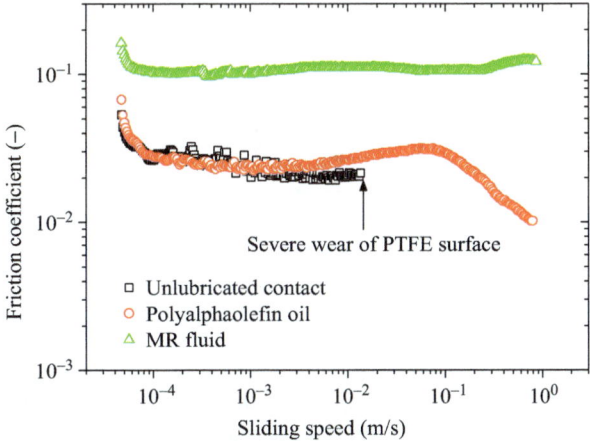

Figure 6.4 Friction coefficient *versus* sliding speed curves using a compliant ball-on-three plates tribometer: (squares) unlubricated, (circles) polyalphaolefin lubricated, (triangles) MR fluid lubricated; OS carbonyl iron particles at 10 vol% in PAO including 0.3 wt% organoclay.
Adapted from ref. 8.

expected to be entrained in the contact. Experiments revealed that the friction coefficient initially increases when increasing the speed, reaches a maximum and then decreases (see Figure 6.4). The Shallamach model[34] predicts such a maximum to occur at a critical velocity of approximately 10 m s^{-1} and this is a much larger value than the one we find in the experiments (typically ~ 0.1 m s^{-1}). A more plausible explanation for this maximum is, however, the onset of the mixed lubrication regime after surface indentation.[8]

When micron-sized magnetic particles were added to the base oils, the particles became entrapped in the contact and the lubrication regime was dominated by a solid–solid boundary friction, as particles formed boundary films of a few times the typical particle size (see Figure 6.4). Optical observations demonstrated the appearance of a blackish wear scar after running the tests and the presence of iron particles indented in the PTFE matrix under sliding. These iron microparticles formed a third body film and hence prevented the rubbing faces from direct contact. As a consequence, the friction coefficient is expected to depend on the grade of the iron employed (size, roughness, surface chemistry and composition). A question naturally arises: which is the best CIP grade to keep friction at the lowest possible value?

The average friction coefficient and wear scar diameters for a range of carbonyl iron grades employed are shown in Figure 6.5. The effect of surface coating on the tribological performance can be ascertained by comparing OM, OS and OX powders since they have very similar size and only differ in the surface chemistry. As demonstrated in Figure 6.5, OS carbonyl iron particles (that are coated with a thin silica shell) provide the best friction-reduction behavior and also the smaller wear scar. On the other hand, OX carbonyl iron

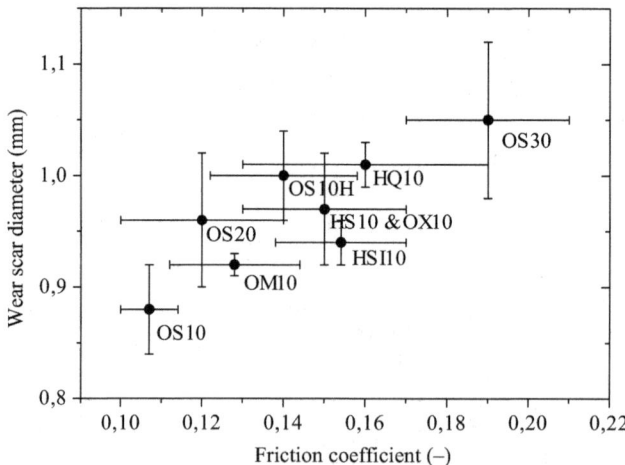

Figure 6.5 Tribological performance of MR fluids in the boundary lubrication regime. First two letters correspond to the grade of the carbonyl iron employed in the formulation of the MR fluid. Numbers correspond to the particle concentration in vol%.
Adapted from ref. 35.

particles give the largest friction coefficient and wear scar diameters, in this case probably due to the larger surface roughness of the particles. By comparing HS and HIS powders it is found that particles coated with phosphate groups do not significantly reduce friction and/or wear. Finally, the effect of particle size can be explored by comparing OM, HS and HQ powders. The results obtained reveal that friction and wear are reduced by employing larger particles. In all cases investigated, frictional results were reasonably well correlated to the wear scar diameter; the larger the friction coefficient, the larger the wear scar diameter.

Also shown in Figure 6.5 are results for MR fluids at different particle content. Results are shown for OS particles because of the lower friction and smaller wear scar diameters if compared to OM and OX grades. The friction coefficient was found to be very sensitive to the particle concentration. In fact, the friction coefficient increased up to 95% upon increasing the particle content from 10 to 30 vol%. The friction mechanism seemed to change as the concentration increased because particles accumulated at the inlet resisting entrapment and obstructing the replenishment of the contact with lubricant. The application of magnetic fields gave also a larger friction coefficient and wear scar diameter. Again, a significantly large concentration of particles was found at the inlet of the contact.

6.3.2 Full Film Lubrication Regime: Dispersing Medium Effects

The aim of this section is to interrogate the full film-forming abilities of MR fluids in the absence of magnetic fields (*i.e.* the full film lubrication regime). To do so, the viscosity of the dispersing phase had to be increased. Newtonian fluids employed in this section were silicone oils of different viscosities; 9.3 mPa s (SO10), 98.0 mPa s (SO100) and 340 mPa s (SO350) (Sigma-Aldrich). The tribopairs employed were again PTFE and steel.

In the isoviscous-elastic regime, the pressures developed within the contact are large enough to cause an extensive deformation in the compliant surface but at the same time they are quite low to cause a substantial increase in the lubricant viscosity.[33] As a result, in this regime, the elastic deformation of the PTFE is a very significant part of the thickness of the fluid film.

To a sufficient accuracy, the elastic deformation of the PTFE plates produces a circular Hertzian contact region of approximately constant film thickness h_C and radius,

$$a = \left[\frac{\sqrt{2}}{4} R \left(\frac{1 - \nu_1^2}{E_1} + \frac{1 - \nu_2^2}{E_2} \right) F_N \right]^{1/3} \tag{6.2}$$

and within this region, most of the friction originates from Couette flow. Consequently, the friction force in the contact F_C can be estimated by integration of the Couette shear stress $\tau = \eta V / h_C$ over the circular Hertzian contact according to:

$$F_C = \int_{r=0}^{r=a} \tau 2\pi r \mathrm{d}r = \eta \pi a^2 \frac{V}{h_C} \tag{6.3}$$

Therefore, the friction coefficient, μ_{EHL} is given by:

$$\mu_{EHL} = \frac{F_C}{F_{Li}} = \frac{3}{\sqrt{2}} \eta \pi a^2 \frac{V}{h_C F_N} \tag{6.4}$$

where F_{Li} is the normal load on surface i.

Many equations exist in the literature for the central film thickness h_C in the isoviscous elastohydrodynamic regime, which are obtained by regression-fitting computed results over a wide range of conditions of entrainment speed, viscosity, load, modulus and radius.[33,36-40] The Biswas and Snidle equation[39] together with eqn (6.4) will be used later in this work to estimate the friction under full film EHL lubrication.

Apart from this contribution - subtracting the plastic deformation of PTFE - to the friction coefficient, the deformation term contribution must be included. This deformation term arises from the incomplete recovery of the energy dissipated by subsurface viscoelastic deformation. Greenwood *et al.*[41] calculated that for a sphere on a flat compliant surface, the friction coefficient due to deformation energy losses is given by:

$$\mu_{def} = \alpha \frac{3}{16} \frac{a}{R} \tag{6.5}$$

where the loss factor α, is proportional to the damping factor $\tan \delta$, through $\alpha = k\pi \tan \delta$, where k is typically between 1 and 3. For PTFE surfaces, $\tan \delta \approx 0.07$ at 25 °C.[42] Assuming a value of $k = 1$ and $\tan \delta \approx 0.07$, a value of $\mu_{def} = 0.002$ is obtained.

The prediction of the friction coefficient in full film lubrication regimen calculated as described above, $\mu = \mu_{def} + \mu_{EHL}$, is shown as a straight line in Figure 6.6 and compared to experimental data on Newtonian fluids.

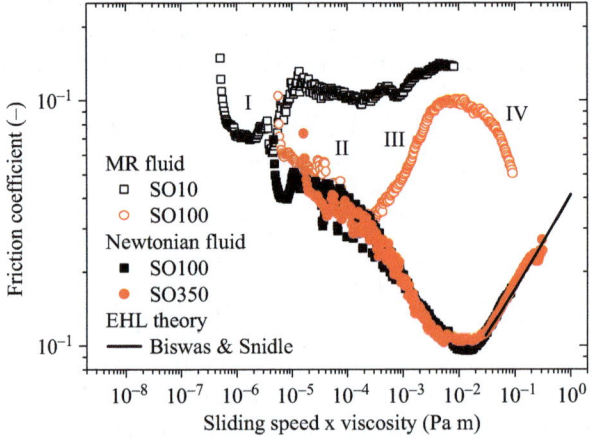

Figure 6.6 Friction coefficient *versus* sliding speed ×viscosity curves; open symbols, MR fluids (HQ 6 vol%); closed symbols, silicone oils; SO10, 9.3 mPa s; SO100, 98.0 mPa s and SO350, 340 mPa s; solid line, EHL theory prediction.[39] Adapted from ref. 8.

As observed, the approximate prediction of assuming only Couette friction in the theoretical Hertzian contact satisfactorily estimates the experimental results for Newtonian fluids and captures quite well the expected $(\eta V)^{0.4}$ dependence in the full film regime.[33]

The friction mechanism dramatically changes when iron microparticles are dispersed in the base silicone oils. In all cases investigated, friction increases when adding magnetic microparticles if compared to experiments for particle-free liquids (see Figure 6.6). At low speeds (region I), solid–solid friction dominates independently of the oil viscosity; in this regime, the characteristics of the particles are important (see section before). At intermediate speeds (region II), the effect of oil viscosity manifests. In general, the friction coefficient quickly starts decreasing due to a three-body abrasive wear that likely occurs where hard iron microparticles are free to roll and slide between the sliding surfaces ("ball bearing" effect) of the microparticles (see Figure 6.6). The carrier fluid seems to be entrained in the contact through the thin packed bed of particles and it is hypothesized that the microparticles may act as the combination of rolling and sliding bearings between the frictional surfaces leading to a decrease in the friction coefficient. Next, at a critical sliding speed ($\eta V \approx 10^{-4}$ Pa m), microparticles are likely to penetrate into the PTFE surfaces in the rubbing process giving way to a two-body abrasive wear region (region III). As a result, the friction coefficient increases, presumably because the rolling motion of the microparticles is hampered. At a sufficiently large speed, the oil may be entrained in the contact (region IV). At this stage, the theoretical base oil film thickness h_C is larger than the typical size of the iron particles and dispersed particles are expelled out of the contact hence reducing the friction coefficient. This critical speed correlates well with the onset of the full film lubrication regime in Newtonian fluids (see Figure 6.5) and increases with increasing the particle size.[8]

6.4 Conclusions

The tribological behavior of magnetic suspensions has been revisited. First, magnetic nanoparticles are demonstrated to reduce friction up to one order of magnitude when employed to separate interacting surfaces under non-uniform magnetic fields. The same suspensions can also be used to delay starvation in those cases where the amount of lubricant is restricted. Next, the tribological mechanism involved in the lubrication of MR fluids in compliant (soft) contacts has been investigated. Friction increases at low speeds because of the formation of a compact mattress of agglomerated iron particles. As a consequence, this regime is sensitive to the grade of carbonyl iron employed in the formulation. In particular, silica coated iron particles exhibit lower friction levels if compared to naked or oxide coated iron particles. At large sliding speeds, the viscosity of the dispersing medium employed plays a significant role. Initially friction decreases at low speeds because of the so-called "ball bearing" effect and later friction increases as particles become embedded in the PTFE matrix. At exceedingly large speeds particles are expelled out of the contact. These results are believed to provide guidance to developers of MR formulations as we emphasize that the formulation of the magnetic suspensions has an important impact on the tribological performance under application.

Acknowledgements

This work was supported by MICINN MAT 2010-15101 project (Spain), CAPES BEX 0834/11-4 and FAPEMIG APQ-00463-11 (Brasil), by the European Regional Development Fund (ERDF) and by Junta de Andalucía P10-RNM-6630 and P11-FQM-7074 projects (Spain).

References

1. J. M. Ginder, *MRS Bulletin*, 1998, August 26.
2. J. de Vicente, D. J. Klingenberg and R. Hidalgo-Álvarez, *Soft Matter*, 2011, **7**, 3701.
3. D. J. Carlson and J. Intell., *Mat. Syst. Str.*, 2002, **13**, 431.
4. D. J. Carlson and J. Int., *Vehicle Des.*, 2003, **33**, 207.
5. J. de Vicente, J. R. Stokes and H. A. Spikes, *Tribol. Lett.*, 2005, **20**, 273.
6. G. T. Y. Wan and H. A. Spikes, *Tribol. Trans.*, 1987, **35**, 12.
7. K. Lee, Y. Hwang, S. Cheong, Y. Choi, L. Kwon, J. Lee and S. H. Kim, *Tribol. Lett.*, 2009, **35**, 127.
8. A. J. F. Bombard and J. de Vicente, *Tribol. Lett.*, 2012, **47**, 149.
9. G. Liu, X. Li, B. Qin, D. Xing, Y. Guo and R. Fan, *Tribol. Lett.*, 2004, **17**, 961.
10. X. Tao, Z. Jiazheng and X. Kang, *J. Phys. D Appl. Phys.*, 1996, **29**, 2932.
11. G. K. Nikas and I. Proc., *Mech. Eng. J*, 2010, **224**, 453.
12. Z. S. Hu, J. X. Dong and G. X. Chen, *Tribol. Int.*, 1998, **31**, 355.
13. Z. S. Hu and J. X. Dong, *Wear*, 1998, **216**, 87.
14. C. Cusano and H. E. Sliney, *Tribol. Trans.*, 1981, **25**, 190.
15. F. Chiñas-Castillo and H. A. Spikes, *Tribol. Trans.*, 2000, **43**, 387.
16. S. Lingard, W. A. Bullough and W. M. Shek, *J. Phys. D: Appl. Phys.*, 1989, **22**, 1639.
17. P. L. Wong, W. A. Bullough, C. Feng and S. Lingard, *Wear*, 2001, **247**, 33.
18. W. A. Bullough, P. L. Wong, C. Feng and W. C. Leung, *J. Intell. Mat. Struct*, 2003, **14**, 71.
19. W. C. Leung, W. A. Bullough, P. L. Wong and C. Feng, *Proc. I. Mech. Eng. J.*, 2004, **218**, 251.
20. V. R. Iyengard, A. A. Alexandridis, S. C. Tung and D. S. Rule, *Tribol. Trans.*, 2004, **47**, 23.
21. C. H. Lee, D. W. Lee, J. Y. Choi, S. B. Choi, W. O. Cho and H. C. Yun, *J. Trib. – Trans. ASME*, 2011, **133**, 031801.
22. W. Li-jun, G. Chu-wen, Y. Ryuichiro and W. Yue, *Trib. Int.*, 2009, **42**, 792.
23. Z. D. Hu, H. Yan, H. Z. Qiu, P. Zhang and Q. Liu, *Wear*, 2012, **278**, 48.
24. W. Kordonski and S. D. Jacobs, *Int. J. Mod. Phys. B*, 1996, **10**, 2837.
25. W. Kordonski and A. Shorey, *J. Intell. Mat. Struct.*, 2007, **18**, 1127.

26. W. Kordonski and S. Gorodkin, *Appl. Optics*, 2011, **50**, 1984.
27. D. K. Singh, V. K. Jain, V. Raghuram and R. Komanduri, *Wear*, 2005, **259**, 1254.
28. S. Jha and V. K. Jain, *Wear*, 2006, **261**, 856.
29. R. E. Rosensweig, *Ferrohydrodynamics*, Dover, New York, 1997.
30. R. C. Shah and M. V. Bhat, *Tribol. Int.*, 2004, **37**, 441.
31. E. Andablo-Reyes, R. Hidalgo-Alvarez and J. de Vicente, *Soft Matter*, 2011, **7**, 880.
32. E. Andablo-Reyes, J. de Vicente, R. Hidalgo-Alvarez, C. Myant, T. Reddyhoff and H. A. Spikes, *Tribol. Lett.*, 2010, **39**, 109.
33. B. J. Hamrock and D. Dowson, *Ball Bearing Lubrication – The Elastohydrodynamics of Elliptical Contacts*, Wiley-Interscience, New York, 1981.
34. A. Shallamach, *Wear*, 1963, **6**, 375.
35. A. J. F. Bombard and J. de Vicente, *Wear*, 2012, **296**, 484–490.
36. I. M. Hutchings, *Tribology: Friction and Wear of Engineering Materials*, Edward Arnold, Great Britain, 1992.
37. M. Esfahanian and B. J. Hamrock, *Tribol. Trans.*, 1991, **34**, 628.
38. A. D. Roberts and D. Tabor, *Proc. R. Soc. Lond. A*, 1971, **325**, 323.
39. S. Biswas and R. W. Snidle, *J. Lubr. Technol.*, 1976, **98**, 524.
40. W. E. Jamison, C. C. Lee and J. J. Kauzlarich, *ASLE Trans.*, 1978, **21**, 299.
41. J. A. Greenwood, H. Minshall and D. Tabor, *Proc. Royal Soc. Lond. A*, 1961, **259**, 480.
42. J. Blumm, A. Lindemann, M. Meyer and C. Strasser, *Int. J. Thermophys.*, 2010, **31**, 1919.

Coated Magnetorheological Composite Particles: Fabrication and Rheology

YING DAN LIU AND HYOUNG JIN CHOI*

Department of Polymer Science and Engineering, Inha University, Incheon
402-751, Korea
*Email: hjchoi@inha.ac.kr

7.1 Introduction

Magneto-responsive magnetorheological (MR) systems, which can be controlled by an applied external magnetic field strength, include MR fluids,[1–3]
MR gels[4] and MR elastomers,[5] depending on the medium materials used and
how those are prepared. Among these, this chapter will focus on MR fluids
even though coated MR composite particles can be also applied to MR gel and
MR elastomers.

MR fluids are generally a class of particle suspensions consisted of
micron-sized (micron to several tenths of microns) magnetizable particles
dispersed in a non-magnetic fluid. Micron sized particles of ferromagnetism
or ferrimagnetism are suitable for MR fluids as suspended particles, which are
able to be polarized by an external magnetic field and connect to adjacent
particles by induced strong dipole–dipole interactions. This phenomenon
results in particulate chain-like or column-like architectures parallel to the
applied magnetic field direction with non-Newtonian behaviors of strongly
field-dependent yield stress and shear viscosity.[6–8] However, as the size of

RSC Smart Materials No. 6
Magnetorheology: Advances and Applications
Edited by Norman Wereley

magnetic particles becomes in nanoscale, the suspension drops into another category of ferrofluid,[9–11] in which the Brownian motion of the dispersed particles dominates the actions of the nanoparticles rather than field-induced interaction or gravity-generated settling, resulting in small magnetoviscous effect and no sedimentation phenomena. The size scale of the particles in the suspension gives rise to not only distinct physical performance but also different applications. Advanced applications of MR fluids require large field-induced yield stress, low off-state shear viscosity, stability against chemical corrosion and settling, which are all closely dependent on the constituting materials (carrier liquids, additives, and magnetic particles) used in the MR system. From this point of view, better understanding of the MR materials is necessary and crucial for its further development and commercial applications. Even though the viscous carrier liquid has significant effect on the flow and field-responsive behavior of the MR fluid,[12,13] magnetizable particles, as the main active body of MR suspension, attract considerable attention mainly in two directions of research: one is to synthesize special particles of this kind; the other is to modify conventionally adopted particles, for example soft magnetic carbonyl iron (CI) microparticles for the purpose of exploring new material sources, or achieving higher MR performance, better settling or chemical stability and so on.

On the other hand, composite particles with tailored morphologies and components on the size-scale of colloid or even larger regimes in general are the subject of rapidly developing interest in research recently, offering great promise in a variety of scientific studies and technological applications.[14–18] The incorporation of diverse materials containing organics or inorganics, as well as both categories, makes them attractive in materials science due to the improved physical and chemical properties in the composite particles, such as optical, electrical, magnetic, thermal, mechanical or biological properties and so forth.[19–21] This thus enables a broad range of applications in many fields with a corresponding requirement of material properties. However, it is not only applications that are of interest, the combination of diverse materials are also attractive in their design and synthesis processes including dispersing, surface modification, coating, polymerization, *etc.*[22–25] Different processes and strategies affect the morphological, physical and chemical properties of the composite particles. In addition, microscope representation of the colloidal particles is closely related to their physical state when detected, such as suspension, film and 3D pattern of particulate building blocks.[18,26–29]

Based on these synergistic effects of composite material fabrication, various coated MR composite particles have been developed. As the core materials in magnetic composite particles, magnetizable particles used in MR fluids include iron, iron alloys (such as those with nickel or cobalt),[30,31] and iron oxides of diverse shapes and size distributions. It is not only the particle composition, but also the particle shape and size distribution, that will affect the performance of the MR suspension; for example, the settling rate of a bidisperse MR fluid using nanometre-sized particles was reduced because the nanoparticles fill pores which are created between the larger particles.[32]

Concurrently, the materials research in electrorheological (ER) fluids, an electric analogue of MR suspensions, has been widely developed over the last ten years. It has widened from hydrous inorganics to anhydrous inorganics and organic conducting polymers.[33–37] In particular, to date, synthesizing electro-responsive inorganic-organic composite particles is a leading topic in the material research of ER systems, which has become a focus because of the particular components, shapes, sizes, or structures of the fabricated particles, as well as their attractive ER performances.[38–44] In addition, methods and strategies for synthesizing colloidal composite particles are applied when designing specific ER particles such that more new materials, including one dimensional carbon nanotubes (CNTs)[45,46] and two dimensional layered silicates and graphenes,[47–51] have been adopted in preparing naonocomposite ER materials. Particles possessing core-shell type architectures, containing polystyrene (PS),[42,51] poly(methyl methalcrylate) (PMMA)[43,46,52] or silica[53,54] spheres as cores and conducting polyaniline (PANI), polypyrrole (PPy), poly(3,4-ethylenedioxy-lthiophene),[41] CNTs or graphene oxide, *etc.* as shell materials, are frequently reported. A similar concept has been applied when preparing magnetic MR composite particles but for a different motivation: coated magnetic composite particles are favored because the magnetic properties of the pure magnetizable particles are sustained, lower specific gravity and higher anti-corrosion stability can be achieved, and so on. Furthermore, particles of this kind can be superior in specific applications, for example optical finishing.[55]

Even though materials used in MR systems are not as rich as that applied in ER fluids, their higher mechanical response to magnetic fields advantageously enables the commercial application of MR fluids; for example, to primary suspensions in ground vehicles. Therefore, studies aiming to prepare more practically available MR composite particles are of interest. In this chapter, coated composite MR particles will be discussed, including their fabrication and magnetorheological performance, especially when used in a suspension. Both polymeric (*e.g.* PMMA, PS, poly(vinyl butyral) (PVB))[56] and inorganic (*e.g.* silica, zirconia)[57,58] materials have been applied as encapsulating shell materials for magnetic CI particles, which results in a substantial decrease in the particle density and excellent resistance to chemical oxidation. These coatings are well-suited to micron sized CI particles but are also well-suited to coating nano-scale iron oxide particles. Magnetic nanoparticles can also be loaded on the surface of polymer spheres in order to prepare magnetic particles with low density.

7.2   Carbonyl Iron-based MR Particles

As the commercially adopted magnetic particles in the MR system, the carbonyl iron (CI) particle enjoys its particular favor in both academia and industries. CI is a high purity iron powder, which is 97% pure and has unique spherical morphology. BASF Corp. produces CI powder (CIP), which is synthesized by thermal decomposition of iron pentacarbonyl ($Fe(CO)_5$) and yields powders with high chemical purity, unique spherical shape and high magnetic saturation. The BASF CIP has a wide selection of powder grades;

hard and soft, as well as coated and uncoated, products. Among the CIP grades, several popular candidates currently used in MR systems are the grades denoted as CC, CM, HQ[59] and HS, which are summarized in Table 7.1. Table 7.1 tabulates their average particle size (D_{50}), chemical composition and surface treatment (with or without coating), all of which have relative low hysteresis and high magnetic saturation of more than 200 emu g^{-1}.[60] In addition, the CC grade of CIP is found to have the highest Fe content and correspondingly the lowest carbon and nitrogen content. Figure 7.1 shows the SEM images of the aforementioned CIPs: all have spherical shape but different size distributions.

However, another physical parameter we should consider is their density, which is about 7.9 g cm^{-3} and extremely higher than common carrier liquids such as silicone oil, mineral oil, or other aqueous solutions. Thus, the particles readily sediment in the absence of frequent agitation. Various strategies have

Table 7.1 Comparison of several different kind of BASF CI powders.

CIP Grade	Average Particle Size $D_{50}/\mu m$	Coating (or not)	Chemical Composition (% w/w)			
			Fe	C	N	O
CC	4.50	0.1% SiO_2	99.6	0.016	0.01	0.24
CM	9.00	No	99.5	0.05	0.01	0.20
HQ	2.00	No	97.5	0.79	0.99	0.47
HS	2.50	No	97.3	1.00	1.00	0.50

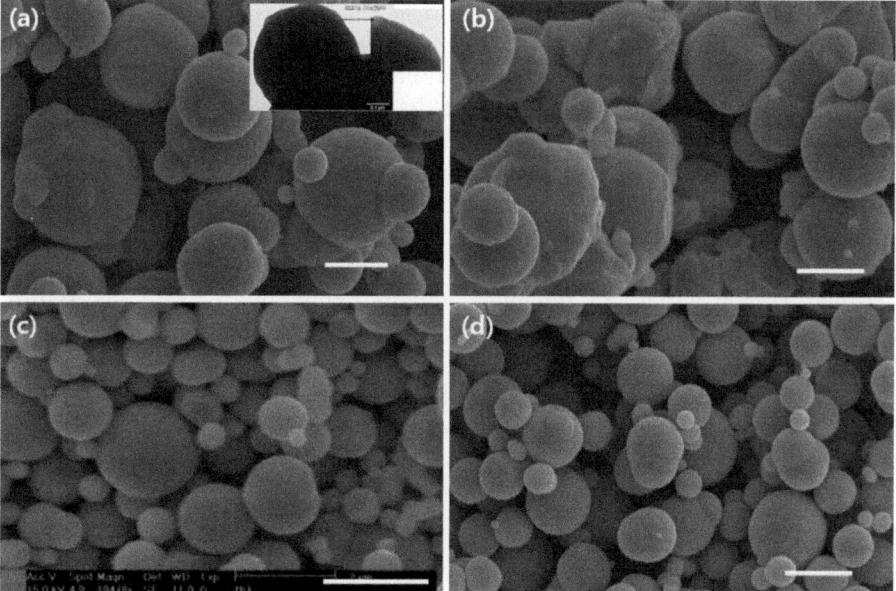

Figure 7.1 SEM images of several BASF CIPs: (a) CC (inset: TEM image, reprinted from ref. 64), (b) CM, (c) HQ,[59] and (d) HS (scale bar: 2μm).

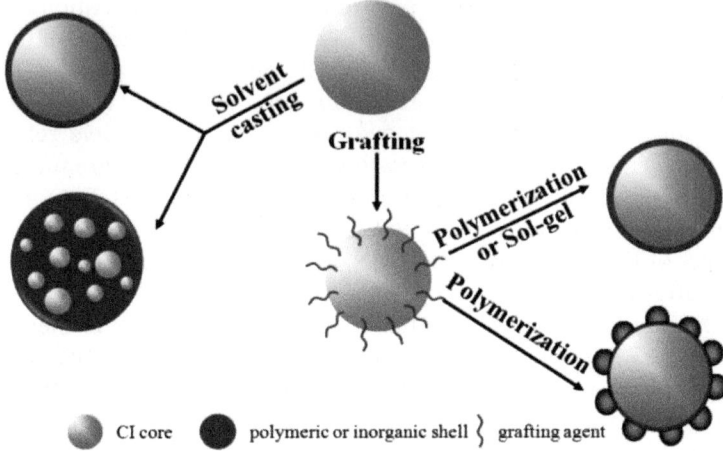

Figure 7.2 Schematic representation of synthesis of coated CI particles by polymers or inorganics *via* different methods.

been reported to solve this problem, the main ideas of which could be clarified into two approaches: one is to decrease the density mismatch between the CI particles and carrier liquids by particle modification, and the other is to increase the resistance of the particles to settling and/or prevent them from coming in contact with each other (steric separation). For the latter one, high viscous dispersing media are used instead of pure liquids, for example grease-mixed oil,[61] ionic liquid[12] or nanoparticles (fumed silica, organoclay, carbon nanotube, *etc.*)[62–64] added liquids. Nevertheless, these media of high viscosity induce another problem of high off-state shear viscosity, which may limit the potential applications of MR fluids in specific fields. Coating of the CI particles is the most favored technique to prepare magnetic MR particles with low density and designed surface properties for typical applications of MR fluids. Furthermore, the strategies for coating themselves and the resulted improved morphologies[6,56] interest the researchers in this field. The unique spherical shape and tailored chemistry on the surface of the CI particles make the coating process feasible: a wide variety of polymers and inorganics have been successfully coated onto the surface of CI particles. A schematic diagram for preparing coated CI particles is shown in Figure 7.2. Based on the first coated layer, a second layer can be introduced onto the coated CI particle. Therefore dual-coated CI particles are fabricated and employed as MR materials, indicating diverse surface property and MR behaviors.

7.2.1 Single-coated CI Particles

7.2.1.1 Inorganic Modification

As shown in Figure 7.1a, the BASF CI powder CC grade was the first to apply CI particles with a coated surface: a thin layer of silica. It can be observed from the TEM image (inset of Figure 7.1(a)) that the whole particle surface is covered by silica even though its thickness is not completely uniform.[64] The silica layer may affect wettability of the particles, dispersing state and shear viscosity of the MR fluid. To improve the chemical stability of the iron particles, polysiloxane encapsulated 4 μm sized CI particles were prepared by Pu *et al.*[65] in a hydrolysis-condensation polymerization procedure. The coated polysiloxane layer was in the nanoscale ($\sim$60 nm) but showed good resistance to thermal oxidation and acidic corrosion. However, the above research did not mention the MR behaviors of the encapsulated CI particles. In addition, it can be considered that a coated thin layer could not result in a significant reduction in the density of the particles. Recently, we[66,67] reported a coated CI particle with a thick silica shell prepared by a sol-gel process using a modified Stöber method because of its easy control in the coating process and the large amount of available precursors. Pre-grafting to the pristine CI particles with methacrylic acid and vinyltrimethoxylsilane was applied in order to generate a uniform coating layer. Figure 7.3 shows the scanning electron microscopy (SEM) image and energy dispersive X-ray spectroscopy (EDS) analysis of the silica coated CI particles. The silica coated particles look much smoother than pure CI particles and the interconnected parts between particles become gentle due to the filling of silica. The EDS spectra indicate the appearance of the element silicon from silica as a certain mass fraction that also confirms the successful coating by silica of CI particles. In this situation, the density of the particles was reported to be reduced from 7.90 to 6.35 g cm^{-3} with the potential of having better settling stability than pure CI particles. The particles also showed better resistance to both acid and heating-oxidation. Figure 7.4a indicates the thermal gravimetric analysis (TGA) curves of the CI particles before and after encapsulation. The encapsulated CI particles were not oxidized until the temperature reached higher than 350 °C. Similarly, as shown in Figure 7.4b, the pH value of the HCl solution containing polysiloxane encapsulated CI particles increased more slowly than that of the HCl containing pure CI particles, implying that the resistance of the CI particles to corrosion by acid was improved by the encapsulation with silica.

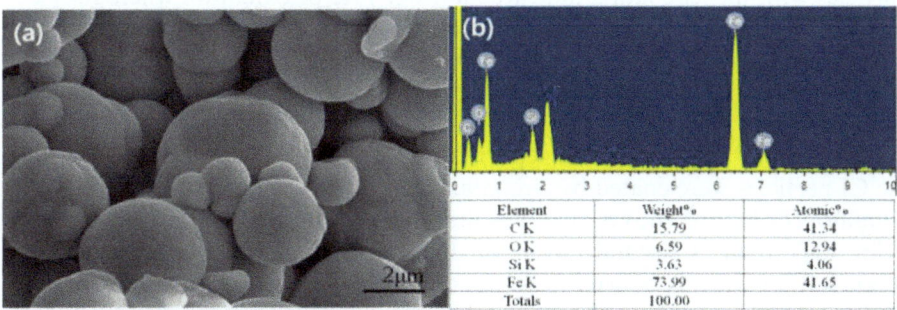

Figure 7.3 SEM image of silica-coated CI particles (a) and EDS analysis (b) of the surface of coated particles.[58]

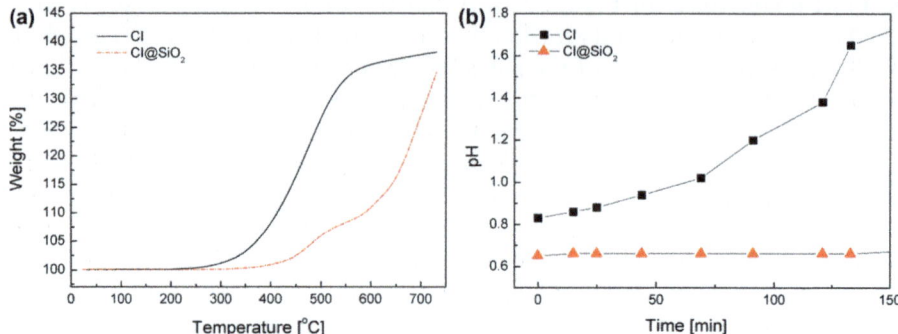

Figure 7.4   TGA curves (a)[67] and resistance to acidic corrosion (b)[58] of the CI particles before and after encapsulation by polysiloxane. The resistance to acidic corrosion was measured by dispersing the two kinds of CI particles in HCl of same concentration respectively, and then testing the pH value of each as a function of time.

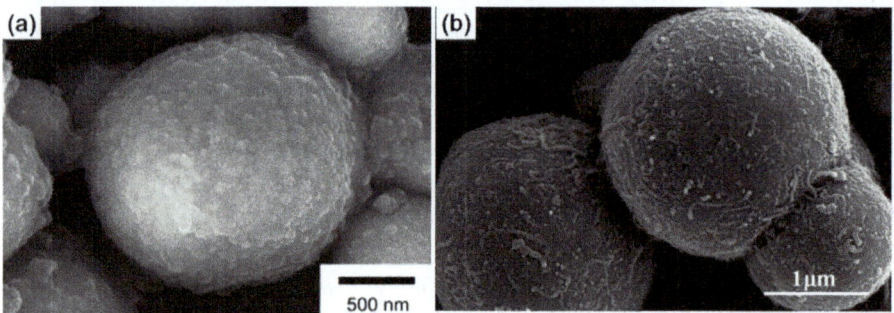

Figure 7.5   SEM images of coated CI particles by zirconia (a)[69] and MWCNT (b).[72]

On the other hand, nickel-coated CI (HS grade) particles prepared by an electrode-less deposition process, were also introduced as MR materials.[68] The particles with nickel coating indicated a lower rate of oxidation when anaysed by TGA. Meanwhile, it also induced an undesirable decrease in the saturation magnetization of the CI particles, which was also the main problem that appeared in other situations when the CI particles are modified with non-magnetic materials.

Another well done work was reported by Jacobs and his group[55,69] in which the CI particles were coated by zirconia: a thin silica adhesive layer made from 3-aminopropyl trimethoxysilane was first generated on the surface of CI particles and then a second layer of highly faceted zirconia nanocrystals (Figure 7.5a) was introduced from zirconium butoxide both using the sol-gel method. The prepared particles were able to enhance the chemical stability and the lifetime of the MR fluid in aqueous solution for the application of polishing of optical materials. Moreover, the particles were totally hydrophilic, so the viscosity of the MR fluids could be reduced by adding a mixture of glycerol with a polyelectrolyte.

Multi-walled carbon nanotubes (MWCNT) are also a frequently applied material in coating techniques, especially on the surface of micro- or nano-particles due to their excellent self-assembling property.[70,71] For the application as a coating material on the CI particles, the rough shell of a thick MWCNT (usually carboxylic acid functionalized) nest is not only good for enhancing the dispersion stability and redispersion ability but also shows less reduction in saturation magnetization than the CI particles coated by other non-magnetic materials, as the MWCNT nest possesses a weak magnetic property due to the small amount of residue iron catalyst within the walls.[72,73] The facile way to prepare the CI/MWCNT particles (Figure 7.5b) includes two steps (Figure 7.6): the first is to modify the CI particles by 4-aminobenzoic acid and the second is to coat the particles in the aqueous suspension of MWCNT by ultrasonication and vigorous stirring. When followed by the self-assembly with sonication, a solvent casting process in the w/o emulsion is developed that generates better-coated MWCNT shell.

7.2.1.2 Polymer-incorporated Composite Particles

Magnetic particles with a core-shell structure, particularly polymer encapsulated magnetic CI particles, are a class of the most attractive magnetic materials being applied in MR systems owing to their greatly reduced specific gravity. A series of CI/polymer core-shell particles have been prepared using various methods. *In situ* dispersion polymerization was one of the first techniques applied to coat CI particles with PMMA by means of grafting agents, such as acrylic acid or methacrylic acid, to enhance the compatibility between the CI particles and methyl methacrylate monomers.[74] A cross-linked PMMA shell was created by adding a cross-linking agent, ethylene glycol dimethacrylate, with the monomers to enhance chemical resistance and surface hardness of the particles.[75] In another case, the coating process by PS was conducted in a similar manner, which did not produce a uniform PS layer but a layer of nano-scaled PS spheres sprinkling on the surface of the CI particles.[76,77] Figure 7.7a–c shows the morphology of the aforementioned core-shell particles, where the particles sustained their spherical shapes and possessed significantly diverse surface morphologies. Compared to coarse pure CI particles (Figure 7.1), the PMMA coated ones show a smoother surface, whereas bulgy points are observed in the cross-linked PMMA coated sample. The presence of localized PS nanospheres rather than free PS particles off the surface of the CI particles is caused by the high speed stirring and the application of a graft agent and stabilizer.

Also, conducting PANI and PPy are being applied widely as ER materials owing to their large available polarizability for a high ER effect and easy synthesis.[37] To search for an improved ER effect and its correlation with particulate ER materials, smart core-shell particles containing PANI as either the core or shell matter have attracted considerable interest from researchers.

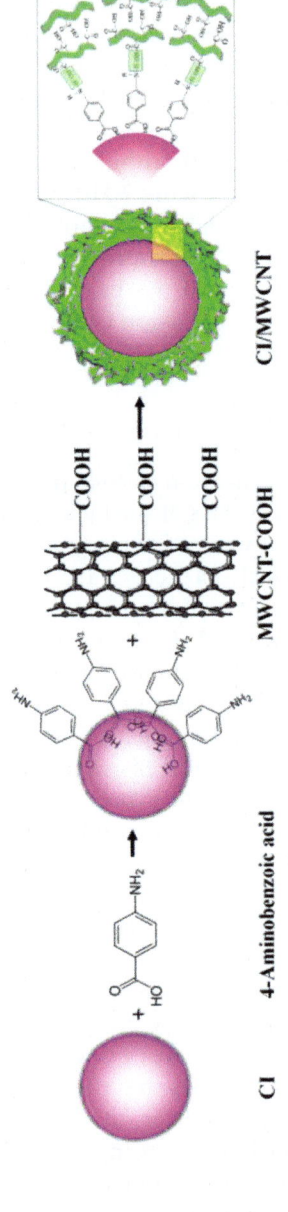

Figure 7.6   Schematic diagram of the synthesis route for CI/MWCNT core-shell particles.

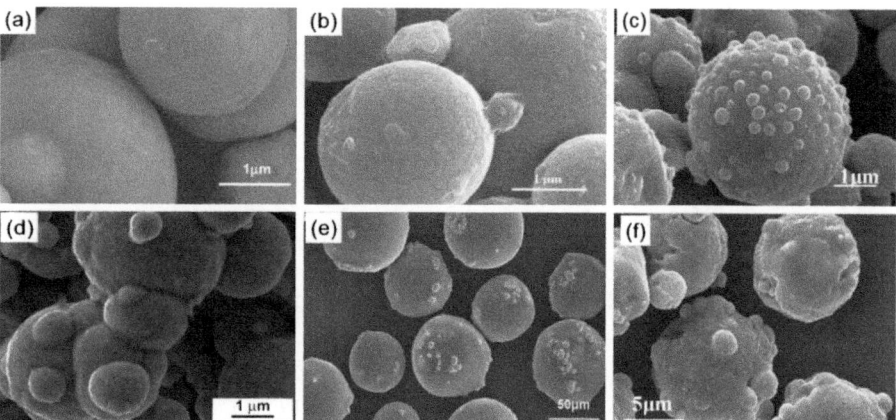

Figure 7.7   SEM images of polymer coated CI particles: (a) CI/PMMA, (b) CI/cross-linked PMMA, (c) CI/PS by dispersion polymerization,[76] (d) CI/PANI,[80] (e) CI/PS by solvent casting[82] and (d) CI/PC.[84]

Based on the methods developed in preparing core-shell ER particles, PANI- or PPy-coated CI particles were synthesized for MR fluids. In contrast to the aforementioned polymers (*e.g.* PMMA, PS), which are sensitive to the stirring rate during polymerization, PANI forms a reproducible shell on the 4-aminobenzoic acid-grafted core sphere in the procedure of *in situ* oxidation polymerization.[78] Similarly, PPy-coated CI particles were prepared by Sedlacik *et al.*[79] using surfactant (cetryl trimethylammonium bromide) modified CI particles. They also reported a different method to prepare PANI-coated CI particles by dispersing CI particles in a PANI colloid dispersion in chloroform. Figure 7.7d shows an SEM image of the prepared PANI-coated CI particles. In contrast to the CI/PMMA particles, the CI/PANI particles demonstrated a rather rough surface according with the essential property of PANI[80]. This suggests that the morphologies (or shapes) of the obtained polymer-coated CI particles are obviously different from pristine CI particles, which also corresponds to the coating method, coated materials and mass. These characteristics are believed to affect their MR performance when applied as dispersed solid particles in the MR fluid. On the other hand, another significant factor in influencing the MR behaviors is the magnetic properties of the particles: the incorporation of CI particles with non-magnetic polymer coatings can reduce responsive to magnetic fields.

In addition to direct polymerization, a solvent casting or evaporation method is possibly the simplest way of preparing general polymer composites or polymer-coated composite particles. This has been used frequently to prepare polymer/clay or other intercalated nanocomposites by dispersing layered nanoparticles and polymers in small molecular organic solvents, such as chloroform and toluene.[81] To fabricate the CI embedded polymer composite particles, the CI particles are first dispersed in an organic polymer solution and transferred to the non-solvent of an aqueous solution containing both the

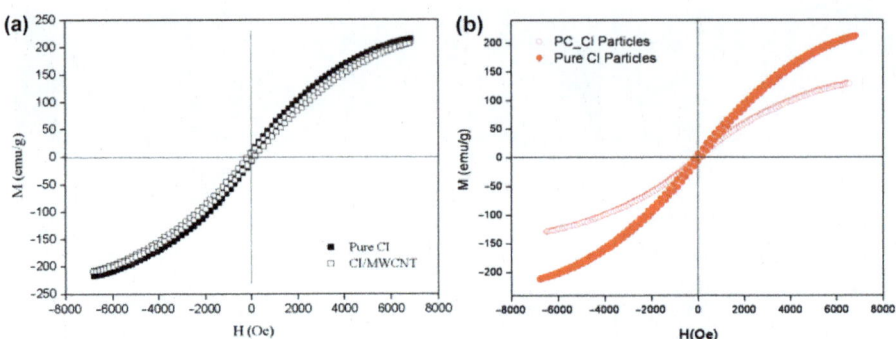

Figure 7.8   Vibrating sample magnetometer (VSM) data comparison between pure CI
              and CI/MWCNT (a),[73] and pure CI and CI/PC (b).[84]

stabilizer and emulsifier with vigorous stirring. Therefore, the CI particles
containing polymer solution droplets are formed and stabilized by the surfactants. By using this method, Choi *et al.* prepared CI composite particles with
PS,[82] poly(vinyl butyral) (PVB),[83] and polycarbonate (PC).[84] Figure 7.7e and
7.7f present the SEM image of the composite particles of CI/PS and CI/PC
prepared using this method, respectively. The particles obtained are spherical
but with a bumpy surface due to the embedded CI particles. In contrast to the
core-shell coated particles, these particles are much larger (~ 10–50 μm) than
the original CI particles. In addition, the density of the magnetic composite
particles by this method are considerably lower (CI/PS: 2.51 g cm^{-3}, CI/PC:
3.28 g cm^{-3}) than that of the core-shell particles due to the large amount of
polymer incorporated, which on the other hand reduced the magnetic properties of the particles much more than the pure CI particles. As shown in
Figure 7.8b, the saturation magnetization of the CI/PC particles is only
130 emu g^{-1} which is much lower than the pure CI particles (200 emu g^{-1}), while
the CI/MWCNT particles exhibit a similar value to pure CI particles
(Figure 7.8a). It was reported that bio-polymeric materials, such as guar gum,
were able to coat the CI particles by the solvent casting method.[85]

7.2.2   Dual-coated CI Particles

As has been mentioned above, MWCNTs were recently applied to wrap CI
particles *via* a simple ultrasonication method with a grafting reagent, generating a considerably rough surface. However, the reduction in density of the
MWCNT-coated CI particles was not as high as that caused by coating with
inorganics or polymers, even though the rough surface prevented sedimentation of the particles. Therefore, to combine the effects by different coatings, a
two-step (dual) coating process was developed to introduce a polymer or inorganic layer first and then a MWCNT nest layer on the CI particles. Therefore, after the initially deposition of PS nanobeads on the surface of CI particles
through conventional dispersion polymerization, densely piled MMCNT nests

are constructed over the surface of CI/PS particles by a simple solvent casting method in a water/oil emulsion system. CI particles dual-coated by PANI and MWCNT in sequence were prepared in a similar way.[78] Thereafter, densely constructed MWCNT nests were prepared on the surface of silica-coated CI particles. In this study, a layer-by-layer self-assembly procedure was applied by placing the silica-coated CI particles alternatively in poly(diallyldimethyl-ammonium chloride) and poly(sodium 4-styrenesulfonate) polyelectrolytes, and finally in a carboxylic acid functionalized MWCNT suspension.[86] The dual-coated particles result in sequential decrease in density and a considerably rough surface that are expected to be beneficial to the dispersion stability of the corresponding MR fluids.

7.3 Iron Oxide Wrapped Particles

Iron oxides (IO) including Fe_3O_4 and γ-Fe_2O_3 have been reported to be good MR candidates. Compared to the mostly used CI particles, which have saturation magnetization of approximately 200 emu g^{-1}, the IO particles exhibit lower saturation magnetization (possibly several dozens of emu g^{-1}) with some hysteresis. On the other hand, the IO particles possess much lower density, *e.g.* 4.32 g cm^{-3} for Fe_3O_4. Therefore, an IO particle-based MR fluid will present better sedimentation stability than CI or even CI/polymer composite-based MR fluids.

Modifications to the particles should be applied to promote not only their dispersion stability against settling and coagulation but also chemical stability against corrosion. For example, a polymer coating of IO particles is one suggested method, which also includes being loaded onto other matrices, such as CNTs or polymer spheres, and being incorporated with inorganic materials (*e.g.* Fe_3O_4-TiO_2, Fe_3O_4-SiO_2).[87,88] By applying PS matrix spheres, magnetite nanoparticle-coated core-shell particles (PS/Fe_3O_4) were adopted as MR particles,[89] which were prepared *via* the controlled precipitation of inorganic precursors onto the core particles of highly monodisperse polystyrene-actoacetoxyethyl methacrylate nanoparticles. These nanoparticles with low density, however, showed low magnetic response and sequentially low MR responses under an applied magnetic field. Later, Fang *et al.*[90] fabricated Fe_3O_4 particle-coated PS microspheres for applications to MR system. The microporous PS particles were first prepared by etching PS/silica composite particles obtained *via* suspension polymerization. Figure 7.9 shows the Fe_3O_4 nanoparticles coated PS particles on the nano (Figure 7.9a) and micron (Figure 9c) scale. A difference in size and surface morphology between the two situations was observed. The microporous PS spheres had a smooth surface after being etched in a HF bath, where the micropores were generated by the removal of loaded silica particles. On the other hand, the Fe_3O_4-loaded PS spheres in Figure 7.9b exhibited a very rough morphology, in which all the micropores were blocked with Fe_3O_4 particles. Figure 7.9d represents the morphology of micron PS particles after loading Fe_3O_4 particles. The inset is a magnified view of one PS/Fe_3O_4 particle. Compared with the smooth surface of the PS particles

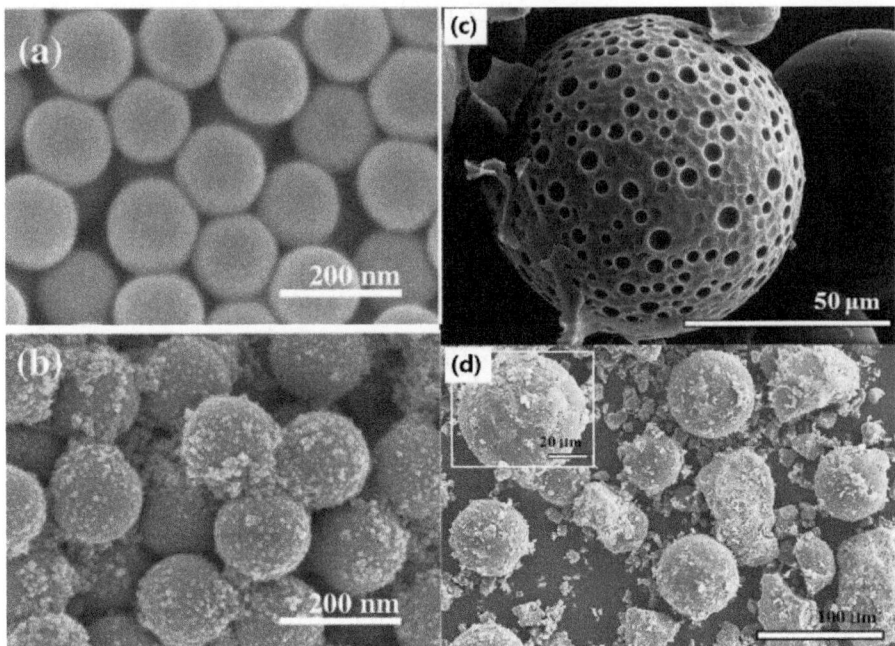

Figure 7.9 SEM images of etched pure PS and Fe_3O_4-coated PS nanoparticles (a, b),[89] microprorous PS and Fe_3O_4-coated PS microspheres (c, d).[90]

with a microporous structure, the Fe_3O_4 loaded PS spheres exhibit a very rough appearance, in which all the micropores are choked up with Fe_3O_4 particles.

7.4 Rheological Characterization

MR fluids based on the as-fabricated magnetic composite particles are prepared by dispersing the particles in a range of oils through shaking and/or sonication. Their MR performances are then detected by loading the well-dispersed MR fluids between a parallel-plate cell of a rotational rheometer equipped with an electromagnet. A controlled and stable magnetic field can be applied when the samples are measured under either a rotation or oscillation test at magnetic field strengths ranging from 0 to 343 kA m^{-1}.

7.4.1 Steady Shear Test

The rotation test using a rotational rheometer is carried out in either controlled shear rate (CSR) or controlled shear stress (CSS) mode. The flow curves of the MR fluid obtained from CSR mode indicate the increase of shear stress (τ) and shear viscosity (η) with the increase of shear rate. Similar to its electrical analogue ER fluid, a general MR fluid behaves like a Newtonian fluid without a magnetic field applied by showing no yield stress and no shear viscosity dependence on the shear rate. However when a magnetic field is applied, the MR fluid changes to a Bingham fluid ($\tau = \tau_y + \eta\dot{\gamma}$) with a yield stress. Note that

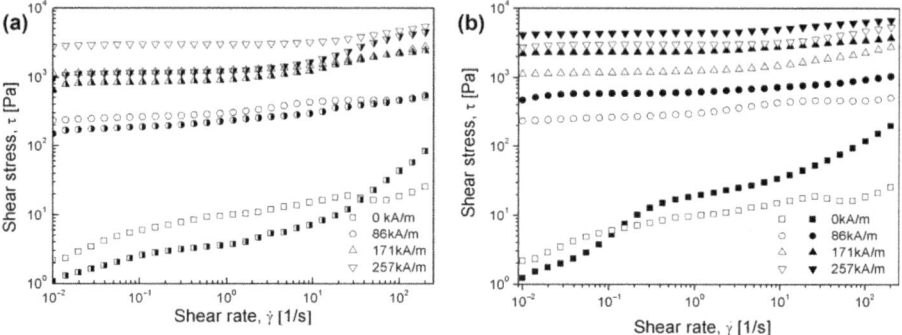

Figure 7.10   Shear stress *vs.* shear rate of the pure CI MR fluid (open) compared with CI@SiO$_2$ MR fluid (half right) (a)[58] and the comparison between pure CI (open) and silica/MWCNT dual-coated CI (closed) MR fluids (b).[86]

compared to MR fluids, ER fluids often exhibit deviation from a Bingham model equation for their flow curve characteristics and thereby various rheological equations of state have been introduced.[91,92] Moreover, the yield stress and shear viscosity of the MR fluids are found to be closely related to the intensity of the magnetic field. In addition, the shear stress shown by the coated composite particles is generally lower than that of the pure CI particles due to the incorporation of a non-magnetic polymeric component. Figure 7.10 represents the flow curves of the MR fluids containing pure CI, silica-coated CI and silica/MWCNT dual-coated CI particles. In Figure 7.10a, steady shear stresses with plateau behavior over the whole shear rate range are observed due to the filed induced chain structures of dispersed particles. The silica-coated CI particles shows lower shear stress at each magnetic field strength indicating that the non-magnetic coating brings down the yield value of the MR fluid.[58] For another situation, the silica/MWCNT dual-coated CI particles display relatively higher shear stress (Figure 7.10b) regardless of whether the magnetic field is present or not. In the absence of a magnetic field, it is easy to understand that particles in a MR suspension with rough and soft surfaces suffer more contact and friction during shear flow, which will result in an increase in shear stress and apparent viscosity.[86] An increase in coating roughness tends to increase the shear stress when a magnetic field is applied due to the increased friction. The reason is that the coated MWNT nests are more loosely packed than the rigid and closely packed coatings in the previous study.

In addition to the flow curves, dynamic yield stresses obtained by extrapolating the shear stress to a zero shear rate limit are also important for analyzing both MR and ER fluids. There are several constitutive equations which have been developed to analyze the yield stress.[93–95] When analyzing these models, we found that, regardless of the applied external fields, the induced yield stress changes in an analogous pattern. Ginder *et al.*[96] who examined the yield stress (τ_y) dependence on the applied magnetic field strength (H_0), divided the range of yield stress into two different regimes based on H_0. At low H_0, τ_y is proportional to H^2, due to the local saturation of the magnetized particles, while

$\tau_y \propto H_0^{3/2}$ at an intermediate range of H_0. For field strengths high enough to achieve complete saturation, the particles can be treated rigorously as dipoles and the stresses and moduli are then independent of field strength and scale as saturation magnetization (M_S) approaches. In this case, under a moderate magnetic field strength, there may exist a critical magnetic field strength (H_c) satisfying $\tau_y \propto H_0^{3/2}$ ($H < H_c$) and $\tau_y \propto H_0^2$ ($H > H_c$). A universal yield stress equation has been previously proposed by adopting critical electric field strength (E_c) to examine ER fluids.[93] By assuming the similarity between ER and MR fluids, the existence of a critical magnetic field strength (H_c) was hypothesized for MR fluids by busing the yield stress data of a MR fluid containing CI particles and single-walled carbon nanotubles (CI/SWCNT),[97] and a new universal correlation was proposed as:

$$\tau_y(H_0) = \alpha H_0^2 \left(\frac{\tanh \sqrt{H_0/H_c}}{\sqrt{H_0/H_c}} \right) \tag{7.1}$$

Here, α is related to the susceptibility of the fluid and volume fraction or other analogous physical parameters. τ_y possesses two limiting behaviors with respect to H_0,

$$\tau_y = \alpha H_0^2, \quad \text{for } H_0 \ll H_c;$$
$$\tau_y = \alpha H_0^{3/2}, \quad \text{for } H_0 \gg H_c; \tag{7.2}$$

Figure 7.11a shows the dynamic yield stress as a function of magnetic field strength for CI/SWCNT suspension, which shows excellent agreement with eqn (7.1). H_c is about 151.74 kA m^{-1}. Then, a generalized scaling relationship can be obtained by scaling eqn (7.1) via H_c and $\tau_y(H_c) = 0.762\alpha H_c^2$ as:

$$\hat{\tau} = 1.313\hat{H}^{3/2} \tanh \sqrt{\hat{H}} \tag{7.3}$$

It can be seen that the dynamic yield stresses collapse onto a single curve by using eqn (7.3), as demonstrated in Figure 7.11b.

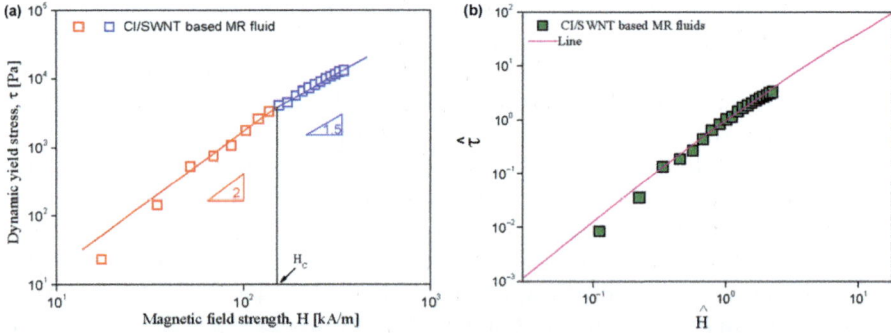

Figure 7.11   (a) Replotted dynamic yield stress *vs.* magnetic field strengths for CI/SWCNT suspension and (b) Plot of $\hat{\tau}$ *vs.* $\hat{H}$ for CI/SWCNT suspension. The solid line is obtained from eqn (7.3).[97]

7.4.2 Oscillation Test

As the MR fluid is always considered as a solid-like material in the applied magnetic field, its viscoelastic properties, represented by the dynamic modulus of both storage or elastic (G') and loss (G'') moduli, are necessary and significant for exhibiting its solid properties. The modulus is always obtained in an oscillation test, during which operation the particles undergo chain formation by the application of magnetic field and then deformation by the applied oscillation strain. The dynamic modulus of a MR fluid based on silica-coated CI (CI@SiO$_2$) particles is shown in Figure 7.12, compared with the pure CI MR fluid. To make the same measurement conditions, a fixed strain value of 0.001% was selected for the dynamic oscillation test in an angular frequency range of 1–100 rad s^{-1}. In Figure 7.12a, the storage (elastic) modulus of the CI MR fluid at each magnetic field remains stable over the whole frequency range, while the storage modulus increases at first and eventually reaches a plateau for CI@SiO$_2$. Considering the polarization and movement of the particles to form chain-like structures is time-dependent,[98] this could be interpreted as the silica shell enhancing the affinity and interaction between particle and polysiloxane (silicone oil) chains, therefore dragging the movement of the particles. In addition, a higher storage modulus in the CI MR fluid is observed at the magnetic field of 86 and 171 kA mm^{-1}, which is easy to understand considering the reduced magnetic properties of the coated particles. However, at a higher magnetic field ($H = 256$ kA m^{-1}) the plateau value of the storage modulus in CI@SiO$_2$ MR fluid is nearly as high as that in CI MR fluid. This is also considered to be caused by the surface features and chemical properties of the CI@SiO$_2$ particles.[55] In Figure 7.12b, the loss moduli of the MR fluids are compared. In contrast to the storage modulus, the loss modulus of CI@SiO$_2$ MR fluid at each magnetic field is much higher than that of CI MR fluid, indicating the significant viscous properties of the CI@SiO$_2$ MR fluid. However, in general, all the storage moduli expressed by the two MR fluids are

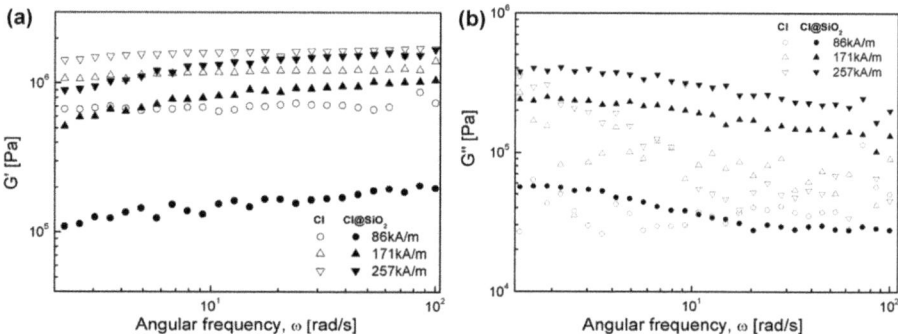

Figure 7.12 (a) Storage (G') and (b) loss modulus (G'') of the CI@SiO$_2$ MR fluid under various magnetic field strengths over the frequency range of 1–100 rad s^{-1}.[67]

all much higher than their corresponding loss moduli, which evidences the more solid-like properties shown by the MR fluids.

The relaxation modulus, indicating the time dependent change in stress when maintaining a constant strain $(G(t) = \tau(t)/\gamma 0)$, is an another way to detect the viscoelastic characteristics of MR fluids. The relaxation modulus can be determined by applying a step-strain (γ). However, it is possible only when a rheometer can perform the γ step quickly enough. As reported, there are different methods[99] for converting rheological functions, for example, G' and $G''(\omega)$ into the relaxation modulus. One of the converting methods was introduced by Schwarzl[100] *via* an empirical equation as follows:

$$G(t) = G'(\omega) - 0.560G''(\omega/2) + 0.200G''(\omega) \qquad (7.4)$$

The calculated relaxation modulus by using eqn (7.4) is plotted in Figure 7.13. In the short relaxation time interval, both CI and CI@SiO$_2$ MR fluids show plateau relaxation moduli implying their highly solid-like (elastic) properties under the applied magnetic field. In the CI@SiO$_2$ MR fluid, a slight decrease in relaxation modulus from $t = 0.1$ s is observed at the three different magnetic field strengths due to the reduced particle interactions by coating, from which point of view we could conclude that the coated silica shell has somewhat negative effect on the magnetic field-induced stress between particles.

The MR effect with time dependent reversibility or durability of a MR fluid is of great importance for its commercial application. Therefore, a long-term MR effect was studied by using a designed rheometer built in a mechanical pulsator.[101] The rheometer design allows measurement of the rheological properties of MR fluids and its exposure to a long-term loading simultaneously, without any manipulation of the measured sample. The experiments were performed under conditions which correspond to real conditions in an automotive MR

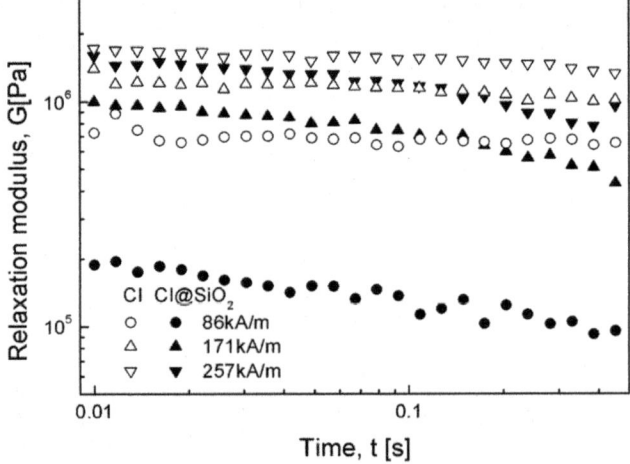

Figure 7.13   Stress relaxation modulus of the CI@SiO$_2$ MR fluids based on pure CI and CI@SiO$_2$ particles.[67]

damper. The results of long term operation have shown that during the durability test the yield stress increases up to five times in the fluid off-state due to in-use-thickening. The same increase was identified in the on-state. These results disprove the hypothesis that a measurable decrease of MR effect arises during the long term operation due to the degradation of magnetic properties. On the other hand, Ulicny *et al.*[102] investigated the chemical changes and a methodology to detect the changes of the CI particles in the MR fluid which had been subjected to a durability test in a MR fan clutch. They indicated that, in a long-term testing, the particles were oxidized, finally forming a porous magnetite layer on the surface of the particles that resulted in loss of fan clutch torque capacity.

7.5 Sedimentation Stability

MR fluids always experience settling problems due to the large particle size and the large difference in density between the magnetic particles and carrier medium. Therefore, it is important to observe the settling performance of an MR fluid. The settling behavior is represented by the sedimentation ratio, which is defined as follows:

$$\text{Sedimentation ratio} = \frac{\text{Volume of the supernatant liquid}}{\text{Volume of the entire suspension}} \times 100\% \qquad (7.5)$$

The volume of the supernatant liquid and entire suspension is substituted by the height of each part when the MR fluid samples are contained in glass vials. Figure 7.14 shows the recorded sedimentation ratio of the pure CI and PS coated CI particle-based MR fluids as a function of time, respectively.[77] With time, the

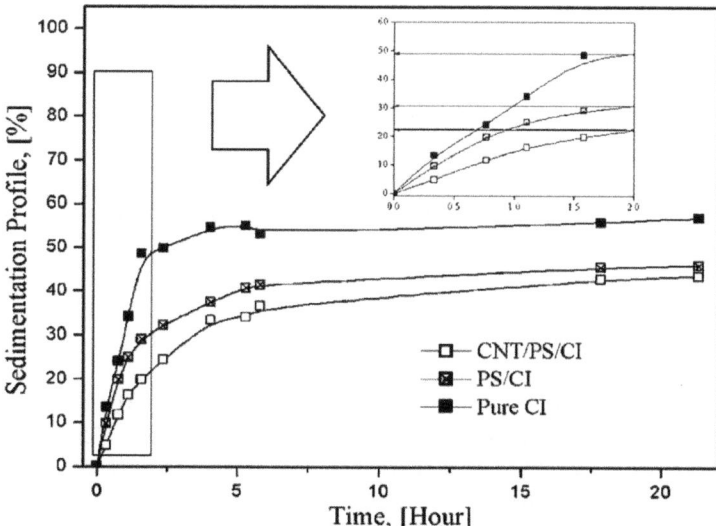

Figure 7.14 Sedimentation profile recorded as a function of time for pure CI, CI/PS, and CI/PS/MWCNT suspensions. The inset is a magnified view of the sedimentation ratio tested at the initial 2h.[77]

particles settled and the sedimentation ratio increased. On the other hand, the sedimentation ratio of the composite particle-based MR fluid increased more slowly than that of the pure CI particle-based MR fluid. Clearly, the CI/PS MR fluid exhibited better sedimentation stability. Moreover, slower settling occurred when a second layer of multiwalled carbon nanotubes was introduced to the CI/PS particles, due to the considerably rough surface of the particles.

7.6   Conclusion

A series of magnetic composite particles are introduced together with their unique applications as MR colloidal particles. The particles, prepared from different magnetic sources and methodologies, show a range of morphologies, magnetic properties and MR performance. The incorporation of polymers and inorganics with magnetic carbonyl iron or iron oxide particles results in a large decrease in particle density and magnetic response. Nevertheless, their properties are considered suitable as MR particles. Both the surface morphology and mass ratio of the incorporated polymers affect the MR performance and sedimentation stability of the composite particle-based MR fluid. On the whole, the coating technique for preparing composite particles is significantly important to develop long-term stable magnetic particles for MR fluids, which is also applicative for producing modified particles applied in other areas. Especially, iron (or iron alloy) nanoparticles are fabricated and applied as MR materials, in which systems inorganic or polymeric modification to the nanoparticles are also essential to the dispersibility and long-term stability of the nanoparticles in the MR system.

References

1. P. J. Rankin, J. M. Ginder and D. J. Klingenberg, *Curr. Op. Colloid Interf. Sci.*, 1998, **3**, 373–381.
2. I. Bica, *J. Ind. Eng. Chem.*, 2006, **12**, 501–515.
3. D. T. Zimmerman, R. C. Bell, J. A. Filer, J. O. Karli and N. M. Wereley, *Appl. Phys. Lett.*, 2009, **95**, 014102.
4. B. Hu, A. Fuchs, S. Huseyin, F. Gordaninejad and C. Evrensel, *J. Appl. Polym. Sci.*, 2006, **100**, 2464–2479.
5. W. Zhang, X. Gong, S. Xuan and W. Jiang, *Ind. Eng. Chem. Res.*, 2011, **50**, 6704–6712.
6. B. J. Park, F. F. Fang and H. J. Choi, *Soft Matter*, 2010, **6**, 5246–5253.
7. J. D. Carlson and M. R. Jolly, *Mechatronics*, 2000, **10**, 555–569.
8. W. H. Li, C. Lynam, J. Chen, B. Liu, X. Z. Zhang and G. G. Wallace, *Mater. Lett.*, 2007, **61**, 3116–3118.
9. K. Raj, B. Moskowitz and R. Casciari, *J. Magn. Magn. Mater.*, 1995, **149**, 174–180.
10. J. Roger, J. N. Pons, R. Massart, A. Halbreich and J. C. Bacri, *Eur. Phys. J. Appl. Phys.*, 1999, **5**, 321–325.

11. J. Ramos, D. J. Klingenberg, R. Hidalgo-Alvarez and J. d. Vicente, *J. Rheol.*, 2011, **55**, 127–152.
12. C. Guerrero-Sanchez, T. Lara-Ceniceros, E. Jimenez-Regalado, M. Raşa and U. S. Schubert, *Adv. Mater.*, 2007, **19**, 1740–1747.
13. P. J. Rankin, A. T. Horvath and D. J. Klingenberg, *Rheol. Acta*, 1999, **38**, 471–477.
14. D. G. Shchukin, G. B. Sukhorukov and H. Möhwald, *Angew. Chem., Int. Ed.*, 2003, **42**, 4472–4475.
15. Q. Wu, Z. Wang and G. Xue, *Adv. Funct. Mater.*, 2007, **17**, 1784–1789.
16. J. A. Balmer, A. Schmid and S. P. Armes, *J. Mater. Chem.*, 2008, **18**, 5722–5730.
17. J. Pyun, *Polym. Rev.*, 2007, **47**, 231–263.
18. E. Bourgeat-Lami, *J. Nanosci. Nanotechnol.*, 2002, **2**, 1–24.
19. S. Stankovich, D. A. Dikin, G. H. B. Dommett, K. M. Kohlhaas, E. J. Zimney, E. A. Stach, R. D. Piner, S. T. Nguyen and R. S. Ruoff, *Nature*, 2006, **442**, 282–286.
20. S. Pavlidou and C. D. Papaspyrides, *Prog. Polym. Sci.*, 2008, **33**, 1119–1198.
21. M. Sanles-Sobrido, V. Salgueiriño-Maceira, M. A. Correa-Duarte and L. M. Liz-Marzán, *Small*, 2008, **4**, 583–586.
22. F. Caruso, A. S. Susha, M. Giersig and H. Möhwald, *Adv. Mater.*, 1999, **11**, 950–953.
23. P. Singh, D. D. Joseph and N. Aubry, *Soft Matter*, 2010, **6**, 4310–4325.
24. J. Wang and X. Yang, *Langmuir*, 2008, **24**, 3358–3364.
25. M. M. Gudarzi and F. Sharif, *Soft Matter*, 2011, **7**, 3432–3440.
26. X. Deng, L. Mammen, Y. Zhao, P. Lellig, K. Müllen, C. Li, H.-J. Butt and D. Vollmer, *Adv. Mater.*, 2011, **23**, 2962–2965.
27. T. Ding, K. Song, K. Clays and C. H. Tung, *Adv. Mater.*, 2009, **21**, 1936–1940.
28. T. C. Halsey, *Science*, 1992, **258**, 761–766.
29. W. Ming, D. Wu, R. van Benthem and G. de With, *Nano Lett.*, 2005, **5**, 2298–2301.
30. A. J. Margida, K. D. Weiss and J. D. Carlson, *Int. J. Mod Phys B*, 1996, **10**, 3335–3341.
31. Y. Yang, L. Li, G. Chen and E. Liu, *J. Magn. Magn. Mater.*, 2008, **320**, 2030–2038.
32. N. M. Wereley, A. Chaudhuri, J. H. Yoo, S. John, S. Kotha, A. Suggs, R. Radhakrishnan, B. J. Love and T. S. Sudarshan, *J. Intell. Mater. Syst. Struct.*, 2006, **17**, 393–401.
33. X. P. Zhao and J. B. Yin, *Chem. Mater.*, 2002, **14**, 2258–2263.
34. J. H. Sung, D. P. Park, B. J. Park, H. J. Choi and M. S. Jhon, *Biomacromolecules*, 2005, **6**, 2182–2188.
35. Q. Wu, B. Y. Zhao, L. S. Chen, C. Fang and K. A. Hu, *J. Colloid Interface Sci.*, 2005, **282**, 493–498.
36. S. Zhang, W. T. Winter and A. J. Stipanovic, *Cellulose*, 2005, **12**, 135–144.
37. H. J. Choi and M. S. Jhon, *Soft Matter*, 2009, **5**, 1562–1567.

38. W. J. Wen, X. X. Huang, S. H. Yang, K. Q. Lu and P. Sheng, *Nat. Mater.*, 2003, **2**, 727–730.
39. K. Mimura, Y. Nishimoto, H. Orihara, M. Moriya, W. Sakamoto and T. Yogo, *Angew. Chem., Int. Ed.*, 2010, **49**, 4902–4906.
40. Y. D. Liu, F. F. Fang and H. J. Choi, *Soft Matter*, 2011, **7**, 2782–2789.
41. Y. D. Liu, J. E. Kim and H. J. Choi, *Macromol. Rapid Commun.*, 2011, **32**, 881–886.
42. Y. D. Liu, B. J. Park, Y. H. Kim and H. J. Choi, *J. Mater. Chem.*, 2011, **21**, 17396–17402.
43. M. S. Cho, Y. H. Cho, H. J. Choi and M. S. Jhon, *Langmuir*, 2003, **19**, 5875–5881.
44. J. B. Yin, X. A. Xia, L. Q. Xiang and X. P. Zhao, *J. Mater. Chem.*, 2010, **20**, 7096–7099.
45. H. J. Choi, J. Y. Lim and K. Zhang, *Diamond Relat. Mater.*, 2008, **17**, 1498–1501.
46. K. Zhang, Y. D. Liu and H. J. Choi, *Chem. Commun.*, 2012, **48**, 136–138.
47. K. P. S. Parmar, Y. Meheust, B. Schjelderupsen and J. O. Fossum, *Langmuir*, 2008, **24**, 1814–1822.
48. Y. Meheust, K. P. S. Parmar, B. Schjelderupsen and J. O. Fossum, *J. Rheol.*, 2011, **55**, 809–833.
49. W. L. Zhang, B. J. Park and H. J. Choi, *Chem. Commun.*, 2010, **46**, 5596–5598.
50. W. L. Zhang and H. J. Choi, *Chem. Commun.*, 2011, **47**, 12286–12288.
51. W. L. Zhang, Y. D. Liu and H. J. Choi, *J. Mater. Chem.*, 2011, **21**, 6916–6921.
52. F. F. Fang, Y. D. Liu, I. S. Lee and H. J. Choi, *RSC Advances*, 2011, **1**, 1026–1032.
53. N. Kuramoto, M. Yamazaki, K. Nagai, K. Koyama, K. Tanaka, K. Yatsuzuka and Y. Higashiyama, *Thin Solid Films*, 1994, **239**, 169–171.
54. J. Y. Hong, E. Kwon and J. Jang, *Soft Matter*, 2009, **5**, 951–953.
55. C. Miao, R. Shen, M. Wang, S. N. Shafrir, H. Yang and S. D. Jacobs, *J. Am. Ceram. Soc.*, 2011, **94**, 2386–2392.
56. B. J. Park, F. F. Fang, K. Zhang and H. J. Choi, *Korean J. Chem. Eng.*, 2010, **27**, 716–722.
57. S. N. Shafrir, H. J. Romanofsky, M. Skarlinski, M. Wang, C. Miao, S. Salzman, T. Chartier, J. Mici, J. C. Lambropoulos, R. Shen, H. Yang and S. D. Jacobs, *Appl. Opt.*, 2009, **48**, 6797–6810.
58. Y. D. Liu, F. F. Fang and H. J. Choi, *Colloid Polym. Sci.*, 2011, **289**, 1295–1298.
59. A. J. F. Bombard, M. Knobel, M. R. Alcantara and I. Joekes, *J. Intell. Mater. Syst. Struct.*, 2002, **13**, 471–478.
60. J. Park, M. Kwon and O. O. Park, *Korean J. Chem. Eng.*, 2001, **18**, 580–585.
61. B. O. Park, B. J. Park, M. J. Hato and H. J. Choi, *Colloid. Polym. Sci.*, 2011, **289**, 381–386.
62. S. T. Lim, M. S. Cho, H. J. Choi and M. S. Jhon, *Int. J. Mod. Phys. B*, 2005, **19**, 1142–1148.

63. F. F. Fang, I. B. Jang and H. J. Choi, *Diamond Relat. Mater.*, 2007, **16**, 1167–1169.
64. A. Gómez-Ramírez, M. T. López-López, F. González-Caballero and J. D. G. Durán, *Smart Mater. Struct.*, 2011, **20**, 045001.
65. H. Pu, F. Jiang and Z. Yang, *Mater. Lett.*, 2006, **60**, 94–97.
66. Y. D. Liu, F. F. Fang and H. J. Choi, *Colloid Polym. Sci.*, 2011, **289**, 1295–1298.
67. Y. D. Liu, H. J. Choi and S. B. Choi, *Colloid Surf., A*, 2012, **403**, 133–138.
68. J. C. Ulicny and A. M. Mance, *Mater. Sci. Eng. A*, 2004, **369**, 309–313.
69. R. Shen, S. N. Shafrir, C. Miao, M. Wang, J. C. Lambropoulos, S. D. Jacobs and H. Yang, *J. Colloid Interf. Sci.*, 2010, **342**, 49–56.
70. H. J. Jin, H. J. Choi, S. H. Yoon, S. J. Myung and S. E. Shim, *Chem. Mater.*, 2005, **17**, 4034–4037.
71. S. H. Lee and C. M. Liddell, *Small*, 2009, **5**, 1957–1962.
72. F. F. Fang and H. J. Choi, *Colloid. Polym. Sci.*, 2010, **288**, 79–84.
73. F. F. Fang and H. J. Choi, *J. Appl. Phys.*, 2008, **103**, 07A301.
74. M. S. Cho, S. T. Lim, I. B. Jang, H. J. Choi and M. S. Jhon, *IEEE Trans. Magn*, 2004, **40**, 3036–3038.
75. B. J. Park, M. S. Kim and H. J. Choi, *Mater. Lett.*, 2009, **63**, 2178–2180.
76. F. F. Fang and H. J. Choi, *Phys. Status Solidi A Appl. Mater. Sci.*, 2007, **204**, 4190–4193.
77. F. F. Fang, H. J. Choi and Y. Seo, *ACS Appl. Mater. Interf.*, 2010, **2**, 54–60.
78. F. F. Fang, Y. D. Liu, H. J. Choi and Y. Seo, *ACS Appl. Mater. Interfaces*, 2011, **3**, 3487–3495.
79. M. Sedlacik, V. Pavlinek, P. Saha, P. Svrcinova and P. Filip, *AIP Conf. Proc.*, 2011, **1375**, 284–291.
80. M. Sedlačík, V. Pavlínek, P. Sáha, P. Švrčinová, P. Filip and J. Stejskal, *Smart Mater. Struct.*, 2010, **19**, 115008.
81. P. C. LeBaron, Z. Wang and T. J. Pinnavaia, *Appl. Clay Sci.*, 1999, **15**, 11–29.
82. F. F. Fang, M. S. Yang and H. J. Choi, *Magnetics, IEEE Trans. Magn.*, 2008, **44**, 4533–4536.
83. I. B. Jang, H. B. Kim, J. Y. Lee, J. L. You, H. J. Choi and M. S. Jhon, *J. Appl. Phys.*, 2005, **97**, 10Q912.
84. F. F. Fang, Y. D. Liu and H. J. Choi, *IEEE Trans. Magn.*, 2009, **45**, 2507–2510.
85. W. P. Wu, B. Y. Zhao, Q. Wu, L. S. Chen and K. A. Hu, *Smart Mater. Struct.*, 2006, **15**, N94–N98.
86. Y. D. Liu and H. J. Choi, *J. Appl. Phys.*, 2012, **111**, 07B502.
87. J. Pacull, S. Gonçalves, Á. V. Delgado, J. D. G. Durán and M. a. L. Jiménez, *J. Colloid Interface Sci.*, 2009, **337**, 254–259.
88. J. H. Wei, C. J. Leng, X. Z. Zhang, W. H. Li, Z. Y. Liu and J. Shi, *J. Phys. Conf. Ser.*, 2009, **149**, 012083.

89. H. J. Choi, I. B. Jang, J. Y. Lee, A. Pich, S. Bhattacharya and H. J. Adler, *IEEE Trans. Magn.*, 2005, **41**, 3448–3450.
90. F. F. Fang, J. H. Kim and H. J. Choi, *Polymer*, 2009, **50**, 2290–2293.
91. Y. P. Seo and Y. Seo, *Langmuir*, 2012, **28**, 3077–3084.
92. Y. P. Seo, H. J. Choi and Y. Seo, *Soft Matter*, 2012, **8**, 4659–4663.
93. H. J. Choi, M. S. Cho, J. W. Kim, C. A. Kim and M. S. Jhon, *Appl. Phys. Lett.*, 2001, **78**, 3806–3808.
94. Y. T. Choi, J. U. Cho, S. B. Choi and N. M. Wereley, *Smart Mater. Struct.*, 2005, **14**, 1025.
95. P. Gonon, J. N. Foulc, P. Atten and C. Boissy, *J. Appl. Phys.*, 1999, **86**, 7160–7169.
96. J. M. Ginder, L. C. Davis and L. D. Elie, *Int. J. Mod. Phys. B*, 1996, **10**, 3293–3303.
97. F. F. Fang, H. J. Choi and M. S. Jhon, *Colloid Surf., A*, 2009, **351**, 46–51.
98. M. Kaushal and Y. M. Joshi, *Soft Matter*, 2011, **7**, 9051–9060.
99. M. Baumgaertel and H. H. Winter, *Rheol. Acta*, 1989, **28**, 511–519.
100. F. R. Schwarzl, *Rheol. Acta*, 1975, **14**, 581–590.
101. J. Roupec and I. Mazůrek, eds. *R. Jabloński and T. Březina*, Springer Berlin Heidelberg, 2012, pp. 561–567.
102. J. C. Ulicny, M. P. Balogh, N. M. Potter and R. A. Waldo, *Mater. Sci. Eng. A*, 2007, **443**, 16–24.

CHAPTER 8

Microstructures and Physics of Super-Strong Magnetorheological Fluids

R. TAO

Department of Physics, Temple University, Philadelphia, PA 19122, USA
Email: rtao@temple.edu

8.1 Introduction

Magnetorheological (MR) fluids, which change from a liquid state into a solid state and *vice versa* in a magnetic field, have many industrial applications.[1] A typical MR fluid is a suspension of micrometre-sized magnetic particles in a base liquid, such as silicon oil. Surfactants are usually added to alleviate the settling problem. In the absence of an external magnetic field, no net magnetization exists within the particles and MR fluids have a relatively low viscosity. When a magnetic field is applied, the particles are polarized and align in the field direction to form chains and columns. The induced solid structure produces a respectable yield shear stress, exceeding the requirement for several mechanical applications, 40 kPa, set by manufacturing engineers.[2] For example, iron based MR fluids at 40–50% volume fraction have a yield stress about 100 kPa under a magnetic field of 1 T.[3] As the magnetic field further increases, the strength of the solidified fluid is further strengthened. This process is reversible. Once the magnetic field is removed, MR fluids return back to their original liquid state. The fluids' response time to a magnetic field is of the order of milliseconds.

RSC Smart Materials No. 6
Magnetorheology: Advances and Applications
Edited by Norman Wereley
© The Royal Society of Chemistry 2014
Published by the Royal Society of Chemistry, www.rsc.org

MR fluids were first discovered by Rabinow in 1948,[4] following the discovery of electrorheolgical (ER) fluids.[5] After 60-years of research and development, MR fluids began to have industrial applications. Commercial products, such as shock absorbers, linear dampers, and MR polishing machines, are now in production and applied in automobile industry and other industrial sectors.[1,6] Currently, MR fluids are considerably stronger than ER fluids, which have a yield stress around or below 10 kPa. This makes MR fluids very attractive. While the MR applications are still at an early stage, their impact on the automobile industry, railroad transportation, and bridge construction has already become notable.

Since the discovery of MR fluids, there has been some expectation that MR fluids may revolutionize certain industrial sectors. This possibility will become reality only if we can enhance the strength of MR fluids. The current strength of MR fluids, for example, is not sufficient for manufacturing flexible fixtures. Stronger MR fluids are needed for these applications. In addition, it is especially desirable if these strong MR fluids only require a moderate magnetic field. MR applications need electromagnets inside the MR devices. If the required magnetic field is strong, the electromagnets will be heavy and bulky, making the MR device bulky. The electromagnets are also well known for their delay time since the magnetic coils are an RL circuit. To produce a strong magnetic field, the magnetic coils must have a low DC resistance R and a high inductance L. The stronger is the magnetic field, the longer is the delay time. For a big electromagnet, the delay time L/R can be as long as several seconds. While MR fluids themselves respond to an external field quickly, the delay time of electromagnets lengthens the response time of MR devices. The size and delay time of electromagnets will not become an issue and the MR devices will remain agile only if the required magnetic field is moderate.

This, however, is not an easy task. Efforts in searching for new MR materials in the past decades came with limited results. For example, under a magnetic field of $H = 398$ kA m^{-1} (or $B = 0.5$ T), a new iron-cobalt MR fluid at 25% volume fraction produces a yield shear stress of 80 kPa,[3] while old carbonyl iron MR fluid at 25% volume fraction has yield shear stress of 60 kPa.[4] Ginder and Davis made a finite element calculation for a single-chain structure of MR fluids.[7] After taking the magnetic saturation of iron, 2.1 T, into account, they predicted that an iron-based MR fluid at 50% volume fraction with a single-chain structure had a yield shear stress limit of 200 kPa. From their results, to produce a yield stress of 100 kPa requires a magnetic field around 1T, quite high indeed. This implies that improving the physical properties of magnetic particles alone is unlikely to bring the strength of MR fluids to a required level.

In this chapter, we will report a completely different and novel approach that makes MR fluids super-strong. The key is to improve the microstructure of MR fluids. We do not seek new materials, but employ a quick compression-assisted-aggregation process to force MR fluids to form a microstructure that is much stronger than the single-chain structure. After doing so, the MR fluids well exceed the theoretical strength predicted from the single-chain structure. For example, our experiments with an iron-based MR fluid find that at a moderate magnetic field, $H = 458$ kA m^{-1} ($B = 0.576$ T), this structure-enhanced yield shear stress exceeds 800 kPa, ten times the value without the compression-assisted aggregation. In addition, since it is dealing with the fluid's microstructure, this approach is general and applicable to all kinds of MR fluids.

We will first review the physical mechanism and conventional microstructure of MR fluids in section II and discuss the rheological property and weak points of the single-chain structure in Section III. The experimental data to identify the weak points will be in Section IV. We will then describe our compression-assisted-aggregation experiment and rheological properties of the super-strong MR fluids in Section V. The new MR microstructure of these fluids after the compression process, evidence of the improved microstructure, is in Section VI. Finally, we will give some phenomenological formula in Section VII and discuss related issues in Section VIII.

8.2   Microstructure of MR Fluids

When there is no additional force, the conventional microstructure formation in MR fluids is determined by the field-induced polarization, the same as ER fluids.[8] Thus, MR and ER fluids should have the same microstructure. Experiments also confirm this conclusion.[9,10]

Let us consider a model of MR fluids consisting of N identical spherical magnetic particles of magnetic susceptibility μ_p in a base liquid of susceptibility μ_f, and $\mu_p > \mu_f = 1$. The fluid is placed between two magnetic poles, which are denoted as two planes $z = 0$ and $z = d$. The particles have radius a. The applied magnetic field is $\vec{H}_0$ along the z direction. We assume that $\vec{H}_0$ is below the particles' magnetic saturation. Then in the field, the magnetic particles are polarized, obtaining an induced magnetic dipole moment, $\vec{m} = \vec{H}_{loc} a^3 (\mu_p - \mu_f)/(\mu_p + 2\mu_f)$, where $\vec{H}_{loc}$ is the local field. Two magnetic dipoles at $\vec{r}_i$ and $\vec{r}_j$ have an interaction

$$U(\vec{r}_{ij}) = m^2 \mu_f (1 - 3\cos^2\theta_{ij})/r_{ij}^3 \qquad (8.1)$$

where $\vec{r}_{ij} = \vec{r}_i - \vec{r}_j$ and $0 \leq \theta_{ij} \leq \pi/2$ is the angle between the z direction and the joint line of the two dipoles. The dipolar force acting on the particle at $\vec{r}_i$ by the particle at $\vec{r}_j$ is given by

$$\vec{f}_{ij} = (3m^2\mu_f/r_{ij}^4)[\vec{e}_r(1 - 3\cos^2\theta_{ij}) - \vec{e}_\theta \sin(2\theta_{ij})] \qquad (8.2)$$

where $\vec{e}_r$ is a unit vector parallel to $\vec{r}_{ij}$ and $\vec{e}_\theta$ is a unit vector parallel to $\vec{e}_r \times (\vec{e}_r \times \vec{H}_0)$. The motion of the ith particle is described by a Langevin equation

$$M d^2\vec{r}_i/dt^2 = \vec{F}_i - 6\pi a\eta d\vec{r}_i/dt + \vec{R}_i(t) \qquad (8.3)$$

where $\vec{F}_i$ is the electric force on the ith dipole from all other dipoles and all images, η is the viscosity of the base liquid, $-6\pi a\eta\vec{v}_i$ is the Stokes drag force, the leading hydrodynamic force on the particle, $\vec{R}_i(t)$ is the random Brownian force, and M is the particle's mass. Under the dipolar approximation, it is easy to derive analytic expressions for $\vec{F}_i$ from eqn (8.2).[11] The Brownian force has a white-noise distribution: the averages $\langle \vec{R}_i(t) \rangle = 0$ and $\langle R_{i,\alpha}(0)R_{i,\beta}(t) \rangle = 12\pi ak_BT\eta\delta_{\alpha\beta}\delta(t)$, where k_B is the Boltzmann constant and T is the temperature.

The molecular dynamics simulation of the Langevin equation (8.3) provides a clear picture of the dynamic structure formation of MR fluids. As shown in Figure 8.1, the particles are randomly distributed before a magnetic field is applied. At this stage, the MR fluid is a suspension. At $t = 0$ a magnetic field is turned on and the particles begin to move according to eqn (8.3). As shown in Figure 8.2, after 10 ms (milliseconds) single chains are formed. While at this stage these chains are not straight and the microstructure has almost no lateral ordering yet, the fluid already has a yield shear stress and its effective viscosity is significantly increased. With time, the microstructure develops into a three-dimensional (3-D) ordering structure. As shown in Figure 8.3, at $t = 0.5$ s, the chains are more straight, some chains aggregate together, and a lateral order begins to build up. After 2.2 seconds, chains move to form thick columns. While the column has a limited thickness in the x-y direction and maintains the field direction as its preferred direction, it is clear from Figure 8.4 that this

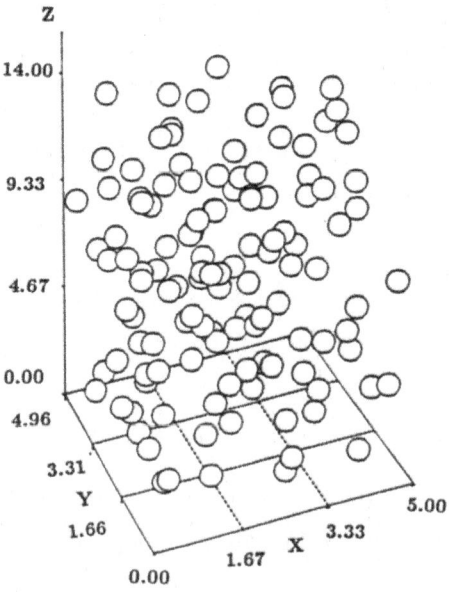

Figure 8.1   Before a magnetic field is applied, the magnetic particles are randomly distributed in the base liquid.

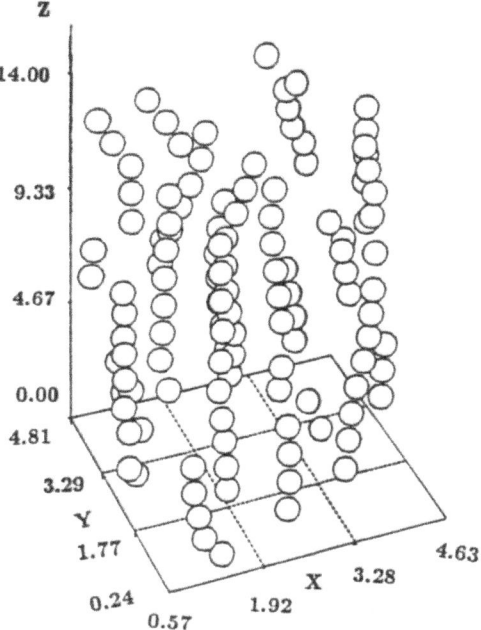

Figure 8.2 The magnetic particles form chains in about 10 ms after a magnetic field is applied

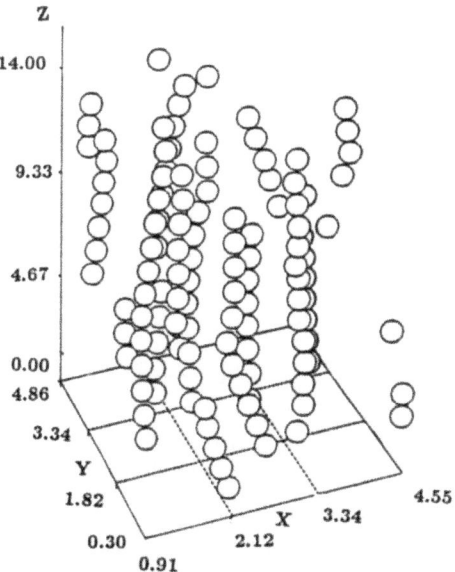

Figure 8.3 The chains become more straight and some chains aggregate together about 0.5 seconds after the magnetic field is applied.

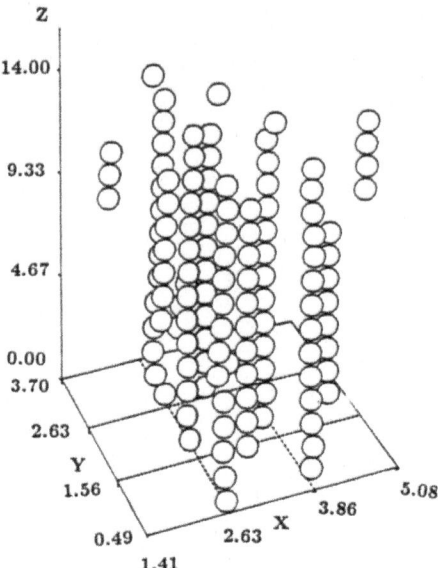

Figure 8.4   The chains aggregate together to form thick columns, which have a body-centered tetragonal lattice structure.

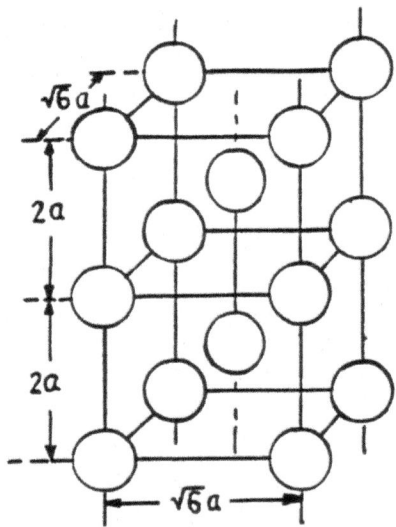

Figure 8.5   The ground state of MR structure, body-centered tetragonal lattice. The particles have radius a and are not shown to scale.

structure is a crystalline structure. In fact, the columns have a body-centered tetragonal (bct) lattice structure (Figure 8.5).[8]

The above simulation confirms the experimental finding that the formation of the conventional MR microstructure has two development stages. The first stage is the chain-formation process, which only takes several milliseconds to complete after the magnetic field is applied. The second stage is a relatively slow process, involving the aggregation of chains into thick columns. In the simulation this process takes several seconds. When the MR system is big, as in the experiment, this process may take a couple of minutes.

The above MR microstructure and the dynamic process can be understood from the induced dipolar interactions. It is clear from eqn (8.1) that the interaction energy is minimized if $\theta_{ij} = 0$ and $r_{ij} = 2a$. Therefore, once we apply a magnetic field, the particles quickly form chains. A finite chain between the two poles, combining with their infinite images, becomes an infinite chain. For the formation of 3-D structure, we need to examine the chain–chain interaction. Let us consider one infinite chain along the z axis with its particles at $z = (2j + 1)a$ $(j = 0, \pm 1, \pm 2, ...)$. If the second infinite chain parallel to the z axis is a distance ρ away from the first one and its particles are at $z = (2j + 1)a + \varsigma$, the leading term of the interaction between these two infinite chains is given by

$$U_{cc}(\rho, \varsigma) = (m^2 \mu_f \pi^2 / a^3)(2\rho/a)^{1/2} \exp(-\pi\rho/a) \cos(\pi\varsigma/a) \qquad (8.4)$$

When $\varsigma = 0$, the interaction is repulsive. When $\varsigma = a$, all particles in the second chain slide a distance of a along the z direction, then the interaction energy becomes negative and minimized. The force between such two mismatched chains is attractive. At the final 3-D structure, to minimize the energy, every chain must have four mismatched chains as its nearest neighbors. This leads to the bct lattice structure.

In comparison with the dipolar interaction, the Brownian force is very weak. For most MR fluids, the ratio of the interaction energy of two magnetic dipoles to the thermal energy, $\alpha = m^2 / (8a^3 k_B T)$, is quite large, $\alpha > 100$. Therefore, at the first stage of chain formation, the Brownian force is negligible in comparison with the interaction between two dipoles. However, once the chains are formed, the dipolar interaction is screened. As noted from eqn (8.4), the interaction between two straight chains is weak and short-ranged. Then the Brownian force plays an important role. For example, if $\alpha = 100$, when the distance between the two straight chains $\rho \sim 3a$, the Brownian force is already comparable with the chain–chain interaction. Then, the Brownian force easily makes the chains bend or randomly fluctuate, as seen in the simulation (Figure 8.2 and 8.3). This phenomenon is associated with the instability of one-dimensional solids. The attracting force between curved chains is stronger than that of straight chains.[12] Therefore, the chains' fluctuations under Brownian motion help their aggregation together. However, once several chains get together to form a column, the three-dimensional column is not easy to bend or fluctuate (Figure 8.4). The Brownian force has little effect on columns. Thus, the column–column aggregation is difficult in the unassisted aggregation process and the developed columns are of a limited thickness.

The rheological properties and strength of MR fluids are related to the field-induced microstructure. For example, thick column structure is stronger than single-chain structure.[13] However, most agile MR applications require a quick response and cannot wait long enough for the columns to form. Moreover, the unassisted aggregation process only produces columns consisting of 10–15 chains.[9] The relevant MR microstructure in most MR applications is the single-chain structure.[7]

8.3 The Weak Points of MR Microstructure

To develop strong MR fluids, we need to know the rheological properties of the induced MR structures. In particular, we want to identify the weak points of these structures and find a way to strengthen them. To begin with, we examine the single-chain-structure, which is the base of all other MR structures.

We consider applying an external shear force on the chain. Under the force, the chain deforms, becomes slanted, then breaks into two broken chains if the shear force exceeds a threshold. The breaking point should be the chain's weak point under the shear. The conventional wisdom assumes that the breaking point is in the middle of chain.[12,13] Why is it at the middle? The system is symmetric around the middle between the two poles; based on this symmetry, it is natural to guess that the chain could break at the middle.

However, this is not the case. There is a symmetry breaking.[14] After a magnetic field is applied, the space is no longer isotropic. To form a long chain along the field direction is energetically favored in this space. Let us assume that the shear process is quasi-static, that the chain relaxes into the lowest energy state at each shearing step. Under such a deformation, the configuration with the minimum energy is no longer symmetric. The symmetric configuration becomes unstable. In particular, during the initial deformation, a gap emerges at one end. The gap widens as the strain increases. When the strain exceeds a critical value, the chain breaks. The breaking point is at either end, but not at the middle. In another words, the weak points of a MR chain under shear are at its two ends.

In order to demonstrate the above points, we apply the Green's function method to find the energy for various configurations. The calculation includes the multipole contributions and the effect of electrodes. A comparison of the energies between different configurations enables us to identify the minimum-energy configuration. We thus find the deformation process and locate the breaking point.

After the fluid is activated, chains are formed. The induced attractive force among the chain particles provides a capacity to resist an external stress. The yield shear stress of a chain is defined as the stress below which a deformed chain can recover after the stress is removed, but above which the chain incurs non-recoverable damage even if the stress is released. It is clear that quantitative estimation of the yield stress depends on how the chain deforms in response to an external stress.

The chain is originally along the field direction, spanning between the electrodes with particles touching each other. Upon application of a pair of shear forces at the chain's two end particles, the chain is strained and elongated.

These two forces are of the same magnitude but in the opposite direction, perpendicular to the field. As mentioned before, the shear process is slow and can be described as quasi-static.

Since the magnetic field satisfies $\nabla \times \vec{H} = 0$, we define a magnetic potential $\Phi(\vec{r})$, $\vec{H} = -\nabla\Phi(\vec{r})$. Then $\nabla^2\Phi(\vec{r}) = 0$ with its boundary conditions as $\Phi(\vec{r})\big|_{z=0} = 0$ $\Phi(\vec{r})\big|_{z=d} = -H_0 d$. Here $\vec{H}_0 = H_0\hat{z}$ is the applied magnetic field. We denote $\hat{z}$ as a unit vector along the z direction. On the particle surface, $\Phi_{out}(\vec{r})\big|_{s_i} = \Phi_{in}(\vec{r})\big|_{s_i}$ and $\mu_f\partial\Phi_{out}(\vec{r})/\partial n_i\big|_{s_i} = \mu_p\partial\Phi_{in}(\vec{r})/\partial n_i\big|_{s_i}$ where $\Phi_{out}(\vec{r})$ and $\Phi_{in}(\vec{r})$ are the potentials outside and inside the ith particle, respectively. S_i is the ith particle's surface and $\partial/\partial n_i$ is the normal derivative at the surface S_i in the outward direction.

The above boundary problem can be simplified by writing the total potential $\Phi(\vec{r})$ of the system with two terms

$$\Phi(\vec{r}) = -\vec{H}_0 \cdot \vec{r} + \Phi^{(1)}(\vec{r}) \tag{8.5}$$

where the second term in eqn (8.5) is the magnetic potential due to the polarization of particles by the applied field. The Laplace equation for $\Phi^{(1)}$ and its boundary condition are given by

$$\nabla^2\Phi^{(1)}(\vec{r}) = 0, \quad \Phi^{(1)}(\vec{r})\big|_{z=0} = \Phi^{(1)}(\vec{r})\big|_{z=d} = 0 \tag{8.6}$$

The Green function, which equals to the potential at $\vec{r}$ due to a unit point magnetic 'charge'' at $\vec{r}'$ and all its images, is given by

$$G(\vec{r}, \vec{r}') = \sum_{k\lambda} \frac{\lambda}{|\vec{r} - \vec{r}'_{k\lambda}|} \tag{8.7}$$

where $k = 0, \pm 1, \pm 2, ..., \lambda = \pm 1, \vec{r} = (x, y, z)$ and $\vec{r}'_{k\lambda} = (x', y', 2kd + \lambda z')$. Application of the Green's function yields

$$\Phi^{(1)}(\vec{r}) = \int_V \rho(\vec{r}')G(\vec{r}, \vec{r}')d^3r' \tag{8.8}$$

where $\rho(\vec{r}')$ is the magnetic 'charge'' at $\vec{r}'$. Since the magnetic 'charge' only appears on the surfaces of particles inside the volume V, Eqn (8.8) becomes

$$\Phi^{(1)}(\vec{r}) = \sum_i \oint_{S_i} \sigma(\vec{r}')G(\vec{r}, \vec{r}')da' = \sum_{ik\lambda} \oint_{S_i} \frac{\lambda\sigma(\vec{r}')}{|\vec{r} - \vec{r}'_{k\lambda}|}da' \tag{8.9}$$

where $\sigma(\vec{r}')$ is the surface charge density to be determined by the boundary conditions.

To apply the boundary conditions, we need to express $\Phi(\vec{r})$ in the coordinate system which has its origin at a particle's center. We therefore make an expansion,

$$\frac{1}{|\,(\vec{r} - \vec{r}_j) - (\vec{r}'_{k\lambda} - \vec{r}_j)\,|} = 4\pi \sum_{lm} \frac{1}{2l+1} \frac{r^l_<}{r^{l+1}_>} Y^*_{lm}(\vec{r}'_{lm} - \vec{r}_j)\, Y_{lm}(\vec{r} - \vec{r}_j) \quad (8.10)$$

where $\vec{r}_j$ is the jth particle's center, and $r_<$ (or $r_>$) is the smaller (or larger) one among $|\,\vec{r} - \vec{r}_j\,|$ and $|\,\vec{r}'_{k\lambda} - \vec{r}_j\,|$. For $m \geq 0$, the spherical harmonic function $Y_{lm}(\vec{r}) = \sqrt{(2l+1)(l-m)!\,/\,[4\pi(l+m)!]}\, P^m_l(\cos\theta) e^{im\varphi}$, where P^m_l is an associated Legendre function, θ and ϕ are the spherical coordinates of $\vec{r}$. If $m < 0$, $Y_{lm} = (-1)^m Y_{l\,|m|}$. With the above expansion, we obtain

$$\Phi(\vec{r}) = - H_0 z + 4\pi \sum_{ik\lambda lm} \frac{\lambda}{2l+1} Y_{lm}(\vec{r} - \vec{r}_j) \oint_{S_i} \frac{r^l_<}{r^{l+1}_>} \sigma(\vec{r}') Y^*_{lm}(\vec{r}'_{k\lambda} - \vec{r}_j) da' \quad (8.11)$$

In particular, the potentials inside and outside the jth particle are given by

$$\Phi_{in}(\vec{r}) = - H_0 z_j - H_0(z - z_j) + 4\pi \sum_{lm} \frac{q_{jlm}}{(2l+1)a^{2l+1}} |\,\vec{r} - \vec{r}_j\,|^l \, Y_{lm}(\vec{r} - \vec{r}_j)$$

$$+ 4\pi \sum_{ik\lambda lm}' \frac{\lambda}{2l+1} |\,\vec{r} - \vec{r}_j\,|^l \, Y_{lm}(\vec{r} - \vec{r}_j) \oint_{S_i} \frac{\sigma(\vec{r}')}{|\,\vec{r}'_{k\lambda} - \vec{r}_j\,|^{l+1}} \, Y^*_{lm}(\vec{r}'_{k\lambda} - \vec{r}_j) da'\,;$$

$$(8.12)$$

$$\Phi_{out}(\vec{r}) = - H_0 z_j - H_0(z - z_j) + 4\pi \sum_{lm} \frac{q_{jlm}}{(2l+1)\,|\,\vec{r} - \vec{r}_j\,|^{l+1}} \, Y_{lm}(\vec{r} - \vec{r}_j)$$

$$+ 4\pi \sum_{ik\lambda lm}' \frac{\lambda}{2l+1} |\,\vec{r} - \vec{r}_j\,|^l \, Y_{lm}(\vec{r} - \vec{r}_j) \oint_{S_i} \frac{\sigma(\vec{r}')}{|\,\vec{r}'_{k\lambda} - \vec{r}_j\,|^{l+1}} \, Y^*_{lm}(\vec{r}'_{k\lambda} - \vec{r}_j) da'$$

$$(8.13)$$

with $a < |\,\vec{r} - \vec{r}_j\,| < |\,\vec{r}'_{k\lambda} - \vec{r}_j\,|$

Here

$$q_{jlm} = \oint_{S_i} \sigma(\vec{r}')a^l Y^*_{lm}(\vec{r}' - \vec{r}_j) da' \quad (8.14)$$

is the magnetic multipole moment of the jth particle with respect to its center. The prime on those summation symbols in eqn (8.12) and (8.13) indicates that the cases with $i = j, k = 0, \lambda = +1$ are excluded in the summations. This notation will be used hereafter throughout the chapter.

To evaluate the above integrals, we introduce the following identity:[15]

$$\frac{Y_{lm}(\vec{r}'_{k\lambda} - \vec{r}_j)}{(2l+1) \mid \vec{r}'_{k\lambda} - \vec{r}_j \mid} = \sum_{l'm'} A^{lm}_{l'm'}(\vec{r}_j - \vec{r}_{ik\lambda}) \mid \vec{r}'_{k\lambda} - \vec{r}_{ik\lambda} \mid^{l'} Y_{l'm'}(\vec{r}'_{k\lambda} - \vec{r}_{ik\lambda})$$

(8.15)

where $\vec{r}_{ik\lambda} = (x_i, y_i, 2kd + \lambda z_i)$ and the expansion coefficient $A^{lm}_{l'm'}(\vec{r}_j - \vec{r}_{ik\lambda})$ is given by

$$A^{lm}_{l'm'}(\vec{r}_j - \vec{r}_{ik\lambda}) = (-1)^{l+m} \frac{Y^*_{l+l',m+m'}(\vec{r}_j - \vec{r}_{ik\lambda})}{\mid \vec{r}_j - \vec{r}_{ik\lambda} \mid^{l+l'+1}}$$

$$\times \left[\frac{4\pi(l+l'+m-m')!(l+l'+m'-m)!}{(2l+1)(2l'+1)(2l+2l'+1)(l+m)!(l-m)!(l'+m')!(l'-m')!} \right]^{1/2}$$

(8.16)

Substituting eqn (8.15) into eqn (8.12), we have

$$\Phi_{in}(\vec{r}) = - H_0 z_j - H_0(z - z_j) + 4\pi \sum_{lm} \frac{q_{jlm}}{(2l+1)a^{2l+1}} \mid \vec{r} - \vec{r}_j \mid^l Y_{lm}(\vec{r} - \vec{r}_j)$$

$$+ 4\pi \sum_{ik\lambda lm} \lambda \mid \vec{r} - \vec{r}_j \mid^l Y_{lm}(\vec{r} - \vec{r}_j) \sum_{l'm'} A^{lm}_{il'm'} \oint_{S_i} \sigma(\vec{r}') \mid \vec{r}'_{k\lambda} - \vec{r}_j \mid^{l'} Y^*_{l'm'}(\vec{r}'_{k\lambda} - \vec{r}_{ik\lambda}) da'$$

(8.17)

For the sake of simplicity, we abbreviate $A^{lm}_{l'm'}(\vec{r}_j - \vec{r}_{ik\lambda})$ as $A^{jlm}_{il'm'}$. Note that in the above equation $\mid \vec{r}'_{k\lambda} - \vec{r}_{ik\lambda} \mid = a$ and $Y^*_{l'm'}(\vec{r}'_{k\lambda} - \vec{r}_{ik\lambda}) = \lambda^{l'+m'} Y^*_{l'm'}(\vec{r}' - \vec{r}_i)$. En (8.17) thus becomes

$$\Phi_{in}(\vec{r}) = - H_0 z_j - H_0(z - z_j) + 4\pi \sum_{lm} \frac{q_{jlm}}{(2l+1)a^{2l+1}} \mid \vec{r} - \vec{r}_j \mid^l Y_{lm}(\vec{r} - \vec{r}_j)$$

$$+ 4\pi \sum_{il'm'} q_{il'm'} \sum_{lm} B^{jlm}_{il'm'} \mid \vec{r} - \vec{r}_j \mid^l Y_{lm}(\vec{r} - \vec{r}_j)$$

(8.18)

where

$$B^{jlm}_{il'm'} = \sum_{k\lambda} \lambda^{l'+m'+1} A^{lm}_{l'm'}(\vec{r}_j - \vec{r}_{ik\lambda})$$

(8.19)

Similarly, the potential outside the jth particle is shown to be

$$\Phi_{out}(\vec{r}) = - H_0 z_j - H_0(z - z_j) + 4\pi \sum_{lm} \frac{q_{jlm}}{(2l+1) \mid \vec{r} - \vec{r}_j \mid^{l+1}} Y_{lm}(\vec{r} - \vec{r}_j)$$

$$+ 4\pi \sum_{il'm'} q_{il'm'} \sum_{lm} B^{jlm}_{il'm'} \mid \vec{r} - \vec{r}_j \mid^l Y_{lm}(\vec{r} - \vec{r}_j)$$

(8.20)

From the boundary condition on the sphere surfaces, we have

$$(\mu_p - \mu_f)\left[\frac{H_0}{\sqrt{12\pi}}Y_{10}(\vec{r}-\vec{r}_j) - \sum_{lm}la^{l-1}\left(\sum_{il'm'}B^{jlm}_{il'm'}q_{il'm'}\right)Y_{lm}(\vec{r}-\vec{r}_j)\right]$$

$$= \sum_{lm}q_{jlm}\frac{l\mu_p + (l+1)\mu_f}{(2l+1)a^{l+2}}Y_{lm}(\vec{r}-\vec{r}_j) \tag{8.21}$$

Applying the orthonormal condition of the spherical harmonics, we find the following relationship among the multipole moments,

$$\sum_{il'm'}\left[\delta^j_i\delta^l_{l'}\delta^m_{m'} + (2l+1)\frac{(\mu_p-\mu_f)a^{2l+1}l}{l\mu_p+(l+1)\mu_f}B^{lm}_{l'm'}\right]q_{il'm'} = \sqrt{\frac{3}{4\pi}}\frac{(\mu_p-\mu_f)a^3}{\mu_p+2\mu_f}H_0\delta^1_l\delta^0_m \tag{8.22}$$

or in matrix notation

$$Cq = b \tag{8.23}$$

where

$$C^{jlm}_{il'm'} = \delta^j_i\delta^l_{l'}\delta^m_{m'} + (2l+1)\frac{(\mu_p-\mu_f)a^{2l+1}l}{l\mu_p+(l+1)\mu_f}B^{lm}_{l'm'} \tag{8.24}$$

q represents the multipole moments, and b is the matrix given by

$$b_{ilm} = \sqrt{\frac{3}{4\pi}}\frac{(\mu_p-\mu_f)a^3}{\mu_p+2\mu_f}H_0\delta^1_l\delta^0_m \tag{8.25}$$

Now let us calculate the interaction energy due to the presence of magnetic particles, ΔW. After assuming that μ_p and μ_f are constants, we have

$$\Delta W = \frac{1}{8\pi}\int(\vec{H}\cdot\vec{B} - \vec{H}_0\cdot\vec{B}_0)d^3x \tag{8.26}$$

where $\vec{H}$ and $\vec{B}$ ($\vec{H}_0$ and $\vec{B}_0$) are the magnetic field and the flux density after (before) the presence of the particles. Eqn (8.26) is transformed into

$$\Delta W = \frac{1}{8\pi}\int(\vec{H}\cdot\vec{B}_0 - \vec{H}_0\cdot\vec{B})d^3x + \frac{1}{8\pi}\int(\vec{H}+\vec{H}_0)\cdot(\vec{B}-\vec{B}_0)d^3x \tag{8.27}$$

It is easy to verify that the second integral in eqn (8.27) vanishes. Consequently, ΔW becomes

$$\Delta W = \frac{1}{8\pi}\int(\vec{H}\cdot\vec{B}_0 - \vec{H}_0\cdot\vec{B})d^3x \tag{8.28}$$

The above integral is taken over the whole region of the system. Only those regions occupied by the particles contribute to the integral since the integrand vanishes outside the particles: as $\vec{B} = \mu_f \vec{H}$ and $\vec{B}_0 = \mu_f \vec{H}_0$ outside the particles, $\vec{H} \cdot \vec{B}_0 - \vec{H}_0 \cdot \vec{B} = 0$. Then eqn (8.28) becomes

$$\Delta W = -\frac{(\mu_p - \mu_f)}{8\pi} \sum_{i=1}^{N} \int_{V_i} \vec{H}_0 \cdot \vec{H} d^3 x \qquad (8.29)$$

where V_i is the volume of the ith particle. Since $\vec{H}_0 = -\nabla \Phi_0$ and $\Phi_0 = -H_0 r \cos \theta$, we have

$$\int_{V_i} \vec{H}_0 \cdot \vec{H} d^3 x = -\int_{V_i} (\nabla \Phi_0) \cdot \vec{H} d^3 x = -\int_{V_i} \nabla (\Phi_0 \vec{H}) d^3 x + \int_{V_i} \Phi_0 \nabla \cdot \vec{H} d^3 x \quad (8.30)$$

The second integral vanishes because $\nabla \cdot \vec{H} = 0$ inside the particle. Applying the divergence theorem to the first integral, we find

$$\int_{V_i} \vec{H}_0 \cdot \vec{H} d^3 x = -\oint_{S_i} \Phi_0 H_n da \qquad (8.31)$$

where H_n is the normal component of the magnetic field on the inside surface of the particle. Note that $\mu_p H_n^{in}|_{S_i} = \mu_f H_n^{out}|_{S_i}$ and the surface 'charge' density $\sigma(\vec{r})$ on the ith particle is given by $\sigma(\vec{r}) = (H_n^{out}|_{S_i} - H_n^{in}|_{S_i})/4\pi$. Hence, $H_n^{in}|_{S_i} = 4\pi\sigma(\vec{r})\mu_f/(\mu_p - \mu_f)$. Substituting $H_n^{in}|_{S_i}$ into eqn (8.31), we have

$$\int_{V_i} \vec{H}_0 \cdot \vec{H} d^3 x = \frac{4\pi H_0 \mu_f}{\mu_p - \mu_f} \int_{S_i} \sigma(\vec{r}) r \cos \theta da = \frac{4\pi H_0 \mu_f}{\mu_p - \mu_f} \sqrt{\frac{4\pi}{3}} q_{i10} \qquad (8.32)$$

The energy is given in the following form,

$$\Delta W = -\frac{1}{2} \sqrt{\frac{4\pi}{3}} H_0 \sum_{i=1}^{N} q_{i10} \qquad (8.33)$$

where we use $\mu_f = 1$.

The procedure of calculation is as follows. We first calculate matrix C from eqn (8.24) for various configurations, then solve q from eqn (8.23). Afterwards, we calculate the energy from eqn (8.33) for these configurations. The energy calculation converges quickly as more multipole moments are included. When a pair of shear forces is applied to a chain through the pole plates, as in most MR applications, we assume that the pair of shear forces is applied to the first particle and the last particle in the chain. It is clear that when the sheared chain is slanted, it is also stretched. In order to locate the breaking point, we first examine where the gap or gaps should be due to this stretch. With various

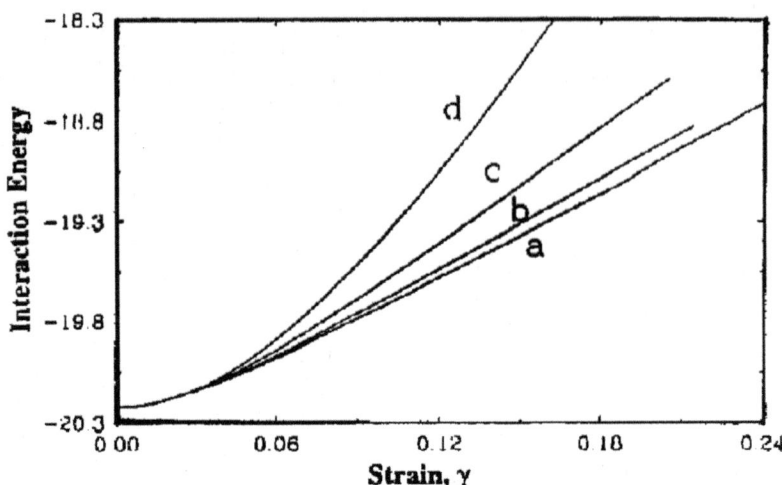

Figure 8.6   A comparison of the interaction energies for different slanted chains: (a) only one gap between the first and second particles, (b) only one gap at the chain middle, (c) two gaps with one between the first and second particles and the other between the last and next last particles, and (d) gaps uniformly distributed along the chain.

proposed configurations, we calculate the energies for all them and, afterwards, make a comparison. Figure 8.6 shows the energies for a couple of these configurations, (1) gaps uniformly distributed along the chain, (2) two gaps, one is between the first and second particles and the other between the last and next last particles, (3) only one gap between the first and second particles, and (4) only one gap in the middle of the chain. It is clear from Figure 8.6 that when the chain has just one gap between the first and the second particles, the energy is the lowest. The configuration with one gap in the middle or the configuration with gaps uniformly distributed along the chain never have the lowest energy. This means that under the external shear force, the slanted chain first has one gap between the first and second particles. Obviously, this gap will get wider and wider if the shear strain becomes larger and larger.

When the shear strain is big enough, the chain breaks. Where is the breaking point? In Figure 8.7, we plot the energy for the broken chains with several different possible broken points. When the broken point is between the first and second particles, the energy is the lowest among them. The broken chain with the break point in the middle does not have the lowest energy. Therefore, when the chain breaks under the shear, the broken point must be at one of its ends. This implies that the weak points of a MR chain are at its two ends.

The above theoretical result is outlined in Figure 8.8. When the shear strain is small, the chain becomes slanted and the gap between the first particle and second particle increases. When the shear strain exceeds a critical value, the chain breaks at one of its ends and the broken chain returns to the vertical position.

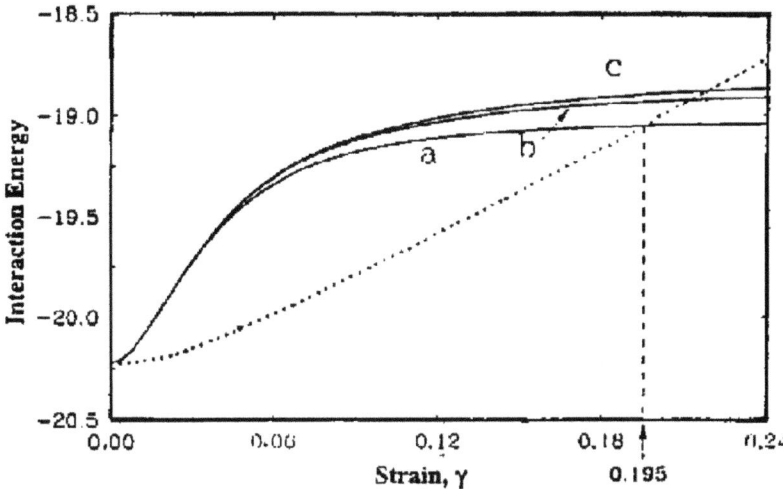

Figure 8.7　A comparison of the interaction energies for different broken chains: (a) the breaking point is between the first and second particles; (b) the breaking point is between the second and third particles; (c) the breaking point is in the middle. The dotted curve is for a slanted chain with a gap between the first and second particles.

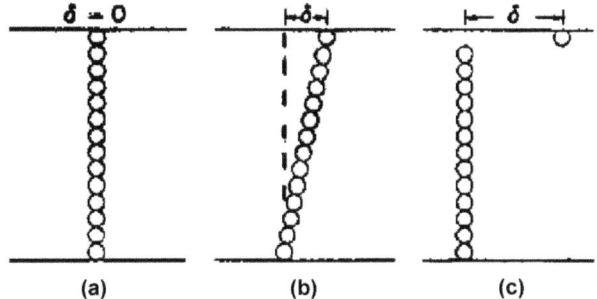

Figure 8.8　(a) A single chain without shear. (b) A slanted chain under a small shear strain. The gap is between the first and second particles. (c) A broken chain when the shear strain exceeds the critical value.

The critical shear strain for the chain to break is just the cross point between the curve for the slanted chain and the curve for the broken chain (Figure 8.7). From our calculation, this critical shear strain is 0.195.

As mentioned earlier, the above discussion assumes that the shear process is slow enough that the chain relaxes into the minimum energy state at each shearing step. Since most MR fluids have a response time in the order of milliseconds, we expect that this assumption could hold if the shear rate $\dot{\gamma}$ is not too high, such as below 100 s^{-1}. Of course, if the dynamic process has a much higher shear rate, the chain will not be able to relax into its minimum energy state during the shear process and thus will eventually break differently.

The experiment described in the subsequent section confirms that the quasi-static assumption is valid if $\dot{\gamma} \leq 100$ s^{-1}. We note that many MR applications have the shear rate within this range.

8.4   Experiment to Determine the Weak Points

The above theoretical result has also been verified by experiments. Since the physics here is the same for both ER fluids and MR fluids, the experiment depicted in Figure 8.9 is carried out with an ER fluid. There is no difficulty to carry out a similar experiment for MR fluids.

The particles used in the experiment are glass beads coated with two layers: a metallic layer and an insulating TiO$_2$ surface layer. The particle diameter is about 40 μm and silicon oil is the base liquid. An electric field close to 1 kV mm^{-1} is applied across the two electrodes. The gap between the electrodes is about 2.8 mm. Therefore, there are about 70 particles in a single chain. Figure 8.9a shows that the chain shears slowly. Figure 8.9b shows that the chain breaks between the first particle and second particle when the shear strain exceeds a critical value. This experiment repeats many times at low shear rate and we always find that the chain breaks at one end, never at the middle. Therefore, it confirms the theoretical result that the weak points of an MR or ER chain are at the chain's ends. In the experiment, we also increase the shear rate and find that the chain breaks at its one end even at a shear rate of 200 s^{-1}.

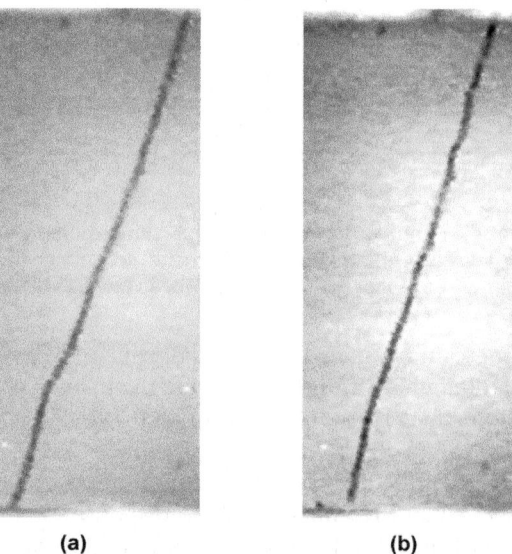

(a)   (b)

Figure 8.9   (a) Under a shear force, the ER chain becomes slanted. (b) The chain breaks between the first and second particles when the shear strain exceeds the critical value.

This experiment also finds that the critical shear strain is 0.25, slightly bigger than the theoretical value 0.195. This difference may be due to a fact that the slanted chain in the experiment is usually curved, while the theoretical calculation assumes that the slanted chain is not curved. While the above discussion is limited to the single-chain structure, the asymmetric breaking behavior should be general. For example, under a magnetic field, MR chains may slowly aggregate into thick columns, whose ideal structure is a bct lattice. Such columns developed from unassisted aggregation, in fact, are a bundle of a dozen chains. Therefore, under a strong shear force, this bundle of a dozen chains may well break at one of its ends.

8.5 Super-Strong MR Fluid

After understanding the microstructure of MR fluids and identifying its weak points, we are in a position to seek a new agile approach to produce super-strong MR fluids. Let us emphasize that the strength of MR fluids solely comes from the induced microstructure. Different microstructures deliver different strengths. For example, both theoretical calculation[13] and experiment[16] have already found that a bct lattice structure has a higher yield shear stress than a single-chain structure. Unfortunately, the unassisted aggregation process, as discussed before, is slow, produces columns of a limited thickness, and is thus not very useful. We need a technology that can rapidly produce thick columns with strong and robust ends, *i.e.*, thick columns with no weak points at their ends.

This technology is a compression-assisted-aggregation process.[17] Immediately after a magnetic field is applied, we compress the MR fluid along the field direction before a shear force is applied. The magnetic field produces chains in milliseconds. The compression pushes these chains to form thick columns with strong and robust ends. In addition, each column consists of at least a couple of hundred chains. Once the weak points of MR microstructure are strengthened and the columns are very thick, the MR fluids become super-strong.

In our experiment, we used a suspension of carbonyl iron particles in silicone oil. The initial experiment had MR fluid of 45% volume fraction for the purpose of making a phase-changeable flexible fixture. We later conducted experiments with low volume fractions and find that this new technology also works well at low volume fractions.

From the scanning electronic micrographic (SEM) image, the carbonyl iron particles were spherical with average diameter 4.5 μm (Figure 8.10). A small amount of surfactant, oleic acid (Sigma), was added into the suspension so that the particles suspended in silicon oil without settling for at least 24 hours. The zero-field viscosity of our MR fluid at 45% volume fraction was 10 poises.

The experimental setup is shown in Figure 8.11, and was originally designed to test the idea of applying MR fluids to manufacturing flexible fixtures. An electromagnet with two water-cooled coils produced a magnetic field in the

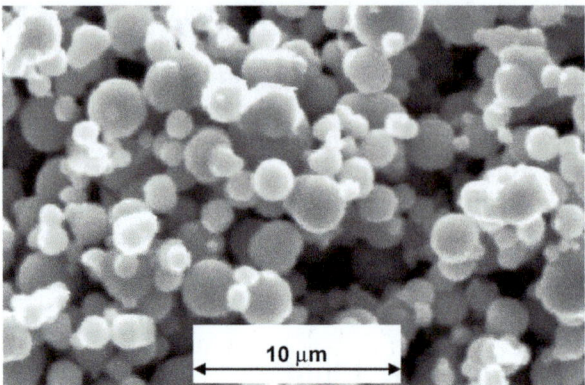

Figure 8.10   The scanning electronic micrographic image of carbonyl iron particles.

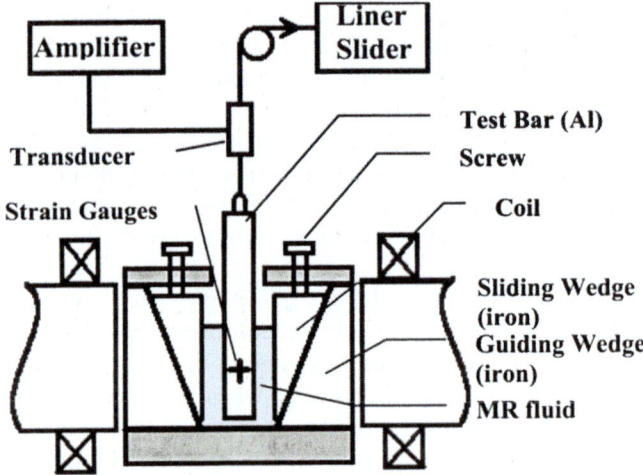

Figure 8.11   The experimental setup.

horizontal direction. The aluminium container between the two magnet poles had one sliding iron wedge and one fixed guiding iron wedge at each side. The interface between the sliding wedge and the guiding wedge had a 12° angle to the vertical direction. As the sliding wedges were pushed down, the MR fluid was compressed in the field direction. The container had height 115 mm and a square horizontal cross section, 89 mm × 89 mm, providing a volume of 200 mL. We poured 120 mL MR fluid into the container. Before application of the magnetic field, we inserted an aluminium bar vertically into the container center. Then, immediately after a magnetic field was applied to solidify the MR fluid, we compressed the MR fluid by pushing the two sliding wedges down symmetrically. As the sliding wedges were down, the MR fluid's level rose and the gap between the two poles was reduced. This inserted aluminium bar simulated a work-piece to be machined.

In order to measure the compression pressure inside the MR fluid, we used four strain gauges (FLA-5-11, Tokyo Sokki Kenkyujo Co., Ltd) on the surfaces of the test bar to form a typical Wheatstone bridge circuit, which enabled us to find the equivalent {*in-situ*} normal stress P_e. To determine the yield stress, we attached a force transducer (Model 3185-500) and a strain gauge conditioner-amplifier (Model 3170, Daytronic Co.) to pull the test bar out. The MR fluid was extremely strong. We had to use a screw-driven linear slider to generate sufficient force to extract the test bar. The yield shear stress depends on the applied magnetic fields and the compression pressure. On many occasions the MR fluid was so strong that we could lift the 170 kg-magnet up with the test bar, but failed to pull the bar out from the MR fluid.

To decide the modulus, we also needed the test bar's vertical displacement under a force. The displacement was very tiny before the MR fluid had yielded. We attached a small mirror to the test bar. A tiny displacement led to a small rotation of the mirror. From the laser beam deflected by the mirror, we could determine the displacement with an accuracy of 1 μm. This displacement is so small that the tensile elongation of the test bar must be subtracted to obtain a correct shear strain.

The test bar's cross section is rectangle. We denote the side perpendicular to the field as w and the side parallel to the field as t. The depth of the bar submerged in the MR fluid is h. The bottom area is $A_b = wt$. The areas perpendicular to the field or parallel to the field submerged in the MR fluid are $A_\perp = 2wh$ and $A_\parallel = 2th$, respectively. The vertical force required to pull the test bar out is mainly determined by the force due to the MR fluid, F_{MR}. The other forces, such as the bar's weight reduced by the fluid's buoyancy, are negligibly small in comparison with F_{MR}. We have

$$F_{MR} = \tau_\perp A_\perp + \tau_\parallel A_\parallel \tag{8.34}$$

where $\tau_\perp$ and $\tau_\parallel$ are the yield stress on a plane perpendicular to the field direction or on a plane parallel to the field direction respectively. By varying the size of $A_\perp$ and $A_\parallel$, we determined $\tau_\perp$ and $\tau_\parallel$. In our experiment, the four aluminium bars used had $t = 1.27$ cm and $w = 2.54, 1.27, 0.635$ and 0.3175 cm. The leading term in F_{MR} is $\tau_\perp A_\perp$.

Since the test bar is non-magnetic, the field around the bar is not uniform. The field H_1 at the front center of the test bar was less than the field H_2 at the side parallel to the field. This was because some magnetic flux, instead of penetrating through the aluminium bar, bypassed it. As it was difficult to measure the field inside the solidified MR fluid and the tangential component of magnetic field was continuous at any interface that had no surface current, we measured H_1 and H_2 at the MR fluid surface. For simplification, we then take the average $H = (H_1 + H_2)/2$ as the mean value of H in our calculation.

Figure 8.12 shows F_{MR} *versus* the compression pressure for a test bar with $t = 1.27$ cm and $w = 2.54$ cm. It is clear that F_{MR} increases linearly with the compression pressure. Hence, the yield shear stress $\tau_y(H)$ increases with the

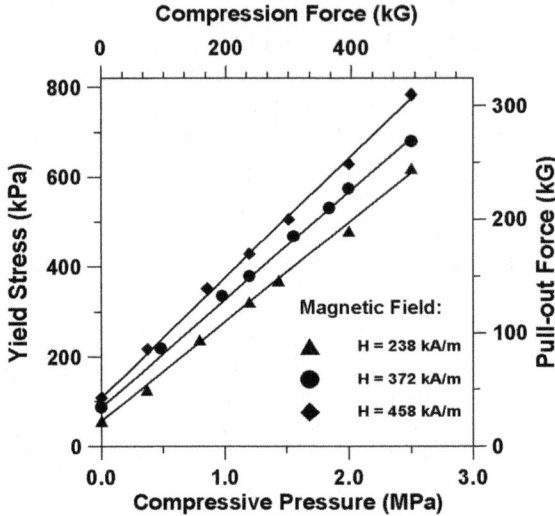

Figure 8.12   The yield shear stress versus the compression pressure at a different magnetic field. The yield shear stress of the MR fluid is linearly increases with the normal stress on it.

normal stress P_e. Through each measurement of F_{MR}, the normal stress P_e and the magnetic field remain fixed. An empirical expression is given by

$$\tau_y(H) = \tau_0 + k_H P_e \tag{8.35}$$

where τ_0 is the yield stress of MR fluid without the compression-assisted aggregation. The slope k_H increases with the field H, from 0.221 for $H = 238$ kA m^{-1}, 0.239 for $H = 372$ kA m^{-1}, to 0.267 for $H = 458$ kA m^{-1}. As shown in Figure 8.12, this linear relationship in eqn (8.34) holds very well. In our experiment, we have obtained a static yield shear stress exceeding 800 kPa.

Figure 8.13 shows the effect of magnetic field on the yield stress. The MR fluid without the compression-assisted aggregation has a static yield shear stress around 80 kPa at $H = 372$ kA m^{-1} and 120 kPa at $H = 514$ kA m^{-1}. The other two curves of compressed MR fluid were obtained as follows. We first applied a magnetic field of 372 kA m^{-1}, then compressed the MR fluid with a normal stress of 1.2 MPa or 2.0 MPa. Afterwards, we varied the coil current to change the magnetic field and measured the pullout force at various fields. During the experiment, we always gave at least 30 seconds for the MR fluid to relax after the compression or change of magnetic field. Figure 8.13 clearly indicates that the yield shear stress is greatly enhanced by the compression.

We note that reducing the magnetic field below 50 kA m^{-1} after the compression led to a sharp drop of the yield shear stress (Figure 8.13). This critical magnetic field can be understood as follows. When we compressed our MR fluid, the top surface was open. Therefore, the top surface of MR fluid has a pressure equal to the atmospheric pressure. The internal pressure inside the MR

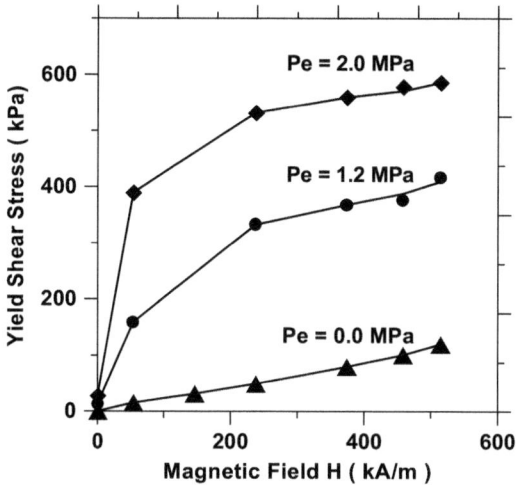

Figure 8.13 The yield shear stress versus the magnetic field with and without compression. All compression-assisted-aggregation processes were performed at a magnetic field of 372 kA/m first. The magnetic field was varied afterwards.

fluid is not uniform under the compression. This requires that the MR microstructure be strong enough to hold the internal pressure difference. When the magnetic field is below 50 kA m^{-1}, the MR microstructure cannot hold the internal pressure difference. During the experiment, we monitored the pressure at the test bar's middle point. The built-up pressure developed by the compression also reduced at the low field, following the sharp drop of the yield stress.

When the magnetic field was off, the MR fluid had a residual yield stress about 20 kPa and a residual field less than 0.5 kA m^{-1}. This hysteresis indicates that the magnetic particles formed a solid structure under the compression-assisted aggregation and the solid structure remained after the external field was removed. However, this hysteresis was so weak that a light stir returned the MR fluid back to its liquid state immediately.

Figure 8.14 shows the relationship between the shear stress and shear strain. The magnetic field was 372 kA m^{-1} for all cases. Without the compression-assisted aggregation, the MR fluid began to yield at a shear stress 40 kPa. The elastic modulus was about 10^7 Pa. After the yield point, the shear stress increases gradually until it reaches a maximum 80 kPa at a shear strain 0.35. With the compression-assisted aggregation, the MR fluid became much stronger and more rigid. The elastic limit, the modulus, and the yield shear stress were all increased dramatically. At $P_e = 2.0$ MPa, the shear modulus is as high as 5.0×10^8 Pa, 2% of aluminium's shear modulus. The overshoot of shear stress occurred at high compression pressure, indicating that the yielding process is sensitive to a structure change. A large shear strain breaks some microstructure of MR fluids and leads to a decrease of shear stress.

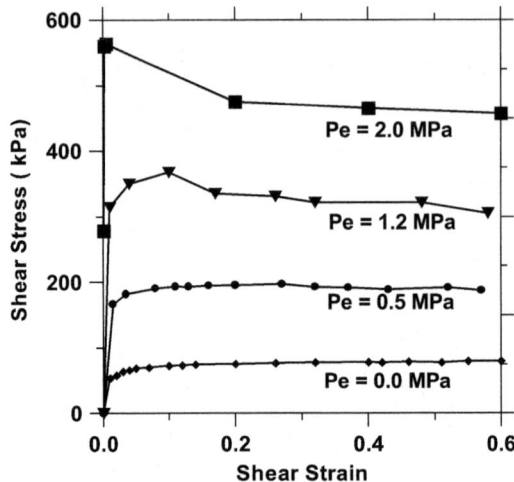

Figure 8.14 Shear stress versus shear strain curves with and without compression. The magnetic field for all curves is 372 kA/m.

It is also worthwhile to mention that the super-strong MR fluid was very stable. The shear modulus did not change in 24 hours after the compression-assisted aggregation, as long as the magnetic field was held. The MR fluid could be re-used many times.

8.6   The Microstructure of Super-Strong MR Fluids

To understand the physical mechanisms underlining this yield shear stress enhancement, we examined the microstructure of MR fluid before and after the compression-assisted aggregation. To do so, instead of silicon oil, we used polymer resins (Epoxy) to mix with the iron particles at 45% volume fraction. Then, we applied a magnetic field of 372 kA m^{-1} on the new irreversible MR fluid. The resin had a cure time of one hour. In one process, we did not compress the fluid and let the resin solidify. In another process, we compressed the MR fluid with a pressure of 1.2 MPa and let the resin solidify under pressure. Afterwards, we cut the cured solid pieces with a diamond saw and conducted SEM analysis. Figure 8.15a shows the microstructure of MR fluid without the compression. Figure 8.15b is the microstructure of MR fluid after the compression-assisted aggregation. It is clear that without the compression, the MR fluid microstructure was dominated by single chains. As our particles were not uniform, the chains were not perfect either, but all of them were not very thick, about 8 μm. As shown in Figure 8.15b, the MR fluid microstructure changed into thick columns after the compression. The average column thickness was over 60 μm, implying that one column had at least 120 chains in its cross section. We paid a special attention to the two ends of these thick columns. As shown in the SEM picture (Figure 8.16), the ends of these columns

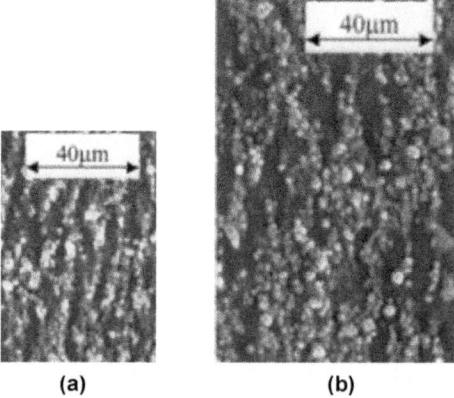

(a)   (b)

Figure 8.15   SEM images of iron-epoxy mixtures cured under magnetic field of 372 kA/m. The field direction is upwards in the diagrams. (a) Without the compression-assisted aggregation, the structure mainly consists of single chains. (b) The compression of 1.2 Mpa forces single chains into a thick column.

Figure 8.16   The column end of the super-strong MR fluid is much thicker than its middle.

were much thicker than their middle. Such structures were robust and strong because the original weak points in the structure were greatly reinforced.

Our compression-assisted-aggregation process can be illustrated in Figure 8.17. When a magnetic field is applied, magnetic particles quickly form chains. As we compress the MR fluid, chains get shorter and bent. When the chains are bent, the attraction between the chains gets stronger and pulls these chains quickly together.[11] Meanwhile, as many particles are pushed by the plates, the ends of the columns are much thicker than their middle. These are the desirable robust structures. The columns produced by our compression process are much thicker than the product of natural aggregation.

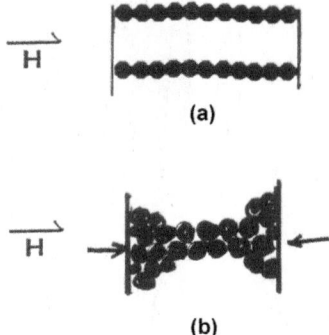

Figure 8.17 Formation of the robust MR microstructure during the compression-assisted aggregation. (a) Chains before the compression. (b) The compression forces the chains to aggregate into thick columns with robust ends.

The robust microstructure is the key to the super-strong MR fluids. In another experiment, we changed the process order: compress the MR fluid before application of magnetic field. Such a process does not produce any yield stress enhancement. The reason is easy to understand. Before application of magnetic field, the magnetic particles can move freely within the base liquid. Compression before the formation of solid structure does not create thick columns. Therefore, there is no change of yield stress.

In addition, we also note that there is a minimum strength of the magnetic chains required for the successful compression. For example, if we apply the compression process when the applied a magnetic field is below 50 kA m^{-1}, then the robust thick columns cannot be produced because the original chains are too weak to sustain the pressure. Under such a situation, the original chains break and the fluid returns to a liquid state.

8.7 Phenomenological Formula

We develop a phenomenological formula here. Let us assume that the MR fluid is uniform everywhere and its stress tensor is given by $\tau_{ij}(i,j=1,2,3)$ that is a combination of the mechanical tensor and Maxwell tensor.[18] The normal stress σ_n and shear stress τ_n on the plane with a normal direction $\vec{n}$ are given as follows.

$$\sigma_n = \sum_{ij} \tau_{ij} n_i n_j$$

$$\sigma_n^2 + \tau_n^2 = \sum_{ijl} \tau_{ij} \tau_{il} n_j n_l \qquad (8.36)$$

If τ_{ij} has three eigenvalues $\sigma_1 \geq \sigma_2 \geq \sigma_3$ and the three principal axes are still denoted by $\vec{n}_i(i=1,2,3)$, then eqn (8.36) can be written as

$$\sigma_n = \sigma_1 n_1^2 + \sigma_2 n_2^2 + \sigma_3 n_3^2$$

$$\sigma_n^2 + \tau_n^2 = \sigma_1^2 n_1^2 + \sigma_2^2 n_2^2 + \sigma_3^2 n_3^2 \qquad (8.37)$$

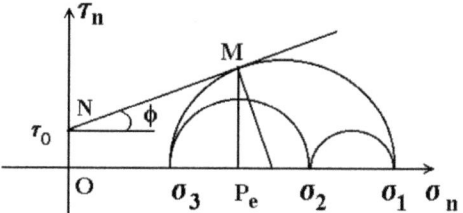

Figure 8.18   Mohr-circle diagram to find the limiting stress. ON represent τ_0 and NM is tangential to the biggest circle.

If we use the points $(\sigma_2 + \sigma_3)/2$, $(\sigma_1 + \sigma_3)/2$, $(\sigma_1 + \sigma_2)/2$ as the centers and $(\sigma_2 - \sigma_3)/2$, $(\sigma_1 - \sigma_3)/2$, $(\sigma_1 - \sigma_2)/2$ as the radii to draw Mohr circles (Figure 8.18), the area confined by the large and two small circles defines all possible values of σ_n and τ_n.[19] As mentioned before, τ_0 is the yield stress without compression, represented by the distance ON. The maximum allowed value of τ_n is represented by a point M, where the line NM is tangential to the biggest circle. Hence, the maximum shear stress τ_y is approximately expressed by

$$\tau_y = \tau_0 + \sigma_n \tan \varphi \tag{8.38}$$

This is the Mohr–Coulomb formula.[19] We note that this phenomenological formula is consistent with our empirical formula eqn (8.35). Also from Coulomb's argument, $\tan \phi$ is the 'internal friction' coefficient and ϕ is the angle of 'internal friction'. If we note that P_e in eqn (8.35) is just σ_n here, then, for example, $\tan \phi = K_H = 0.239$ or $\phi = 13.4°$ at $H = 372$ kA m^{-1}. However, as mentioned before, K_H increases with the magnetic field. Then the 'internal friction' increases with the magnetic field. Therefore, this 'internal friction' is, in fact, mainly the magnetic force.

8.8   Discussion

It is well known that in soil rheology, the soil's viscosity increases if the soil is experiencing a strong normal pressure. A similar phenomenological formula describes the soil rheology, $\tau_y = P_n \tan \phi$, where $\tan \phi$ is the friction coefficient and ϕ is about 30°–50°. In soil rheology, roughly, when P_n is greater than 0.5 MPa, the value of ϕ decreases approximately by 1 degree for an increment of 0.5 MPa. This is attributed to the crushed grains as the pressure increases. In our MR experiment, K_H increases with the field, while it does not appear to be reduced with the increase of P_e. This implies that the surface friction among the MR particles is small, and the field-induced interaction is the main contribution to K_H.

However, the key difference between the soil rheology and our MR experiment is as follows. In the soil rheology, there is a critical dilatancy occurring at volume fraction 50% for spherical particles. Above the critical volume fraction, the soil becomes highly dilatant and its effective viscosity soars and supports large compressive stresses under pressure, while below the critical volume fraction, the soil suspensions flow relatively easily and behave as nearly linear

Newtonian fluids.[19,20] In our MR experiment, we see the soaring yield stress well below 50%. For example, at 25% we already see the yield stress is soaring under the compression. Therefore, the key mechanism of our super-strong MR and ER fluids cannot be the dilatancy as in the soil rheology.

The test bar in our experiment is made of a non-magnetic material (aluminium alloy); the "wall effect" may underestimate the yield stress of MR fluid.[21] However, the average MR particle size 4.5 μm (Figure 8.10) is much smaller than the bar's surface roughness (~ 30 μm). There is no reason to believe that the MR particles could slip on the surface. To verify it, we conducted an experiment with several test bars with different rough surfaces. As expected, there was almost no change in the results of yield shear stress.

From the rising fluid level and the moving wedge position, we noted that our fluid had small compressibility. This was because there were some air bubbles inside the MR fluid. The maximum compressibility ratio of the fluid was 1.5% when the fluid was compressed at $H = 372$ kA m^{-1} and pressure 2.0 MPa. Afterwards, we used a vacuum pump to extract air bubbles from the MR fluids before the experiment. This made the experiment repeatable.

To conclude this chapter, we expect that super-strong MR fluids will have many industrial applications. For example the strength of our super-strong MR fluids exceeds the requirement for manufacturing flexible fixtures and automobile clutches. Since our MR experiment was originally designed for the flexible fixture, the compression speed was not an issue here. There is no difficulty in designing an agile compression-assisted-aggregation process. The technique described in this chapter should be general and applicable to all MR fluids. In fact, the carbonyl iron-oil MR fluid used in our experiment was an old one, discovered 60 years ago.[4] There are several advanced MR fluids on the market. This technique will produce much better results with these advanced MR fluids. We also expect that the current approach is applicable for ER fluids. If the structure-enhanced yield stress of ER fluids can also be 10 times higher than the yield stress without compression-assisted-aggregation, this method will enable ER fluids to have a yield stress strong enough for many industrial applications.

Acknowledgment

This research was supported in part by a grant from NSF DMR-0075780.

References

1. For example, see, *Electrorheological Fluids and Magneto-Rheological Suspensions*, ed. R. Tao, World Scientific Publishing, Singapore, 2011.
2. D. I. Hartsock, R. F. Novak and G. J. Chaundy, *J. of Rheology*, 1991, **35**, 1305.
3. J. D. Carlson and M. J. Chrzan, US Patent 5 277 282, 1994; J. D. Carlson, M. J. Chrzan and F. O. James, US Patent 5 284 330, 1994; O. Ashour,

C. A. Rogers and W. Kordonsky, *J. Intell. Mater. Syst. Struct.*, 1996, **7**, 123.

4. J. Rabinow, US Patent 2575360 (1951); J. Rabinow, *J. AIEE Trans.*, 1948, **67**, 1308.

5. W. M. Winslow, US Patent 2 417 850, 1947; W. M. Winslow, *J. Appl. Phys.*, 1949, **20**, 1137.

6. W. Kordonsky, Magnetorheological Fluids in high precision finishing, in *Electrorheological Fluids and Magneto-Rheological Suspensions*, ed. R. Tao, World Scientific Publishing, Singapore, 2011, pp. 30–39.

7. J. M. Ginder and L. C. Davis, *Appl. Phys. Lett.*, 1994, **65**, 3410.

8. R. Tao and J. M. Sun, *Phys. Rev. Lett.*, 1991, **67**, 398.

9. T. J. Chen, R. N. Zitter and R. Tao, *Phys. Rev. Lett.*, 1992, **68**, 2555.

10. L. Zhou, W. Wen and P. Sheng, *Phys. Rev. Lett.*, 1998, **81**, 1509.

11. R. Tao and Q. Jiang, *Phys. Rev. Lett.*, 1994, **73**, 205.

12. G. L. Gulley and R. Tao, *Phys. Rev. E*, 1993, **48**, 2744.

13. T. C. Halsey and J. E. Martin, *Sci. Am.*, 1993, **10**, 58.

14. R. Tao, J. Zhang, Y. Shiroyanagi, X. Tang and X. Zhang, *Int. J. Modern Phys. B*, 2001, **15**, 918.

15. L. Fu, P. B. Macedo and L. Resca, *Phys Rev. B*, 1992, **47**, 13818.

16. X. Tang, Y. Chen and H. Conrad, *J. Intell. Mater. Syst. Struct.*, 1996, **7**, 517.

17. X. Tang, X. Zhang and R. Tao, *J. Appl. Phys.*, 2000, **87**, 2634.

18. R. E. Rosensweig, *J. Rheol.*, 1995, **39**, 179.

19. For example, see S. S. Vyalov, *Rheological Fundamentals of Soil Mechanics*, Elsevier, New York, 1986, pp. 100–107 and 421–440; J. J. Tuma and M. Abdel-Hady, *Engineering Soil Mechanics*, Prentice-Hall, Inc., New Jersey, 1973, pp. 207–209.

20. C. A. Shook and M. C. Roco, *Slurry Flow*, Butterworth-Heinemann, 1991.

21. E. Lemaire and G. Bossis, *J. Phys. D*, 1991, **24**, 1473; T. Miyamoto and M. Ota, *Appl. Phys. Lett.*, 1994, **64**, 1165.

CHAPTER 9

Magnetorheological Fluids Flowing Through Porous Media: Analysis, Experimental Evaluation, and Applications

NORMAN M. WERELEY,* WEI HU AND
RYAN ROBINSON

Department of Aerospace Engineering, University of Maryland, College
Park, MD, 20742, USA
*Email: wereley@umd.edu

9.1 Introduction

Magnetorheological (MR) fluids are a soft ferromagnetic powder (typically, microscale carbonyl iron) in a silicone oil. Each particle has a natural magnetic dipole. When a magnetic field is applied across the volume of fluid, the magnetic dipoles orient themselves with respect to the applied magnetic field and form chains. It is these chains that induce the yield stress: a local shear stress must be applied that is greater than the yield stress before the chains break and flow is induced. The yield stress can be as high as 100 kPa for commercial MR fluids,[1] and the substantial field-induced yield stresses exhibited by MR fluids make possible many energy absorbing applications, such as rotary brakes or dampers.[2–3] In conventional MR devices, flow channels are straight and electromagnetic coils are configured so that flow streamlines are normal to magnetic field.[2–5] This requires that the magnetic coil be positioned in a bobbin that

RSC Smart Materials No. 6
Magnetorheology: Advances and Applications
Edited by Norman Wereley
© The Royal Society of Chemistry 2014
Published by the Royal Society of Chemistry, www.rsc.org

is concentric to a tubular flux return. The magnetic coil is usually immersed in the MR fluid, which can reduce the life of the coil and lead to difficulties in feeding electrical current to the coil and extracting heat generated by the coil and damping action. This magnetic coil configuration limits the active length of the flow channel to be far less than the valve length. Configuration of the flow channel and flux return is also constrained by magnetic field saturation and precise MR valve flow gaps. Consequently, it is difficult to develop a compact MR damper using the above configurations.

The modern trend of the miniaturization of MR valves requires a more compact MR valve design. One simple flow valve configuration could be obtained with the tortuous channels that naturally exist in porous media as shown in ref. 6. Advantages of this configuration are that, first, the active length of the flow channel can be increased, second, the porous valve can be activated by a magnetic coil or solenoid that is external to the device which prevents all the thermal energy generated through resistance of the solenoid from heating to the damper's hydraulic system. Kuzhir *et al.*[7] developed a hydraulic ram for the investigation of MR fluid flow through porous media in magnetic field parallel to the flow. Their measurements demonstrated that a packed bed of magnetic grains gave a much higher controllable damping range than spiral channels. However, few comprehensive studies have been conducted to evaluate the behavior of the MR fluid flowing through various porous medias.

Three major parameters related to the porous media are used for fluid behavior characterization, that is, porosity, tortuosity and Reynolds number. The porosity, ε, is defined as the ratio of volume of the void space to the volume of the whole space in the valve. The tortuosity, ξ, is defined as the ratio of the length of an imaginary flow path in the porous media to the whole length of the valve. The Reynolds number, Re, is introduced to evaluate the onset condition of a nonlinear viscous force. The interaction between the MR fluid flowing through a cylindrical channel and the applied magnetic field is well described by a chain model,[6,8] and the model can be applied to a porous medium under the assumption that porous channels can be modeled as a network of straight, cylindrical channels.[9–11] It has been shown that the magnetic field applied to the MR fluid is affected by the permeability of the filler materials and the porosity and tortuosity of the porous media[12–13] and the Reynolds number plays an important role in the onset of the nonlinear viscous force.[14] In this study, the effects of utilizing porous media of varying shape and size are examined. The change in maximum force and damping coefficient are analyzed and compared among different porous media (packed beds of cylindrical rods and spheres).

In order to characterize and compare the behavior of the MR fluid in different porous media, a hydraulic testing device is constructed, which consists of a hydraulic cylinder connected to a porous bypass valve. The bypass valve is fabricated around a stainless steel tube and the tube can be packed with filler materials, *i.e.* porous media. A magnetic coil is wrapped around the outside of the bypass valve to induce a magnetic field inside the valve. A piston with a double-rod configuration is used such that the internal volume of the damper remains constant, eliminating the need for a gas accumulator. The hydraulic device is tested in a load frame, and the piston forces MR fluid out of one end of the cylinder and into the bypass valve. Inside the bypass valve, fluid passes through the porous medium. The fluid then exits the bypass valve and reaches the opposite end of the cylinder. Response of the device is measured for steady-state sinusoidal displacements, and equivalent damping and damping coefficient are used to characterize the performance of the porous valve across a range of amplitude and frequency.

Using the developed MR device, the effects of utilizing different porous media and cylinder configurations are observed. First, the effects of varying media porosity and tortuosity are discussed and analyzed. Various types of porous media (packed beds of cylinders and spheres) are tested in a porous-valve-based MR device. The change in maximum force and damping coefficient are analyzed and compared. Secondly, the effects of altering the cross-sectional area of the moving piston inside the hydraulic cylinder are investigated. Damper performance is compared between two differently-sized hydraulic cylinders with the same porous valve configuration. Thirdly, the yield stress produced by each of the porous channels is approximated to determine the magnetic efficiency of each porous medium. A mathematical model based on a multi-channel flow analysis of MR fluid passing through porous channels is employed in order to capture the hysteresis behavior of the MR device and to determine the yield stress produced in the valve. Yield stress is compared between each porous valve configuration. Finally, the effect of the nonlinear viscous force on the performance of the adaptive damper is evaluated. Using the experimental data, the analytical damper behavior with different porous media structures is validated. The effect of the porosity and permeability of the filler material on the field distributions is analytically evaluated. These analytical tools can be explored for performance evaluation and design of MR devices with porous valves.

9.2   Analysis of MR Flow in Porous Media

A porous medium is a material containing voids that can be filled with liquid or gas. With adequate empty space, channels will reach from one end of the medium to the other. In the present study, the porous media are arrays of solid particles constrained in an enclosed valve space as shown in Figure 9.1. An important feature of the porous media is the tortuous fluid channels through

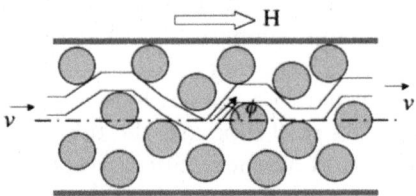

Figure 9.1   Schematic of porous media in a flow channel filled with MR fluids and magnetic field *H* applied as shown.

the porous media. A schematic tortuous channel is shown in Figure 9.1, and at each point of the channel, the inclined angle between the channel axis and magnetic field direction is varied. The characteristics of the fluid flow in the porous media are dependent on the size and shape of the individual particles, the shape of the channels created, and the volume fraction of particles in the medium. If the porous media is used in an MR porous valve, the magnetic properties of the media and the magnetic interactions between particles and between particle and fluid are also important.

Porosity and tortuosity are major factors affecting flow characteristics in the porous media. Porosity, ε, is defined as the ratio of fluid path volume to the total volume of the porous valve, such that:

$$\varepsilon = \frac{V_f}{V} = 1 - \frac{V_p}{V} \tag{9.1}$$

where, V_f and V_p are the volume of liquid and solid phase of the porous media, respectively, and V is the total volume of the porous media. Tortuosity, ξ, refers to the ratio of the true length of the flow channel in the porous media to the length of the porous valve. A straight channel has a tortuosity of one, and a porous valve has tortuosity greater than one. The inherent randomness of packing rods or spheres complicates the calculation of tortuosity, thus empirical measurements are usually used to obtain the tortuosity.

To simplify the analysis of the flow through a porous valve, the flow path in the porous media is considered as a multiple-pipe system, in which a set of identical cylindrical tortuous channels is grouped in parallel as shown in Figure 9.2. The number of the equivalent flow path is assumed as n. The diameter and length of the original porous valve are given as D and L. The cross sectional area and equivalent tortuous length of one single channel are denoted as A_i and L_i, respectively. Notably, the internal surface area of one equivalent channel is a multiplication of the perimeter, C_i, and length, L_i, of the channel. The ratio of the equivalent channel length, L_i, to the porous valve length, L, is the tortuosity of the porous valve. Because the actual flow path in

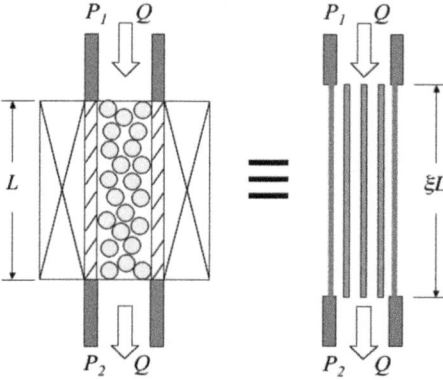

Figure 9.2 Equivalent porous valve flow path.

the porous media is tortuous, $\xi > 1$. Moreover, a hydraulic radius of the equivalent channel, R_h, is introduced as:

$$R_h = \frac{\sum_{i=1}^{n} A_i}{\sum_{i=1}^{n} C_i} = \frac{A_i}{C_i} = \frac{\pi R^2}{2\pi R} = \frac{R}{2} \tag{9.2}$$

where R is defined as the radius of an equivalent cylindrical channel. Clearly, the hydraulic radius is half of the radius of the equivalent circular channel. To relate the hydraulic radius to the porosity, eqn (9.2) can be written as:

$$R_h = \frac{\sum_{i=1}^{n} A_i}{\sum_{i=1}^{n} C_i} = \frac{A}{C} = \frac{A\xi L}{C\xi L} = \frac{V_f}{S_f} = \frac{V_p}{S_p} \frac{\varepsilon}{1-\varepsilon} \tag{9.3}$$

where S_f is surface of the fluid path or the wetted area of the porous medias and S_p is the surface of the porous medias. Notably, due to mutual overlapping of the porous medias, S_f is usually smaller than S_p, except for spherical beads.

The behavior of the MR fluid flow through a cylindrical channel is well described by the chain model.[8] A simulated chain behavior of the MR fluid in a cylindrical channel is shown in Figure 9.3. The uniform magnetic field (of intensity H) is situated in the x-y plane and perpendicular to x-z plane and to the flow direction. While the chain aggregates are much smaller than the channel radius, the analytical yield stress of the MR fluid under the magnetic field is obtained as:

$$\tau_{yz} = \tau_y + \eta \frac{\partial v}{\partial y}$$

$$\tau_{xz} = \eta \frac{\partial v}{\partial x} \tag{9.4}$$

where τ_y is the yield stress when the magnetic field is transverse to the flow, and η is the plastic viscosity of the MR fluid. The momentum equation for the unidirectional flow in a straight channel can be written as:

$$\frac{\partial}{\partial y}\left(\tau_y + \eta \frac{\partial v}{\partial y}\right) + \eta \frac{\partial^2 v}{\partial^2 x} = -\frac{\Delta P}{\xi L} \tag{9.5}$$

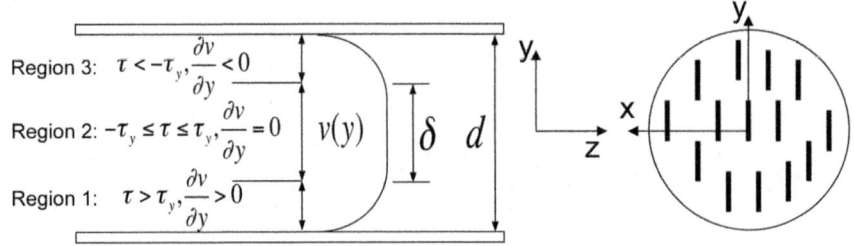

Region 3: $\tau < -\tau_y, \dfrac{\partial v}{\partial y} < 0$

Region 2: $-\tau_y \le \tau \le \tau_y, \dfrac{\partial v}{\partial y} = 0$

Region 1: $\tau > \tau_y, \dfrac{\partial v}{\partial y} > 0$

Figure 9.3   MR fluid flow in straight cylindrical channel.

The solution to eqn (9.5) is obtained using continuity of the stress and ideal flow boundary conditions (*i.e.* no slip at the channel wall), so that the approximate pressure drop of a straight channel, as a function of the yield stress and the flow rate is given by:

$$\frac{\Delta P}{\xi L} = \frac{16}{3\pi} \frac{\tau_y}{R} \mathrm{sgn}(Q) + \frac{8\eta}{\pi R^4} Q \tag{9.6}$$

where Q is the flow rate through the channel and ξL is the length of the channel.

For an equivalent channel in the porous valve, the magnetic field is not perpendicular to the flow direction of the MR fluid. Actually, in a tortuous channel, the inclined angle between the magnetic field and the MR fluid flow path varies as the orientation of the channel relative to valve axis varies. As shown in Figure 9.1, a magnetic field is parallel to the porous valve axis, and at each point of the channel, the inclined angle between the channel axis and magnetic field direction is:

$$\theta = \frac{\pi}{2} - \phi \tag{9.7}$$

The tortuosity of the channel, ξ, can be related to the inclined angel, θ, by:

$$\int_0^L \frac{\mathrm{d}L}{\cos(\phi)} = \int_0^L \frac{\mathrm{d}L}{\sin(\theta)} = \xi L, \quad 0 \leq \theta \leq \frac{\pi}{2} \tag{9.8}$$

For simplicity, the tortuous channel can be approximated as a straight channel with an average inclined angle, $\langle \theta \rangle$, such that

$$\sin(\langle \theta \rangle) = \frac{1}{\xi} \tag{9.9}$$

As the inclined angle is known, the yield stress of the MR fluid in the tortuous channel, denoted as $\langle \tau_y \rangle$, can be determined using an empirical relation with an applied magnetic field as:

$$\langle \tau_y \rangle = \kappa(\theta)\tau_y \tag{9.10}$$

where κ is a correction factor for the yield stress due to the inclination angle of the magnetic field and is a function of θ.

Since the porous valve contains a bed of porous media through which MR fluid flows, the magnetic field applied to the MR fluid is also determined using an averaging process. The magnetic field averaged over the whole volume of the valve is denoted as H, the field averaged over the MR fluids is H_f, and the field averaged over the solid filler medias is H_s. Correspondingly, the effective magnetic permeability for overall valve volume, liquid and solid phase is μ, μ_f and μ_s, respectively. These parameters are related to each other through the following relationship:

$$H = \varepsilon H_f + (1 - \varepsilon)H_s$$

$$\mu H = \mu_f \varepsilon H_f + \mu_s(1 - \varepsilon)H_s \tag{9.11}$$

Thus, the magnetic field applied to the MR fluid is obtained by:

$$H_f = \frac{\mu_s - \mu}{\varepsilon(\mu_s - \mu_f)} H \tag{9.12}$$

In eqn (9.12), the overall permeability is approximated by:

$$\mu = \left[\varepsilon\mu_f^{1/3} + (1 - \varepsilon)\mu_s^{1/3}\right]^3 \tag{9.13}$$

where the relative permeability of the porous media is dependent on the applied field as:

$$\mu_s = 1 + \frac{(\mu_{s0} - 1)M_s}{M_s + (\mu_{s0} - 1)H_s} \tag{9.14}$$

where μ_{s0} and M_s are the material constants of the porous media. From eqn (9.13) to eqn (9.14), the magnetic field applied to the MR fluid can be calculated, and then the yield stress of the MR fluid in the porous valve is obtained using a combination of an empirical equation[15] and eqn (9.10) as:

$$\langle \tau_y \rangle = \kappa(\theta)\left[271700\psi^{1.5239} \tanh(6.33e - 6H_f)\right] \tag{9.15}$$

where ψ is the volume faction of the iron particles in the MR fluid.

Using eqn (9.6) and (9.15), the pressure drop for each equivalent tortuous channel is written as:

$$\frac{\Delta P}{\xi L} = \frac{8}{3\pi} \frac{\langle \tau_y \rangle}{R_h} \text{sgn}(Q_i) + \frac{\eta}{2\pi R_h^4} Q_i \tag{9.16}$$

where Q_i is the flow rate in each channel and can be determined using the total flow rate Q as:

$$Q_i = v_i 4\pi R_h^2 = \frac{Q\xi L}{V_f} 4\pi R_h^2 = \frac{\xi Q}{\varepsilon A_v} 4\pi R_h^2 \tag{9.17}$$

where A_v is the cross-sectional area of the porous valve. Substituting eqn (9.17) into eqn (9.16), the pressure drop across the porous valve can be described as:

$$\Delta P = \frac{8\xi L}{3\pi R_h} \langle \tau_y \rangle \text{sgn}(Q) + \frac{2\eta\xi^2 L}{R_h^2 \varepsilon} \frac{Q}{A_v} \tag{9.18}$$

In eqn (9.18), the first term is a controllable resistance determined by the yield stress of the MR fluid, and the second term includes a linear viscous resistance.

Based on Darcy's law of the flow in a porous media,[14] a nonlinear viscous force exists as the flow rate becomes high. To describe the nonlinear behavior, the Reynolds number, Re, is firstly introduced and defined as:

$$Re = \frac{\rho 4R_h Q}{\varepsilon A_v \eta} \tag{9.19}$$

in which, ρ is the MR fluid density. A nonlinear term as a function of the Reynolds number is introduced into eqn (9.18) and leads to:

$$\Delta P = \frac{8\xi L}{3\pi R_h} \langle \tau_y \rangle \mathrm{sgn}(Q) + \frac{2\eta \xi^2 L}{R_h^2 \varepsilon} \frac{Q}{A_v} \left[1 + \left(\frac{1-\varepsilon}{\varepsilon} + \frac{\varepsilon - \varepsilon_0}{\beta}\right) \frac{Re}{\xi Re_{cr}}\right] \qquad (9.20)$$

where, a constant critical Reynolds number, Re_{cr}, represents the onset of the nonlinear viscous force and nonlinear shape constants ε_0 and β are related to the porous media, and all parameters can be empirically determined. The nonlinear viscous term in eqn (9.20) is equivalent to a nonlinear force derived from inertial terms in the Navier–Stokes equation.[11,14] It is apparent that the flow behavior of the MR fluid in the porous valve is related to the properties of the porous media, such as porosity, tortuosity, shape, and material composition.

9.3 Experimental Characterization and Evaluation

9.3.1 Testing Setup

In order to evaluate the behavior of the MR fluid flowing through the porous media, a hydraulic MR device with a by-pass valve was constructed. The schematic structure and assembly of the device is shown in Figure 9.4. The MR device includes a cylinder that defines a chamber for containing an MR working fluid and a by-pass flow channel that defines a valve for containing porous media. The by-pass valve can be easily detached from the cylinder body, packed with a porous media sample, and reattached. A magnetic coil is wrapped around the outside of the bypass valve to induce a magnetic field inside the valve. A piston with a double-rod configuration was used so that the internal volume of the damper remained constant, eliminating the need for a gas accumulator. In operation, the piston forces MR fluid out of one end of the cylinder and into the bypass valve. Inside the bypass valve, fluid passes through the porous medium. The fluid then exits the bypass valve and reaches the opposite end of the cylinder.

The major components of the MR device are a hydraulic cylinder with a 38 mm (1.5 inch) or 50.8 mm (2 inch) bore and a 152.4 mm (6 inch) stroke and a by-pass valve fabricated using a stainless steel tube with an inner diameter of 9.4 mm and a length of 101.6 mm. The tube fittings are used to mount the

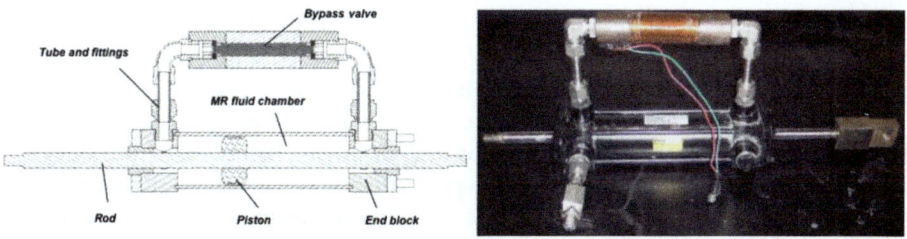

Figure 9.4 MR Damper with a bypass valve filled with porous media.

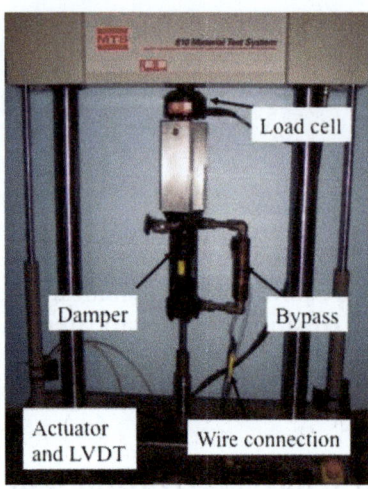

Figure 9.5   Testing setup.

bypass valve to the cylinder as well as the valve used for filling MR fluid. A solenoid with 50 Ohms of resistance is wrapped around the outside of the by-pass valve tube, providing field parallel to valve axis. The wall of the tube is non-magnetic stainless steel, in order to encourage the field to be stronger in the fluid rather than the wall. The damper was filled with Lord Corporation's MR132-DG fluid and pressurized to remove any remaining air.

The testing setup is shown in Figure 9.5. The MR device was subjected to steady-state sinusoidal displacements with amplitude of 12.7 mm (1 in peak to peak) and frequency up to 7 Hz using a 25 kN load frame (MTS 810) with servo-hydraulic actuators. Force and displacement data were recorded using a load cell and displacement sensor (LVDT), respectively. To reduce the noise of the sinusoidal displacement signal, a Fourier series was used to reconstruct the displacement data. The reconstructed displacement signal was then differentiated in the frequency domain to obtain the velocity. A DC power supply provided current control during testing, and the amount of current supplied to the solenoid surrounding the bypass valve was varied between 0 and 1 A.

9.3.2   Characterization of MR Flow in Porous Valve

The representative hysteresis behavior demonstrated by the MR device with a 38 mm cylinder is shown in Figure 9.6. The porous media in the by-pass valve was 3 mm spherical beads, and the damper was applied by a displacement loading at 0.2 Hz and the applied current was varied from 0 to 1.5 A. Figure 9.6(a) shows the force *vs.* piston displacement behavior of the device, and Figure 9.6(b) shows the force *vs.* piston velocity behavior of the device. Apparently, the hysteresis shape of the MR device is similar to that of a conventional MR damper. The total energy dissipated by the MR device is represented by the area within the hysteresis cycles on the force *vs.* displacement

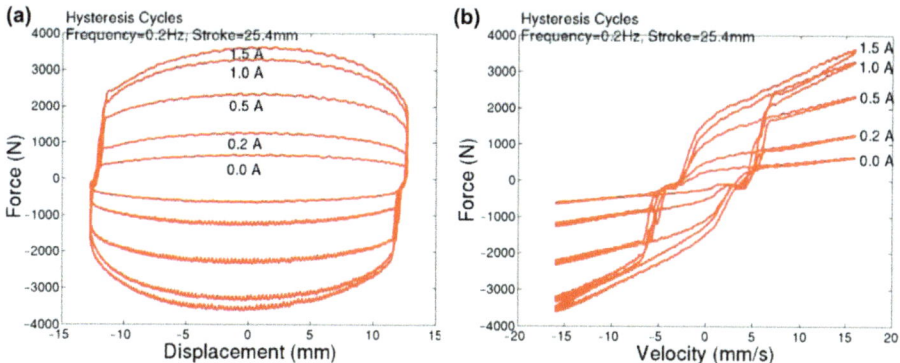

Figure 9.6   Typical measured hysteresis behavior of the MR device with porous valve. (a) Force *vs.* displacement. (b) Force *vs.* velocity

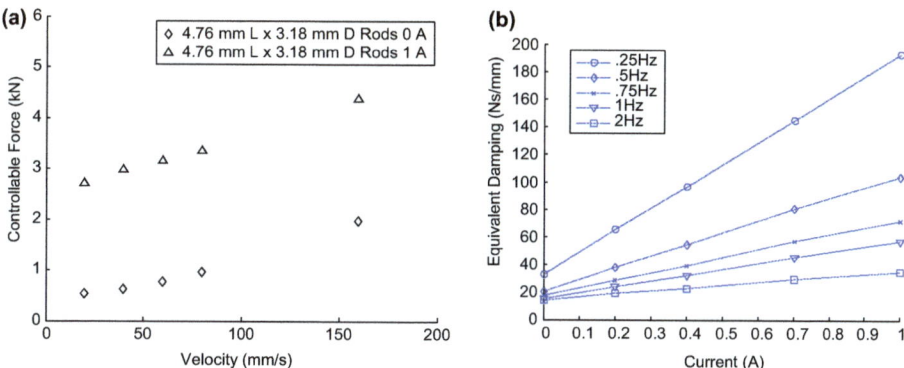

Figure 9.7   Damper characterization: maximum force and equivalent damping. (a) Maximum force as a function of velocity. (b) Equivalent damping with respect to frequency.

diagram in Figure 9.6(a). The typical effect of the applied current shown in the force *vs.* displacement curves is the large change in force near zero velocity (shown at the maximum displacement) as a function of applied current, which in turn increases the enclosed area. As a higher current or magnetic field is applied, the MR device dissipates more energy. However, as the applied current is increased over 1.5 A, the MR device approaches saturation such that no further significant damping force increases.

Maximum force as a function of velocity is used to evaluate the controllability of the MR device with porous valve across various velocity and current ranges. An example of the measured maximum force at 0 A and 1 A was obtained from a porous valve filled with 4.76 mm × 3.18 mm rods and is shown in Figure 9.7a. Apparently, the maximum force of the MR device increases fairly linearly with maximum velocity in the field-off and field-on (1 A) states. This occurs because a viscous fluid such as MR fluid exhibits higher internal resistance as velocity increases.

Equivalent viscous damping, C_{eq}, is used to compare the energy dissipation of different MR device configurations. Equivalent damping is obtained by equating the dissipated energy over a cycle to the energy dissipated by an equivalent viscous damper: equivalent viscous damping is used to quantify the damping augmentation effect demonstrated by the MR device. Equating the energy dissipated by the device to that of a viscous damper, results in an equivalent viscous damping:

$$C_{eq} = \frac{E}{\pi \Omega X_0^2} \tag{9.21}$$

where X_0 is the amplitude of the displacement input. The energy dissipated over one cycle, E, is computed using the trapezoidal rule and given by:

$$E = \oint F(x)\mathrm{d}x \tag{9.22}$$

where, F is the damper force over one cycle.

Damping coefficient is the ratio of between field-on and field-off values of the equivalent viscous damping. In this analysis, the field-on values correspond to experiments at an applied current of 1 A. In order to achieve a high damping coefficient, equivalent damping should be minimized in the field-off state, and maximized in the field-on state.

It is important to note the effects of fluid velocity. Testing was accomplished using sinusoidal excitations, and average piston velocity increases with frequency. As piston velocity increases, equivalent damping in the field-on state decreases considerably. This phenomenon is apparent in Figure 9.7b, in which a large difference between equivalent damping in the field-on and field-off conditions can be seen at 0.25 Hz but a very slight difference in the field-on and field-off conditions at 2 Hz.

9.3.3 Effects of Varying Porosity and Tortuosity

In order to evaluate the effect of various types of porous media on the performance of the MR device, five distinct materials, listed in Table 1, were used to fill the porous valve, *i.e.* 4.76 mm length and 3.18 mm diameter steel rods (1r), 9.52 mm length and 4.76 mm diameter steel rods (2r), 6.35 mm length and 3.97 mm diameter steel rods (3r), 3.5 mm diameter steel spheres (1s), and 5.5 mm diameter steel spheres (2s). The media were packed by dropping the

Table 9.1 Dimensions of packing materials and porous bed parameters.

Bed No.	D/mm	L/mm	No of Particles	ε	ξ_0	ξ	$\pm \sigma$
1r	3.18	4.76	100	0.46	1.48	1.54	0.29
1s	3.50	—	166	0.47	1.26	1.50	0.16
2r	4.76	9.52	16	0.62	1.95	2.40	0.08
2s	5.50	—	30	0.63	2.26	2.66	0.31
3r	3.97	6.35	52	0.42	1.74	1.89	0.15

individual rods or spheres into the valve one at a time until the valve was filled. The media was prevented from escaping the valve by a steel wire mesh placed on each end of the valve. In this study, the diameter of the hydraulic cylinder was 38 mm.

The porosity, ε, was calculated for these filled materials. For a bed of rods, this is given by:

$$\varepsilon = 1 - \frac{Nd^2l}{D^2L} \quad (9.23)$$

where d is the diameter and l is the length of the rod, D is the diameter and L is the length of the valve, and N is the number of rods. For a bed of spheres, porosity is given by

$$\varepsilon = 1 - \frac{2Nd^3}{3D^2L} \quad (9.24)$$

The tortuosity, ξ, is dependent of the applied current. Zero-field tortuosity, ξ_0, was determined empirically, based on the maximum piston force in the zero-field condition. Tortuosity was also determined by estimating post-yield damping at various levels of applied current, and the average tortuosity, ξ, and its uncertainty, σ, can be estimated. In addition, the hydraulic radii, R_h, were calculated using the obtained porosity and known geometry parameters of the porous media . For a bed of cylinders, the hydraulic radius is calculated as:

$$R_h = \frac{d\varepsilon}{2\left(2 + \frac{d}{l}\right)(1 - \varepsilon)} \quad (9.25)$$

For a bed of spheres, it is calculated as:

$$R_h = \frac{d\varepsilon}{6(1 - \varepsilon)} \quad (9.26)$$

The porosity, tortuosity, and hydraulic radius are influenced by the amount of individual cylinders or spheres that were packed inside the bypass valve.

Porosity dictates the ratio between the effective piston area and the effective duct area inside the valve. For a given piston velocity, the volume of fluid that moves through the valve is the same as the volume of the fluid displaced by the piston. If the effective duct area, or the average cross-sectional area of the pores in the valve, decreases, then the equivalent flow velocity in the valve increases for the same piston velocity, such that the force applied by the piston will increase. Hence, decreasing porosity increases damper force. Moreover, beds of rods or spheres with equal porosity tend to exhibit similar hysteresis cycles because the hydraulic amplification ratio between the piston and the duct is the same.

A significant difference, however, can be the manner in which fluid passes through the valve, which is related to tortuosity. Particle size and shape (morphology) influences tortuosity because it dictates how rods and spheres can be packed. The randomness inherent in packing many of these small particles causes tortuosity to vary each time the valve is packed, even if porosity is constant. By forcing localized parts of the flow to move faster, increasing tortuosity increases pressure in the valve, which increases the force necessary to displace the piston. Increasing tortuosity has also been shown to increase the average angle between the flow and the magnetic field inside the solenoid. This leads to greater damper force in the field-on condition. Overall, tortuosity increases both field-on and field-off damper forces.

Between the two sizes of spherical media, different force–displacement cycles are produced, shown in Figure 9.8a. Higher forces are produced with the lower-porosity media, bed 1s. The field-off equivalent damping of bed 1s is greater than that of bed 2s by less than a factor of 2. However, the field-on damping for bed 1s is 3.55 times greater than the field-on equivalent damping for bed 2s. The lower-porosity bed 1s was found to have higher damping coefficients up to 5 Hz. Similarly, the damping coefficients of the three cylindrical media differed greatly because of differences in porosity. While bed 1r produced high damping coefficients, bed 2r produced lower damping coefficients. The force–displacement cycles, shown in Figure 9.8b, indicate a field-on to field-off ratio of maximum force that is higher for the low porosity configuration than the high porosity configuration.

In addition, the porosity of beds 1s and 1r is almost identical. This permits a close analysis on the effects of tortuosity since the tortuosity of the rod-based medium is greater than that of the sphere-based medium. As shown in Figure 9.8, both field-on and field-off damper force of bed 1r is higher than the one of bed 1s. Since the field-off damping increases more than the field-on damping, higher damping coefficients are obtained for the sphere-based medium with a lower tortuosity.

For two sets of media with equal porosity, the higher-tortuosity media will yield higher maximum controllable force than the lower-tortuosity media as shown in Figure 9.8. However, changes in porosity affect maximum controllable forces to a greater extent than changes in tortuosity. In Figure 9.9, it can

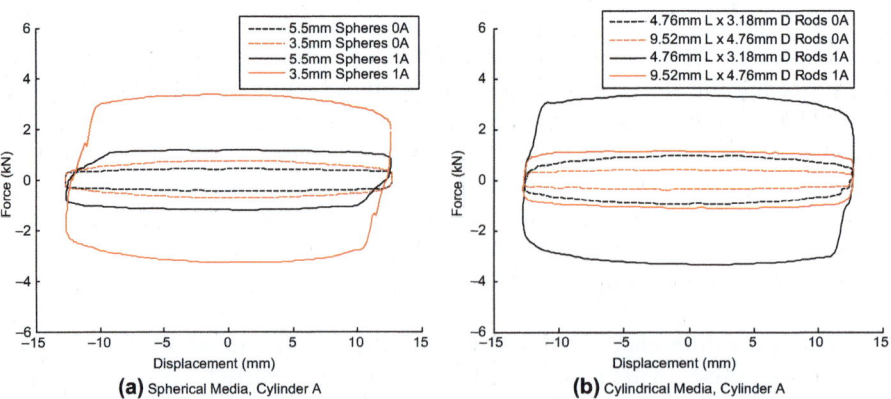

Figure 9.8   Force-displacement hysteresis behavior for different porous media.

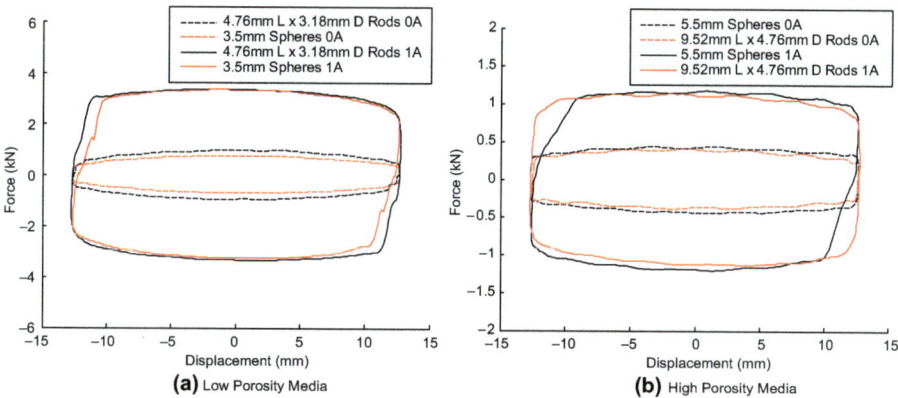

Figure 9.9 Hysteresis behavior of MR device with similar porosity.

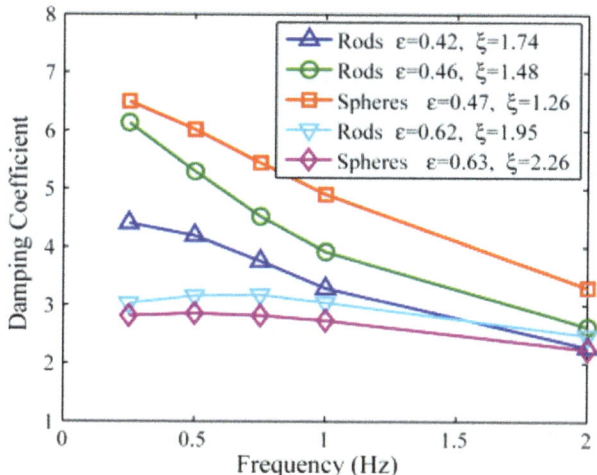

Figure 9.10 Damping coefficient for five different porous media.

be seen that both field-off and field-on hysteresis cycles between similar porosity media are almost identical. On the other hand, altering the porosity of the bed will change the maximum controllable force and damping ratio significantly. In Figure 9.8, there were drastic (greater than 2 kN) differences in force-displacement cycles between media with the same shape and similar tortuosity, but different porosity.

The results also show how changes in porosity and tortuosity of a media-filled valve affect damping coefficients of the system. Figure 9.10 shows the damping coefficient for each porous medium as a function of excitation frequency. The greatest damping coefficients arise for rods and spheres of mid-level porosity, again suggesting an optimal porosity exists within the tested range ($\varepsilon = 0.42$–0.63). This was also observed for packed beds of spheres, and packed beds of rods follow the same trend. Additionally, damping coefficient decreases as tortuosity increases over the tested range of 1.26–2.26. The field-on damping for a valve with a tortuosity of one is negligible, and thus shows a low damping coefficient. Therefore, there exists an optimal tortuosity, which provides a balance between ease of flow in the field-off condition and high yield force in the field-on condition.

For any recorded frequency greater than or equal to 4 Hz, damping coefficients are below 2.0 and there is little difference between the damping coefficients of any of the porous media. This frequency corresponds to a maximum piston velocity of 0.16 m s^{-1}. As viscous effects increase at high frequency, the yield force contribution to the total energy absorption decreases, and field-on damping for all configurations approaches field-off damping.

9.3.4 Effects of Varying Piston Area

To investigate the flow behavior for a higher flow velocity in the porous valve, the cross-sectional area of the moving piston inside the hydraulic cylinder was altered. MR device performance is compared between two differently-sized hydraulic cylinders with the same porous valve configuration. The original hydraulic cylinder, Cylinder A, was then replaced with a second hydraulic cylinder with a larger effective piston area, Cylinder B. A comparison of the properties of each cylinder is included in Table 9.2. One porous bed configuration, bed 1r, was subjected to sinusoidal excitations using the new cylinder configuration.

The piston cross-sectional area for Cylinder B (1829 mm^2) is nearly three times greater than that of Cylinder A (633.4 mm^2). Therefore, Cylinder B displaces a higher volume of MR fluid, producing a higher fluid velocity inside the porous channel. As shown in Figure 9.11, the Cylinder B configuration consistently exhibits greater controllable force than corresponding field-on and field off measurements for Cylinder A. At a velocity of 80 mm s^{-1}, corresponding to an excitation frequency of 1 Hz, the Cylinder B configuration has a maximum field-off force of 6.0 kN and a maximum field-on force of 12.3 kN. In contrast, the Cylinder A configuration has a maximum field-off force of 0.97 kN and a maximum field-on force of 3.4 kN.

Experimental results indicate that increasing piston area causes equivalent damping to increase for each of the assessed values of current and excitation frequency. Equivalent damping values for the two hydraulic cylinder

Table 9.2 Hydraulic cylinder properties.

	Cylinder A	Cylinder B
Cylinder diameter/mm	38.1	50.8
Piston rod diameter/mm	25.4	15.9
Effective area/mm^2	633.4	1829
Duct area/mm^2	69.4	69.4
Piston-duct area ratio	9.1	26.4

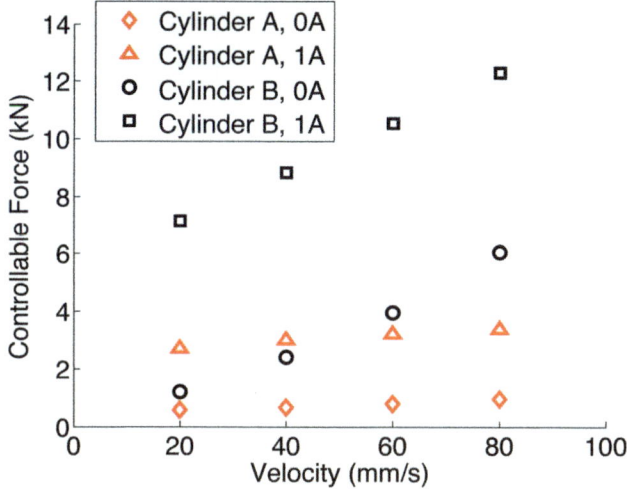

Figure 9.11 Effect of piston area on damper force.

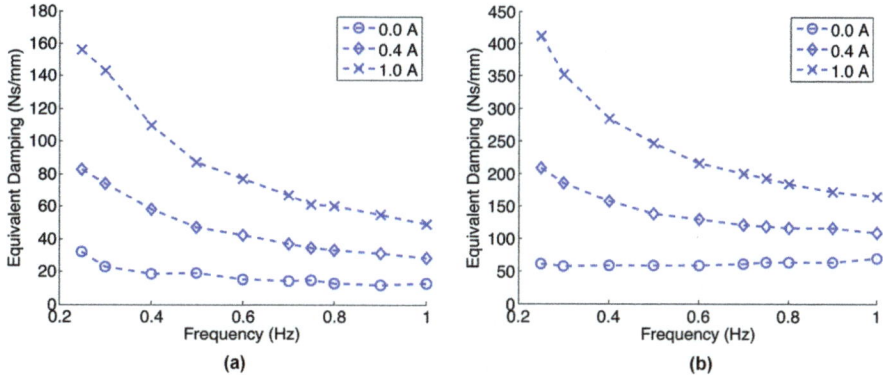

Figure 9.12 Equivalent damping of bed 1r with different piston areas.

configurations utilizing bed 1r are shown in Figure 9.12. The effect this creates on damping coefficients is significant. Damping coefficients for the two different piston configurations are presented in a semi-logarithmic-scaled graph, Figure 9.13. Cylinder B yields the greatest damping coefficients at low frequencies (below 0.5 Hz), but Cylinder A allows for greater damping coefficients at high frequencies. Although the Cylinder B configuration data does not continue past 1 Hz (because the cylinder pressure would have exceeded the maximum allowable pressure), the data trends indicate that the Cylinder A configuration would have superior damping coefficients at least up to 7 Hz. Dampers are typically employed in high frequency applications, which suggests that the piston-area ratio of the Cylinder A configuration is more practical.

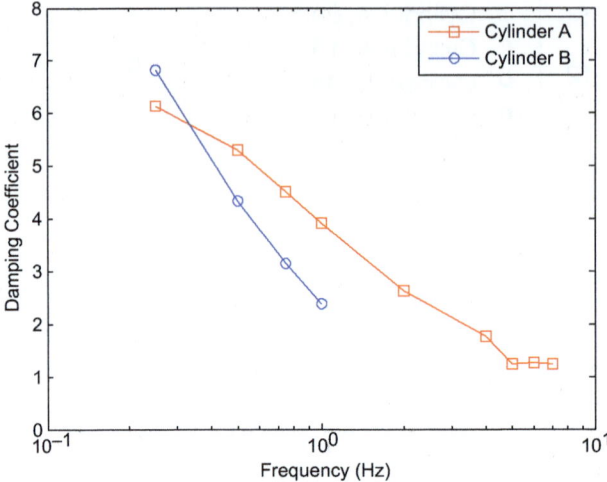

Figure 9.13   Effect of piston area on damping coefficient for bed 1r porous media.

9.4   Modeling of MR Device with Porous Valve

An accurate physical model is desirable in order to predict damper performance of the porous valve based MR devices. The non-Newtonian behavior of MR fluids, coupled with the aspects of flow in porous media, must be taken into account in the model. Data from experimental testing can then be used to validate the model.

The pressure drop across the porous valve due to a flow rate can be described using eqn (9.18). In the MR device with the porous valve, the flow rate in the porous valve is equal to the one displaced by the piston considering conservation of mass and assuming incompressible flow:

$$Q = A_p v_p \tag{9.27}$$

where A_p is the piston area and v_p is the shaft velocity. Since the pressure difference across the piston is equal to the pressure across the porous valve, MR device force as a function the shaft velocity can be obtained as:

$$F = \Delta P A_p = \frac{8\xi L A_p}{3\pi R_h} \langle \tau_y \rangle + \frac{2\eta \xi^2 L A_p^2}{R_h^2 \varepsilon} \frac{A_p^2}{A_v} v_p \tag{9.28}$$

The above force response behavior of the MR device is similar to the Bingham-plastic model[16,17] used for a conventional MR damper.

Since eqn (9.28) can only capture the force–displacement hysteresis of the MR device, a hyperbolic tangent function[18] is introduced to capture the force–velocity hysteresis due to fluid compressibility. Thus, the force

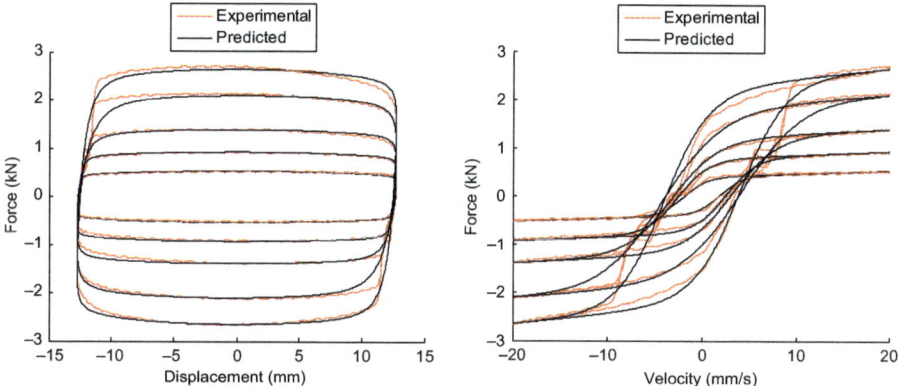

Figure 9.14 Comparison of predicted and experimental force response, bed 1r, 0.25 Hz.

response model of the MR device due to a displacement excitation, x, is described as:

$$F_p = F_y \tanh[(\dot{x} + \lambda_1 x)\lambda_2] + C_{po}\dot{x}$$

$$F_y = \frac{8\xi L A_p}{3\pi R_h}\langle \tau_y \rangle, \quad C_{p0} = \frac{2\eta\xi^2 L}{R_h^2 \varepsilon}\frac{A_p^2}{A_v} \tag{9.29}$$

where λ_1 and λ_2 are characteristic parameters to account for the hysteresis shape.

The predicted force-displacement and force-velocity behavior is shown in Figure 9.14 against the experimental values for bed 1r at 0.25 Hz. By estimating appropriate values of C_{po} and F_y for each damper configuration, the model provides an accurate approximation of damper behavior. The model and experimental values of equivalent damping for bed 1r are compared in Figure 9.15, demonstrating that the model well predicts equivalent damping for magnetic cylinders for frequency less than or equal to 2 Hz.

Since the actual fluid viscosity is a function of the applied magnetic field, C_{po} values increase with the applied current as shown in Figure 9.16, but remain constant at different fluid velocities. Post-yield damping is similar for porous beds with similar porosity. As shown in Figure 9.17, yield force increases linearly with applied current. The yield forces of similar porosity media are also similar. Furthermore, flows through the low-porosity media have yield forces at a current of 1 A that are more than two times that of the yield forces of flows in the high-porosity media. This is expected because decreasing porosity increases yield force by decreasing hydraulic radius and increasing the magnetic field in the fluid.

This trend can be represented by the yield stress of the MR fluid. As shown in Figure 9.18, average fluid yield stress increases as current (applied magnetic field) increases. The yield stress of bed 1s was the highest, despite having the lowest measured tortuosity. This is expected since the high

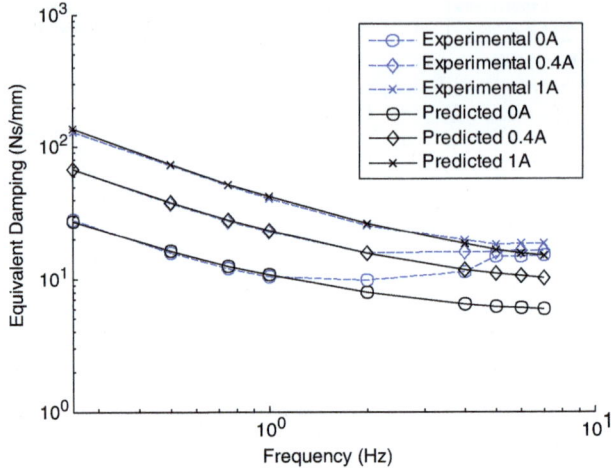

Figure 9.15   Predicted vs. experimental equivalent damping, Bed 1r.

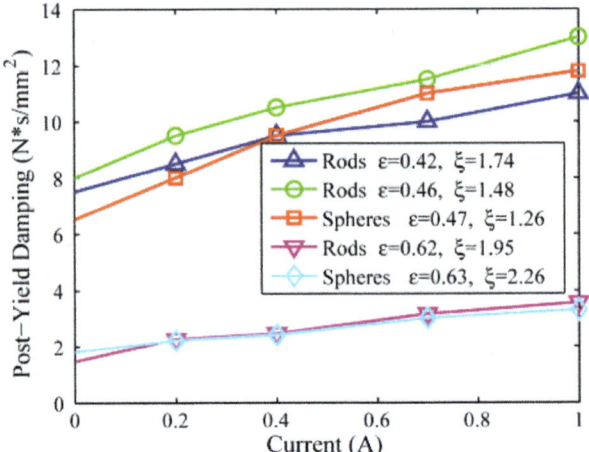

Figure 9.16   Post-yield damping as a function of current.

number of 3.5 mm spheres in bed 1s compared to other porous media created the narrowest tortuous channels, which allowed the magnetic field to permeate these channels more effectively, thus, contributing to higher yield stress levels.

For the modeling of nonlinear viscous behavior of the MR device, eqn (9.20) can be used and the damper force response due to a shaft velocity is described as:

$$F = \Delta P A_{\mathrm{p}} = \frac{8\xi L A_{\mathrm{p}}}{3\pi R_{\mathrm{h}}}\langle\tau_y\rangle + \frac{2\eta\xi^2 L}{R_{\mathrm{h}}^2\varepsilon}\frac{A_{\mathrm{p}}^2}{A_{\mathrm{v}}}v_{\mathrm{p}}\left[1 + \left(\frac{1-\varepsilon}{\varepsilon} + \frac{\varepsilon-\varepsilon_0}{\beta}\right)\frac{Re}{\xi Re_{\mathrm{cr}}}\right] \quad (9.30)$$

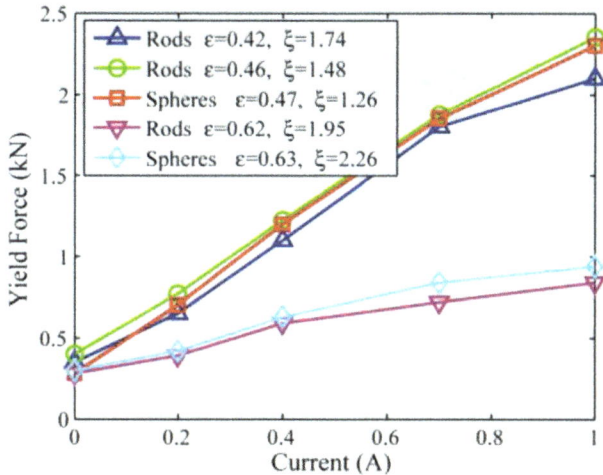

Figure 9.17 Yield force *vs.* current.

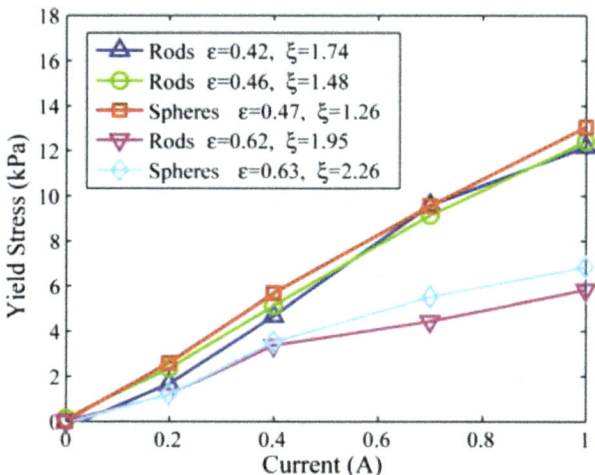

Figure 9.18 Yield stress vs. current.

Figure 9.19a shows an example of the force-velocity hysteresis comparison between analytical and experimental results, in which the experimental data of the MR device with 2.0 mm and 5.5 mm steel beads in the porous valve were compared with the analytical results. Figure 9.19b shows the equivalent damping as a function of the peak velocity for three different beads, *i.e.* 2.0 mm, 3.5 mm and 5.5 mm are grouped together, and the experimental results match the analytical results quite well.

To design an MR device with a porous valve, maximum damper force at a maximum piston velocity and controllable range (the ratio of the total force to the viscous force) are usually given as design objectives, and the MR fluid is chosen in advance. In a design strategy, optimized ε and ξ are determined to

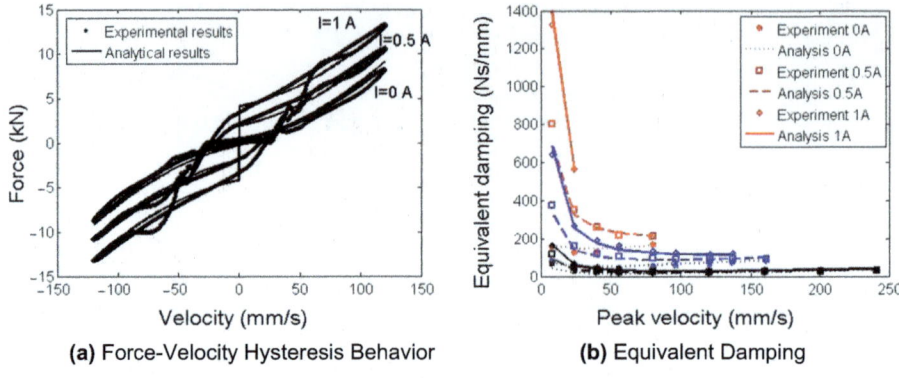

(a) Force-Velocity Hysteresis Behavior   **(b)** Equivalent Damping

Figure 9.19   Modeling results including nonlinear viscous effect.

maximize an achievable yield stress by using eqn (9.11) to (9.15), and then R_h can be found by using eqn (9.3). To reduce the effect of the nonlinear viscous force on the controllable range, the valve area, A_v, will be chosen by maintaining a lower Re than Re_{cr}. Thus, the valve length, L, and the piston area, A_p, can be determined from the required force and controllable range by using eqn (9.28) or (9.30). It should be noted that the efficiency of the magnetic circuits can be evaluated since there is a known relationship between magnetic field strength and the yield stress of one particular type of MR fluid. The magnetic circuit should be designed such that the magnetic field in the fluid can be maximized. These measures include implementing a flux return around the solenoid and use high permeability materials.

9.5   Conclusion

Flow analysis for MR fluids in a porous valve was conducted and it showed that the properties of the porous media, *i.e.* porosity and tortuosity, directly affect the performance of the porous valve. An MR device with a porous bypass valve was tested in order to examine the effects of damper and valve parameters on damper efficiency. The valve was filled with five distinct sets of porous media and subjected to sinusoidal excitations at several applied frequencies and currents. The effects of porosity, morphology, and piston-valve area ratio were investigated. An analytical model to accurately capture the hysteresis patterns of the MR device was validated. A design strategy for an MR damper with a porous valve was developed, in which the effect of nonlinear viscous force on the performance of the MR damper was addressed. Conclusions that were drawn include:

(1) Porosity plays a major role in damper efficiency because it affects the hydraulic amplification between the duct and the hydraulic cylinder. Large differences in damping performance were observed between porous beds with porosity of 0.46–0.47 and 0.61–0.63.

(2) Tortuosity plays a lesser role than porosity in dictating damper performance, but a significant role nevertheless. Dampers configured with similar porosity but different tortuosity produced similar amounts of field-on and field-off damping; however, a higher tortuosity slightly increased field-on and field-off damper forces.

(3) Piston area is a significant factor in damping efficiency. A high piston-valve area ratio provides greater energy absorption and allows for high damping coefficients at low piston velocities, but a low piston-valve area ratio allows for higher damping coefficients at high velocities. Moreover, a damper with a low piston-valve area ratio can accept higher frequency excitations without reaching prohibitively high cylinder pressure.

(4) Nonlinear viscous effects at high velocities on the behavior of the MR device with a porous valve was successfully described using Darcy's equation.

References

1. J. M. Ginder, L. C. Davis and L. D. Elie, Rheology of Magnetorheological Fluids: Models and Measurements, *Int. J. Modern Phys. B*, 1996, **10**, 3293–3303.
2. J. D. Calson and M. J. Chrzan, Magnetorheological Fluid Dampers, U.S. Pat. No. 5277281, 1994.
3. P. C. Chen and N. M. Wereley, Magnetorheological Damper and Energy Dissipation Method, U.S. Pat. No. 6694856, 2004.
4. P. N. Hopkins, J. D. Fehring, I. Lisenker, R. E. Longhouse, W. C. Kruckemeyer, M. L. Oliver, F. M. Robinson, and A. A. Alexandridis, Magnetorheological Fluid Damper, U.S. Pat. No. 6311810, 2001.
5. E. N. Anderfaas, and D. Banks, Magnetorheological Damper System, U.S. Pat. No. 6953108, 2005.
6. Z. Shulman, Magnetorheological Systems and Their Application, *Magnetic Fluids and Applications Handbook*, 1996, 188–229.
7. P. Kuzhir, G Bossis, V. Bashtovoi and O. Volkova, Flow of Magnetorheological Fluid Through Porous Media, *Eur. J. Mech. B, Fluids*, 2003, **22**, 331–343.
8. J. E. Martin and R. A. Anderson, Chain Model of Electrorheology, *J. Chem. Phys.*, 1996, **104**(12), 4814–4827.
9. J. G. Savins, Non-Newtonian Flow Through Porous Media, *Ind. Eng. Chem.*, 1969, **61**(10), 18–47.
10. N. E. Sabiri and J. Comiti, Pressure Drop in Non-Newtonian Purely Viscous Fluid Flow Through Porous Media, *Chem. Eng. Sci.*, 1995, **50**(7), 1193–1201.
11. F. M. White, *Fluid Mechanics*, McGraw-Hill Book Company, 1989.
12. P. Kuzhir, G. Bossis and V. Bashitovoi, Effect of the Orientation of the Magnetic Field on the Flow of a Magnetorheological fluid. I. Pane Channel, *J. Rheol.*, 2003, **47**(6), 1373–1384.

13. P. Kuzhir, G. Bossis and V. Bashitovoi, Effect of the Orientation of the Magnetic Field on the Flow of a Magnetorheological fluid. II. Cylindrical Channel, *J. Rheol.*, 2003, **47**(6), 1385–1398.
14. J. Bear, *Dynamics of Fluids in Porous Media*, New York, Dover Publications, 1988.
15. J. D. Carlson, MR fluids and devices in the real world, Proceedings of the 9th International Conference on Electrorheological Fluids and Magnetorheological Suspensions, 2004, pp. 531–538.
16. R. A. Snyder, G. M. Kamath and N. M. Wereley, Characterization and Analysis of Magnetorheological Damper Behavior Under Sinusoidal Loading, *AIAA J.*, 2001, **39**(7), 1240–1253.
17. N. M. Wereley, L. Pang and G. M. Kamath, Idealized Hysteresis Modeling of Electrorheological and Magnetorheological Dampers, *J. Intell. Mater. Syst. Struct.*, 1998, **9**(8), 642–649.
18. Y. T. Choi, N. M. Wereley and Y. S. Jeon, Semi-Active Vibration Isolation Using Magnetorheological Isolators, *AIAA J. Aircraft*, 2005, **42**, 1244–1251.

CHAPTER 10

MR Devices with Advanced Magnetic Circuits

HOLGER BÖSE,* JOHANNES EHRLICH AND
THOMAS GERLACH

Fraunhofer-Institut für Silicatforschung ISC, Center Smart Materials,
Neunerplatz 2, 97082 Würzburg, Germany
*Email: boese@isc.fraunhofer.de

10.1 Introduction

Magnetorheological (MR) fluids are liquid smart materials, whose rheological
and mechanical properties are controlled by a magnetic field. The reversible
transition from the liquid to a semi-solid state of the material can be varied
continuously, where the degree of stiffening depends on the magnetic field
strength. In strong magnetic fields, the shear resistance of the MR fluid is enhanced by up to three orders of magnitude with respect to the initial liquid
state. Moreover, this transition is fast and occurs within a few milliseconds,
primarily limited by the speed of response of the magnetic field to the rise of
current in the electromagnet of the magnetic circuit.

Due to these outstanding properties, MR fluids have found widespread applications for a variety of mechatronic functions.[1,2] Most attention has been
devoted to adaptive or semi-active vibration damping. First automotive shock
absorbers entered the market about ten years ago. The variability of the
damping force helps to minimize the unsatisfying compromise of passive
suspension systems between safety and comfort. Since its introduction, the
exploitation of semi-active suspension systems has been extended to a

RSC Smart Materials No. 6
Magnetorheology: Advances and Applications
Edited by Norman Wereley
© The Royal Society of Chemistry 2014
Published by the Royal Society of Chemistry, www.rsc.org

multitude of different vehicle models. Further applications of adaptive vibration damping were identified in seats for truck and bus drivers, seismic mitigation of buildings as well as the protection of cable-stayed bridges against wind excitation.

Beside adaptive damping applications, MR fluids can also be beneficially used for electrically controllable clutches and brakes. Such MR clutches are capable of flexibly distributing the torque generated by an engine to several consumers. As an example, MR clutches may control the torque transmission to side aggregates in a vehicle on demand which are usually continuously driven by the combustion engine.[3] A special benefit of MR clutches is the possibility to limit the transmitted torque by the strength of the applied magnetic field. When the input torque reaches a preset limit, the clutch slips and no higher output torque is transmitted. This is especially worthwhile in applications in which safety issues are of crucial importance. Moreover, MR clutches bear the advantage of inherent damping in contrast to conventional clutches which are based on hard disks.

A considerable number of investigations regarding different types of MR dampers have been published so far. They differ in details of their design, but a common feature is the magnetic circuit which is integrated in the moving piston.[4-6] Generally, the magnetic field is generated by a coil and penetrates the MR fluid flowing through an annular gap, whereas the flow resistance and the corresponding damping force are controlled by the current in the coil. Choi *et al.* presented a MR damper with stationary coils for passenger vehicles, where the MR fluid is forced by the piston through an annular duct between two concentric cylinders.[7,8] Due to the crucial importance of the magnetic circuit in MR damping systems, various studies have been performed on the design of conventional magnetic circuits containing electromagnets.[9-11]

Some work on magnetorheological clutches has already been published. The first MR clutch was reported by Rabinow who more than six decades ago was already aware of the strong benefit of MR fluids for the controlled torque transmission.[12] A large variety of careful investigations was performed by Lampe who studied different clutch designs and basic issues of their performance.[13,14] Kieburg *et al.* as well as Gratzer *et al.* described a high torque clutch which was successfully tested in the powertrain of a vehicle.[15,16] A very small and compact MR clutch which can transmit a maximum torque of 6 Nm was developed by Kikuchi *et al.*[17]

Magnetorheological devices like dampers and clutches are mechatronic systems consisting of several components or sub-systems. The main components of the comprehensive MR system are the mechanical part, the magnetic circuit with the power electronics, the control unit and, of course, the MR fluid as the active material (Figure 10.1). The control unit of the MR system generates electric signals which use the input information from internal or external sensors and follow a selected control strategy. According to the signals received from the control unit, the power electronics usually gives rise to a defined current in the coil of an electromagnet, which generates the magnetic field. The magnetic flux is guided by the magnetic circuit from the coil to the

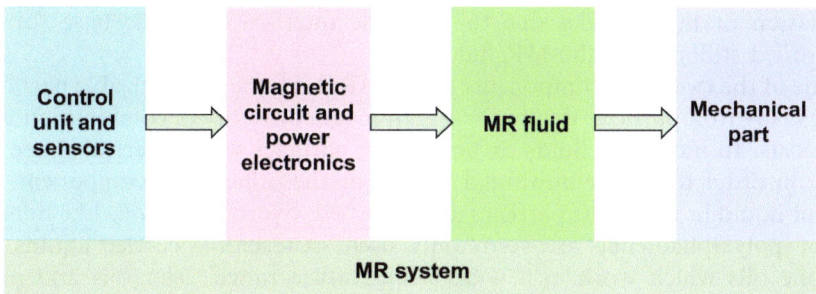

Figure 10.1 Components of a MR device system.

MR fluid. As a result, the controlled variation of stiffening of the MR fluid causes the specific reaction in the mechanical part of the MR device.

It is important to know that all the components or sub-systems of the MR device interact with other components. To give an example, on the one hand the magnetorheological properties of the MR fluid determine the mechanical response of a MR damper or clutch, depending on the composition of the MR fluid. On the other hand, the MR fluid is also a part of the magnetic circuit and has an influence on the magnetic flux in the circuit according to the magnetic properties of the MR fluid and the corresponding magnetic resistance which also depends on the MR fluid composition. As a consequence, in a development of a new MR device the designs of all the components should be adapted with respect to each other. Such a comprehensive proceeding offers a great benefit for the development of optimized MR devices like damping systems or clutches.

In the past, most attention in the literature has been paid to the mechanical part and to convenient control strategies of the MR devices. However, the use of a MR fluid with properties well-adapted to the other system components is also of high importance. Moreover, novel designs of magnetic circuits give the opportunity to realize operational modes of the MR device which are not feasible with conventional magnetic circuits according to the state of the art.

The objective of this chapter is the introduction of novel concepts for advanced magnetic circuits. In conventional magnetic circuits the magnetic field is generated by the current in the coil of an electromagnet. With rising current the magnetic field strength is increased, which leads to an enhanced stiffening of the MR fluid. In novel concepts, hybrid magnetic circuits which contain combinations of different sources of the magnetic field, like permanent magnets and electromagnets, are used in MR devices. The perspectives of such magnetic circuits for the MR technology will be discussed later.

10.2 Magnetorheological Fluids

MR fluids are basically suspensions of magnetizable particles in a carrier liquid. In the magnetic field, the particles become polarized. The magnetic dipoles attract each other and the particles form chains along the magnetic field lines as well as other aggregated structures. This magnetic-field-induced structure formation of the particles due to magnetic interactions is the base for the controlled stiffening of the MR fluid.

One of the two main components of the MR fluid, the magnetizable particles, are usually iron particles due to their high magnetizability in combination with low costs. In most MR fluids carbonyl iron particles with spherical shape are used, in order to reduce unwanted wear. For the other main component, the carrier liquid in which the particles are dispersed, hydrocarbon oils like mineral oil or polyalphaolefine are commonly used. Alternative carrier liquids are silicone oils which work in a wider temperature range, ester oils and poly-glycols. The heavy iron particles in the carrier liquid have to be stabilized against settling by chemical additives. If, however, some separation of the particles due to settling occurs, the sediment must be easily redispersible by agitation of the MR fluid. In a MR damper, this redispersion of the particles can generally be achieved by the motion of the damper piston. Other additives in the MR fluid are responsible for the protection against oxidation and wear. Their purpose is to assure a long lifetime of the MR fluid in use.

In most applications, the relevant properties of the MR fluid are its base viscosity in the absence of a magnetic field and the increase of the shear stress in a magnetic field with increasing field strength. Both features sensitively depend on the concentration of the iron particles in the MR fluid. With increasing particle concentration, the base viscosity and the shear stress in the field are enhanced. In MR fluids for technical applications, the iron particle concentration roughly covers a range between 20 and 50% by volume.

Figure 10.2 gives an overview on how the magnetorheological properties of typical MR fluids with low, medium and high concentration of iron particles vary with rising magnetic flux density of the applied field. The experimental data refer to MR fluids with 20, 35 and 50 vol% iron particles in poly-alphaolefine as the carrier liquid. It becomes apparent that with 50 vol% iron

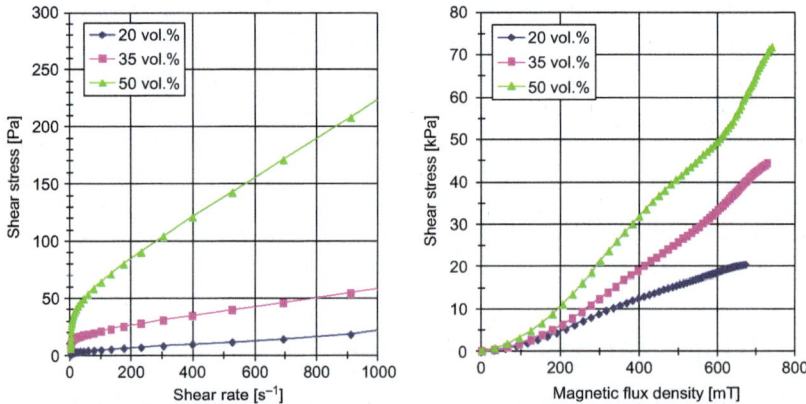

Figure 10.2   Flow curves (left) and dependence of the shear stress on the magnetic flux density (right) for MR fluids containing different concentrations of magnetisable iron particles.

particles a shear stress of 70 kPa at a magnetic flux density of 700 mT can be achieved, whereas the MR fluid with 20 vol% iron particles reaches a shear stress of 20 kPa only. However, the base viscosity of the high-concentrated MR fluid is also enhanced in comparison with the lower-concentrated samples, which can be observed in the flow curves represented in Figure 10.2, left.

Also of high relevance are the magnetic properties of the MR fluids which limit the magnetic flux in the magnetic circuit through the MR fluid. The magnetic properties of the material are represented by the magnetization curves of the magnetic flux density *B* *vs.* the magnetic field strength *H*. Figure 10.3 depicts the magnetization *J* derived from the flux density *B* depending on the field strength *H* for MR fluids with the same iron particle concentrations as already shown in the representation of the magnetorheological properties in Figure 10.2. Obviously, the magnetization increases with rising iron particle concentration and with magnetic field strength, approaching a saturation limit at *ca.* 500 kA m^{-1} field strength.

Shown in Figure 10.3, right, are also the relative permeabilities of the three MR fluids with different iron particle concentrations of 20, 25 and 50 vol%. The relative permeabilities of MR fluids are very low in comparison with bulk material of magnetic steel, whose permeability is above 2000. In contrast to bulk steel, the range of the initial relative permeability of the MR fluids at low magnetic field strength extends from 4 for the MR fluid with 20 vol% iron particles to 12 for the MR fluid with 50 vol%. The reason for these low permeabilities of the MR fluids in comparison to bulk steel is the fine dispersion of their iron particles in the carrier liquid, which causes high magnetic inter-particle resistances for the magnetic flux in the penetration of the MR fluid. This generally low permeability of MR fluids has to be taken into account in

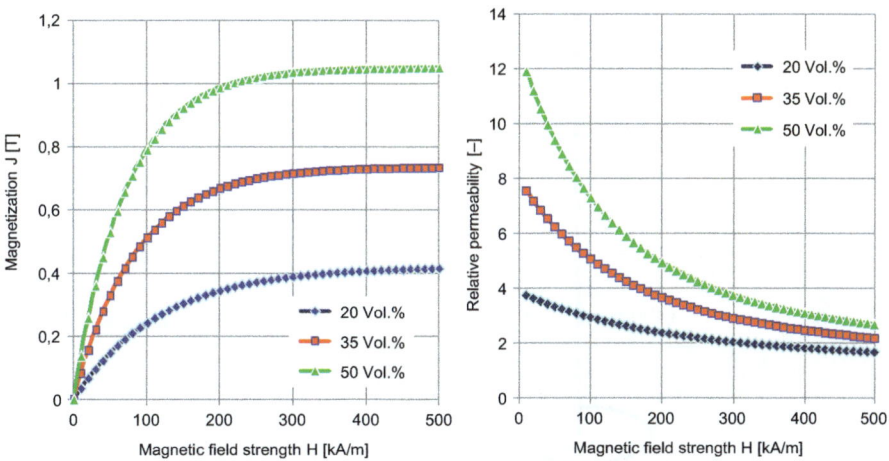

Figure 10.3 Dependence of the magnetization of MR fluids containing different concentrations of magnetisable iron particles on the magnetic field strength (left) and relative permeability derived therefrom (right).

simulations of the magnetic flux density distribution in the magnetic circuit for a MR device.

10.3   Magnetic Circuits in MR Devices

The magnetic circuit as a part of the MR device generates the magnetic field and guides the magnetic flux from the magnetic source to that region in which the MR fluid has to stiffen in order to trigger the intended response of the device. Usually the magnetic flux lines penetrate a gap filled with MR fluid, which weakens the magnetic flux due to its high magnetic resistance, as shown before. Therefore, in the design of the magnetic circuit the width of the MR fluid gap should be made as small as possible, in order to limit the power demand of the electromagnet.

Commonly, the source of the magnetic field in a MR device is an electromagnet with a coil. The current applied to the coil determines the magnetic field strength. The corresponding magnetic flux is guided by the yoke of the magnetic circuit. As an example, Figure 10.4 exhibits a closed magnetic circuit with rotational symmetry. The coil is surrounded by an annular gap filled with the MR fluid. The red lines show the path of the magnetic flux along the yoke parts which are made from magnetic steel with high permeability. The only interruptions of the yoke are the two red MR fluid gap zones above and below the coil. These regions are the active gap zones in which the MR fluid stiffens when the magnetic field is activated.

The design of the magnetic circuit is usually supported by FEM simulations of the magnetic flux density distribution. The goal of these simulations is to optimize the dimensions of the yoke parts of the magnetic circuit such that the magnetic circuit is as small as possible. Moreover, the magnetic flux density in

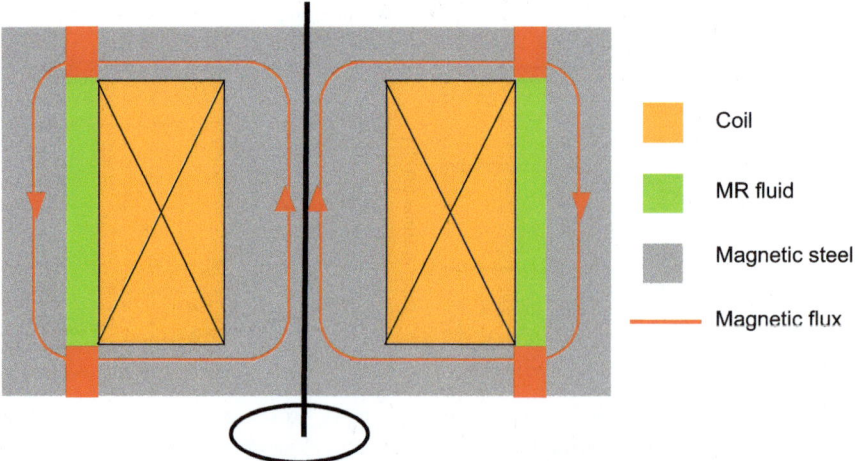

Figure 10.4   Scheme of cross section of a magnetic circuit with a coil generating magnetic flux lines which penetrate and stiffen the MR fluid. The magnetic circuit has rotational symmetry.

the active MR fluid gap zones should be sufficiently high and simultaneously no saturation of the flux density should occur in localized regions of the magnetic circuit. As an example, results of FEM simulations for the magnetic circuit in Figure 10.4 are shown in Figure 10.5. From left to right, the coil current is varied from 0.5 A to 1 A and to 3 A. At a coil current of 3 A, the flux density in the central yoke part of the magnetic circuit approaches the saturation limit of 2.1 T. Saturation can be avoided even with this coil current by thickening the inner yoke part of the magnetic circuit.

The magnetic flux through the active MR fluid gap zones should be as high as possible, in order to activate the MR fluid already with only low effort of electric power. MR fluids with high concentrations of iron particles enhance the magnetic flux due to their higher permeability. This enhancement is considered in the FEM simulations through the introduction of the magnetization curve of the corresponding MR fluid. Hence, such simulations have

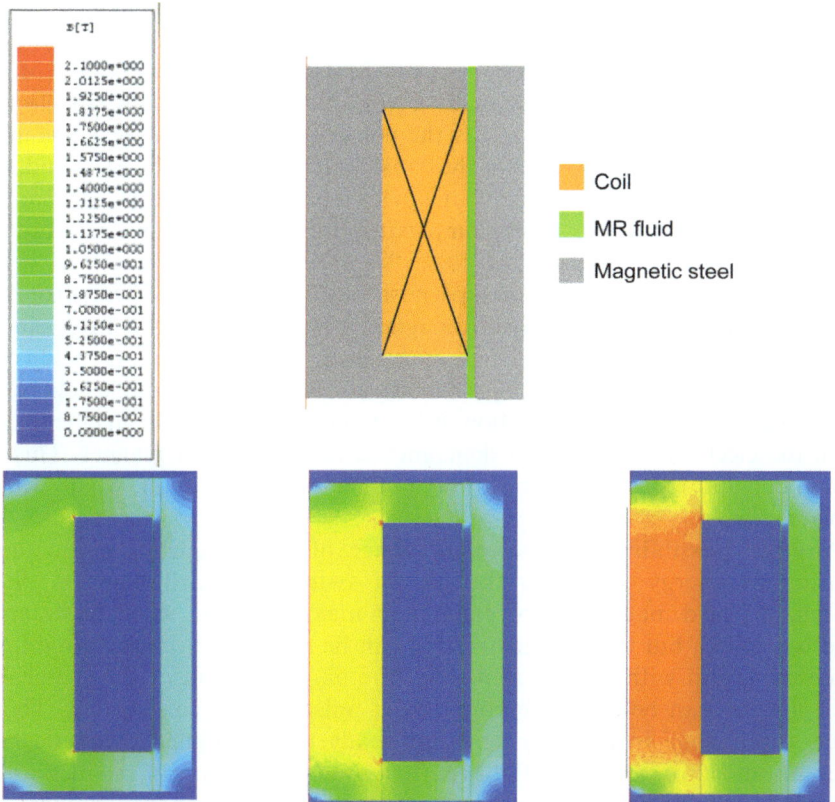

Figure 10.5 Scheme of cross section of a magnetic circuit with a coil (top right) and magnetic flux density in cross section of the magnetic circuit shown in Figure 10.4 (only right half) at different coil currents (0.5 A, 1 A and 3 A) from FEM simulation (bottom).

proved to be a valuable tool for the design of magnetic circuits in a multitude of MR devices.

Up to now, in nearly all MR devices the magnetic circuit is equipped only with an electromagnet as the magnetic field source. This configuration implies that in order to maintain the magnetic field for the stiffening of the MR fluid, electric energy has to be supplied to the electromagnet at any time. For many applications of MR fluids in various devices, this arrangement is not energy-efficient. Furthermore, in case of a power-failure, the magnetic field breaks down immediately and the MR fluid re-liquefies. Such behaviour could eventually lead to severe limitations of fail-safety. To give an example, the damping force of a MR damper would directly drop down to the lowest value in case of a power-failure.

In the following, novel concepts for the improvement of the fail-safe behaviour and the energy-efficiency of MR devices are introduced. For this purpose, so-called hybrid magnetic circuits which contain different sources of the magnetic field generation are integrated in MR devices. Permanent magnets have the inherent capability to generate a magnetic field without the continuous supply of electric energy. However, for the variation of this field at a defined location in the MR fluid, a superposition of the field of the permanent magnet with a variable field caused by an electromagnet is needed. An alternative is the controlled motion of the permanent magnet, which modifies its field at the considered location in the MR fluid, but such motion of the permanent magnet requires a suitable mechanical mechanism. Due to the high complexity, this latter option will not be considered here.

In the more flexible option of superposition of the magnetic fields from the permanent magnet and the electromagnet, the total magnetic field can be electrically varied by the coil current in the electromagnet. The total field is the sum of constant base field of the permanent magnet and the variable field of the electromagnet. In this configuration, it has to be assured that the magnetic field from the electromagnet cannot demagnetize the permanent magnet. This essential demand requires a special design of the magnetic circuit, which will be explained later.

Another possibility for the maintenance of the magnetic field without continuous energy supply is the use of switchable hard magnets. Such switchable hard magnets also retain their magnetization, when no external field is applied, but their magnetization can be modified by an external field. In combination with an electromagnet, the magnetization of the switchable hard magnet can be altered by only short pulses of the coil current. Thereafter, the change in the magnetization leads to a modified magnetic field strength and flux density in the MR fluid. This procedure of magnetic field control by alterations in the magnetization of the hard magnet requires another design of the hybrid magnetic circuit than in the procedure with permanent magnet. These differences in the designs of the magnetic circuits will be represented and discussed in some examples which describe specific adaptive damper and clutch applications of MR fluids.

10.4 Applications

The concept of hybrid magnetic circuits containing different magnetic sources for the field generation offers far-reaching perspectives for a large variety of MR devices. This will be explained in more detail by means of several demonstration models. Since the most relevant fields of applications of MR technology concern dampers and shock absorbers on one side, as well as clutches and brakes on the other side, the following discussion will focus on such controllable MR devices. According to the corresponding applications, they will be distinguished between devices with translational motion and devices in which the MR fluid transmits a rotational motion from the input to the output shaft of the device.

10.4.1 Translational MR Devices

Translational magnetorheological devices are generally used for semi-active dampers and controllable shock absorbers. In these devices, vibrations are attenuated with a variable damping force which is controlled by the magnetic field in real-time. Furthermore, impact energies of moving bodies can be absorbed by the mechanical resistance of the MR fluid in the magnetic field. However, translational devices may also serve as linear force transmission units or for temporarily locking an object in a defined position. In the following, for the sake of simplicity the term damper is used for translational devices.

In a conventional MR damper in which the magnetic field is generated by an electromagnet only, the entire magnetic circuit including the coil is usually integrated in the damper piston. A scheme of such a MR damper with magnetic flux lines in the circuit is depicted in Figure 10.6. The integration of the magnetic circuit in the damper piston allows a more compact damper design in comparison with other dampers like that with an external bypass for the MR fluid flow. This integration will be retained also for the novel hybrid magnetic

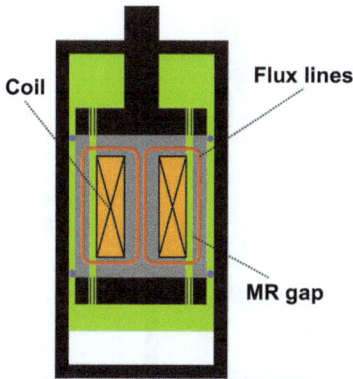

Figure 10.6 Scheme of a conventional MR damper with magnetic circuit containing an electromagnet only.

circuits with additional permanent magnets or switchable hard magnets which are introduced in the following examples describing various demonstration models of MR dampers. Another important issue is the correspondence between the performance of the MR damper and the properties of the MR fluid which is used therein. For this purpose, it will be demonstrated that the achievable damping force can be strongly enhanced with a MR fluid containing a higher concentration of iron particles.

10.4.1.1   *Magnetic Circuit with Permanent Magnet and Electromagnet*

In the first concept of a MR damper with hybrid magnetic circuit, the damper piston contains an electromagnet with a coil in the center and additionally two permanent magnets at the top and the bottom.[18,19] The scheme of the damper with rotational symmetry is depicted in Figure 10.7, wherein the permanent magnets are marked as flat yellow disks. Including the two permanent magnets and the coil, there are in total three sources of the magnetic field. Correspondingly, these three magnetic sources define three magnetic sub-circuits, whose magnetic fluxes may superpose in the active zones of the annular MR fluid gap between the coil and the permanent magnets. The separation of the magnetic circuit into three sub-circuits assures that the permanent magnets cannot be demagnetized by the electromagnet.

Figure 10.8 depicts the three relevant operational modes of the magnetic circuit. In the left scheme, the coil is not activated and the magnetic field is generated merely by the two permanent magnets. The shown flux lines in this operational mode define the upper and the lower sub-circuit. In the middle scheme in Figure 10.8, a strengthening of the magnetic field of the permanent magnets by a positive superposition with the field of the electromagnet occurs. The flux lines around the coil define the medium sub-circuit. Finally, in the right scheme of Figure 10.8, the direction of the coil current is reversed, which

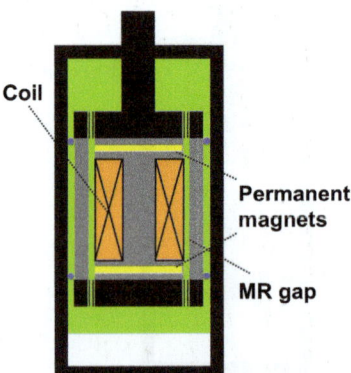

Figure 10.7   Scheme of a conventional MR damper with magnetic circuit containing an electromagnet and two permanent magnets.

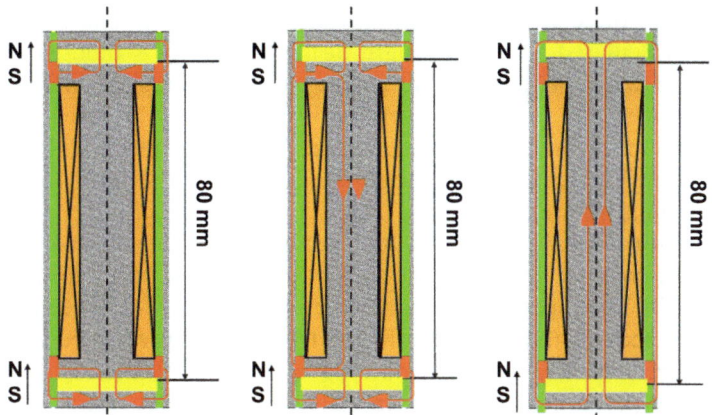

Figure 10.8   Scheme of the magnetic flux in the MR damper piston with electromagnet and permanent magnets in the three different working modes:
magnetic field generated only by the permanent magnets (left), strengthened by the electromagnet (middle) and weakened by the electromagnet
(right).

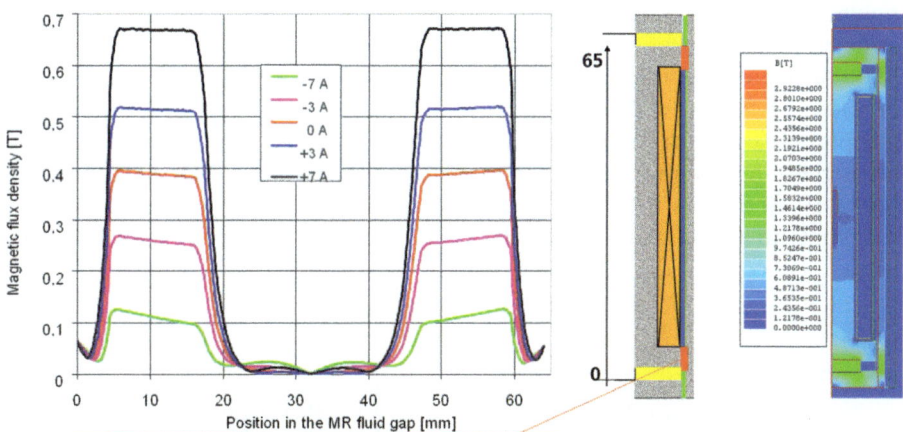

Figure 10.9   Simulated magnetic flux density along the MR fluid gap in the damper
piston for the working modes of field weakening (negative current), field
of the permanent magnets only (0 A) and field strengthening (positive
current), simulation for MR fluid with 35 vol% iron particles (left) and
magnetic flux density distribution for 0 A (right).

leads to a weakening of the magnetic fields of the permanent magnets and the
electromagnet by a negative superposition.

The components in the hybrid magnetic circuit were designed such that the
magnetic field of the two permanent magnets can almost be canceled by that of
the electromagnet in the negative superposition. In the reverse operational
mode, a very high magnetic flux density in the active MR fluid gap zones can be
achieved. Figure 10.9 shows the results of the simulation of the magnetic flux

density for the three operational modes described above. The magnetic flux density was calculated along the length of the annular MR fluid gap from the bottom to the top of the piston. This geometrical line along the MR fluid gap is shown as a green line in the right scheme in Figure 10.9. The red sections on the green line mark the active zones in which the MR fluid is stiffened by the magnetic field. The length of this line is 65 mm and the total length of the damper piston is 80 mm.[20]

The simulation results in Figure 10.9 demonstrate that, when no current is applied to the coil, the magnetic flux density in the active MR gaps amounts to nearly 400 mT. This operational mode determines the fail-safe behavior of the MR damper, because the corresponding damping force is retained without any supply of electric energy. In the operational mode, in which the magnetic fields caused by the permanent magnets and the electromagnet nearly cancel each other (−7 A), the magnetic flux density in the active MR gap zones is about 100 mT only (Figure 10.9). In the reverse operational mode with positive superposition of the magnetic fields (+7 A), the maximum magnetic flux density of nearly 700 mT is achieved.

In the simulations of the magnetic flux density in the magnetic circuit, the magnetization curve of a MR fluid with 35 vol% iron particles was used. As shown in Figure 10.3, the magnetic properties of MR fluids are primarily determined by the concentration of the magnetic iron particles. In order to study the influence of the iron particle concentration of the MR fluid on the performance of the novel MR damper, two in-house-developed MR fluids with iron particle concentrations of 25 and 35 vol%, respectively, in silicone oil as the carrier liquid were used for experimental investigations in the MR damper. Figure 10.10 depicts the measured magnetorheological properties of the two MR fluids in terms of the flow curve and the shear stress *vs.* magnetic flux density. Of course, the MR fluid with 35 vol% iron particles gives rise to

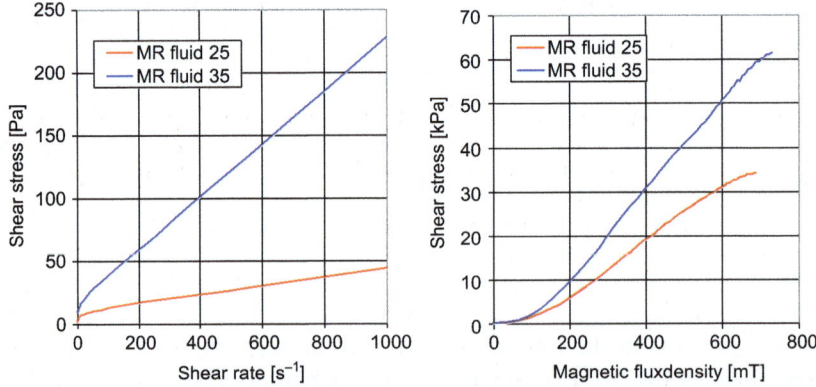

Figure 10.10   Flow curves without magnetic field (left) and shear stress *vs.* magnetic flux density (right) of the two MR fluids 25 and 35 containing 25 and 35 vol% iron particles, respectively, at shear rate 100 s⁻¹.

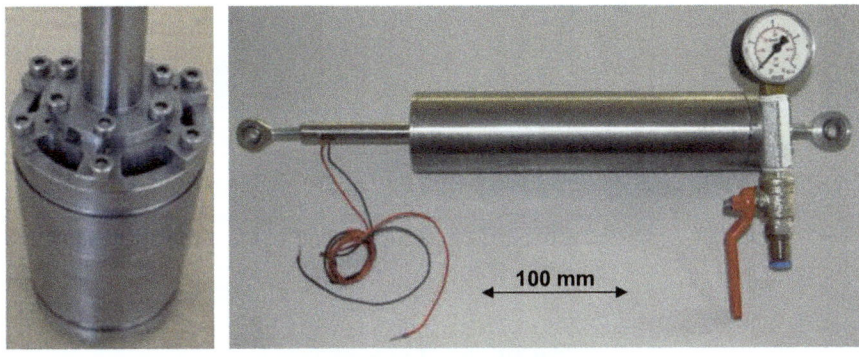

Figure 10.11   Demonstrator model of the MR damper with electromagnet and permanent magnets: piston (left) and assembled damper (right).

considerably higher shear stresses in the magnetic field, but also to a higher base viscosity without field.

In order to prove the simulation results for the magnetic circuit and the behavior of the two selected MR fluids, a demonstration model of the described MR damper was designed, manufactured and assembled. Figure 10.11 shows the constructed damper piston (left) and the assembled damper (right), which contains a gas cushion in order to balance the volume change due to the motion of the piston rod. For the permanent magnets, disks made from NdFeB were used. The MR damper was consecutively filled with the two MR fluids. According to the differing shear stresses in Figure 10.10, the iron particle concentration should sensitively determine also the achievable damping force at a given magnetic flux density.[21] The demonstration model with the selected MR fluid was mounted in a self-designed mechanical testing machine with translational excitation. The motion is generated by a servo motor and converted with a linear gear transmission (Figure 10.12). In these experiments, the performance in terms of the damping force was investigated.

Figure 10.13 exhibits the measured damping force *versus* the displacement of the damper piston at a frequency of 1 Hz with the MR fluid 35 containing 35 vol% iron particles. The damping force varies sensitively with the coil current. As already indicated in the magnetic flux simulation, the damping force without any coil current is in a medium range of about ± 2 kN. This damping force would be relevant in case of an electric power failure and demonstrates the high fail-safe stability of the MR damper.

If a strongly positive current of +7 A is applied to the coil, strengthening the magnetic field of the permanent magnets, the corresponding damping force is increased from ± 2 kN to nearly ± 5 kN. However, if the coil current direction is reversed to −7 A, the measured damping force is lowered to *ca.* ± 0.4 kN (Figure 10.13). These results confirm the intended behavior of the damper and show the wide range of the achievable damping force. The ratio of the highest to the lowest damping force of the MR damper amounts to *ca.* 11 : 1, which represents a very high variability of the MR damper.

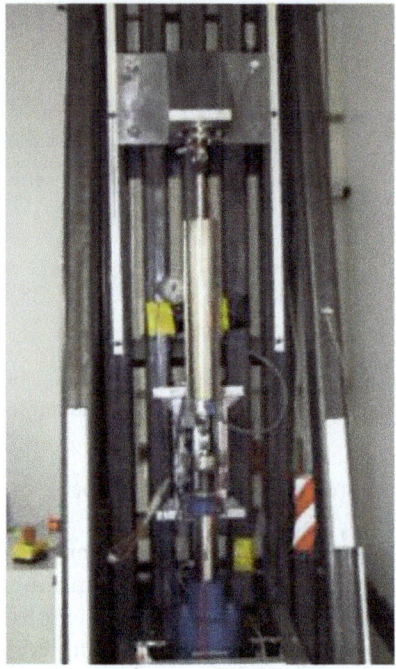

Figure 10.12   MR damper with electromagnet and permanent magnets on the mechanical testing machine.

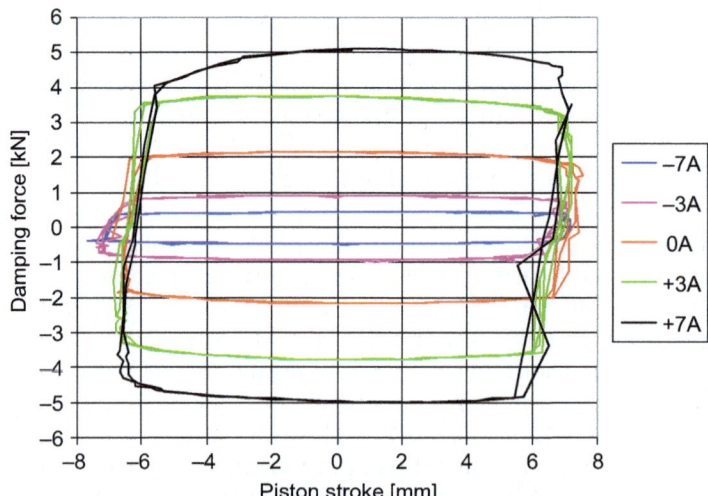

Figure 10.13   Experimental results of force measurements on the MR damper with electromagnet and permanent magnets on the mechanical testing machine at 1 Hz. The MR fluid contains 35 vol% iron particles.

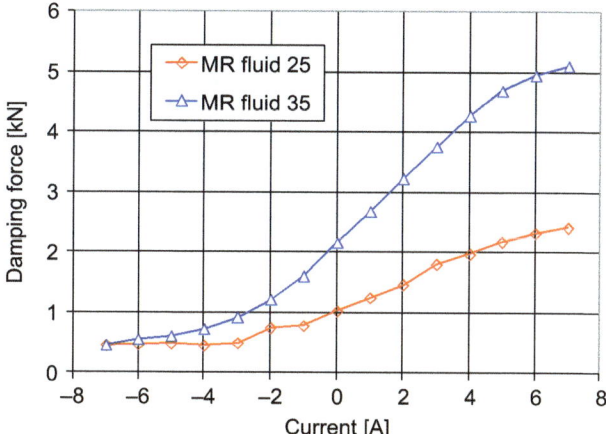

Figure 10.14 Maximum damping force of compression stage *vs.* coil current of the MR damper with MR fluids 25 and 35 containing 25 and 35 vol% iron particles, respectively.

Figure 10.14 shows the damping forces of the damper with the two MR fluids containing 25 and 35 vol% iron particles, respectively, under variation of the coil current in comparison. Obviously, the performance of the damper with the MR fluid 25 with only 25 vol% particles is much lower than for the higher-concentrated MR fluid 35. The highest achievable damping force with the MR fluid 25 at +7 A coil current is less than half of the corresponding damping force with MR fluid 35, which corresponds roughly to the ratio of the shear stresses of the two MR fluids at high magnetic flux density (compare Figure 10.10). Since the lowest damping force at −7 A is about the same for both MR fluids, the maximum factor of force increase by the magnetic field variation is only 5.5 for MR fluid 25, compared to 11 for MR fluid 35. This result is a clear indication that the higher-concentrated MR fluid yields a considerably better performance of the MR damper in terms of the achievable damping force.[21]

In further experiments, the long-term behavior of the MR damper was investigated. In a MR fluid, particle settling occurs due to the high density difference between the iron particles and the carrier liquid. For a good MR damper performance, it is necessary that the particles in the MR fluid completely redisperse in a short time after the restart of operation. For the study of the long-term behavior of the used MR fluids, the damper was stored for 30 days in an upright orientation. Measurements of the damping force with a piston stroke of 6.5 mm and a frequency of 1 Hz in the current-less working mode were performed before and after the storage period.

The results of the tests performed with the damper containing the MR fluid with 25 vol% iron particles are compiled in Figure 10.15. In the diagram, the blue line depicts the initial damping force with the fresh MR fluid and the green line exhibits the damping force after 30 days of storage of the damper. It is

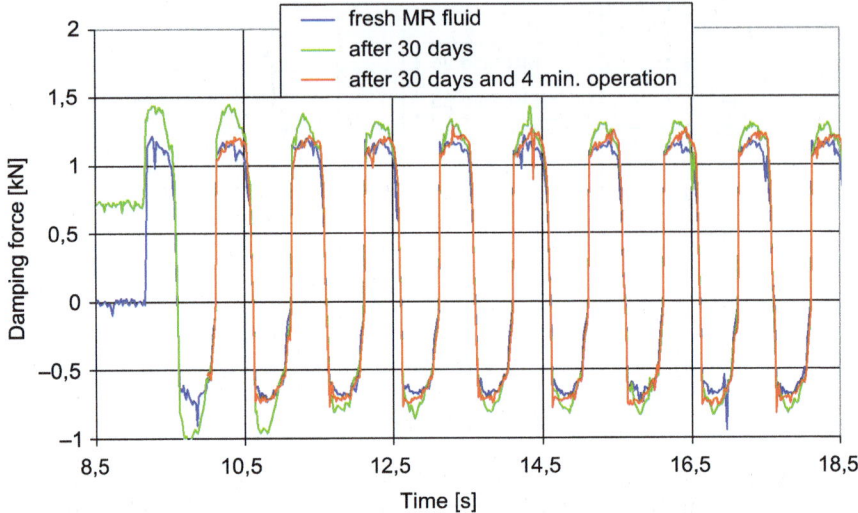

Figure 10.15 Damping forces of the MR damper with MR fluid 25 before and directly after storage for 30 days as well as after 4 minutes of operation.

clearly visible that only at the first few strokes after the storage period, a slightly higher damping force occurs, quickly approaching the original level. The red line depicts the damping force after 4 minutes of operation, which shows constant maxima and minima close to the original damping forces measured with the fresh MR fluid. A corresponding test with the higher-concentrated MR fluid 35 gave similar results. The test results of the long-term behavior demonstrate that particle sedimentation of the used MR fluids in the damper is not a problem, because a fast redispersion of the iron particles in the MR fluid during the operation of the damper occurs. This long-term stability is also supported by the remaining field of the permanent magnets which prevents the iron particles from settling in the active MR fluid gap zones of the damper.

10.4.1.2 Magnetic Circuit with Switchable Hard Magnet and Electromagnet

In the following, another concept for a MR damper with a hybrid magnetic circuit is introduced. The MR damper with the second type of hybrid magnetic circuit contains a switchable hard magnet in addition to the electromagnet. In contrast to the former concept with the permanent magnets, the magnetization of the hard magnet here can be changed by the electromagnet. The possibility to alter the magnetization of the hard magnet is achieved by the design of the magnetic circuit in which both magnetic sources, the hard magnet and the electromagnet, are located in the same circuit.

The main benefit of this configuration is the capability of the magnetic circuit to allow different magnetic flux densities and correspondingly different damping forces to be maintained without permanent electric energy

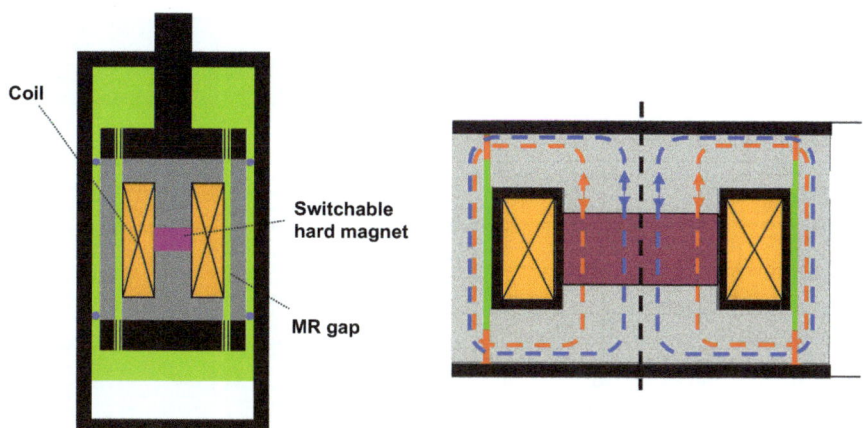

Figure 10.16 Scheme of a MR damper with magnetic circuit containing electromagnet and switchable hard magnet (left) and scheme of damper piston with magnetic flux lines (right).

consumption.[22] Electric energy supply is required solely for the switching of the magnetization of the hard magnet from an initial to a final state. As the most suitable material for the hard magnet, an AlNiCo alloy was selected, because it combines the advantages of high saturation magnetization with low coercive field strength. A high saturation magnetization is required for the generation of high magnetic flux densities by the hard magnet and a low coercive strength limits the energetic effort in the electromagnet to modify the magnetization of the hard magnet.

The possibility to switch the magnetization of the hard magnet requires another design of the magnetic circuit system in comparison to the first damper concept. Here, the hard magnet is located in the same magnetic circuit as the coil. Figure 10.16 depicts the scheme of the MR damper (left) and that of the piston with the magnetic circuit containing the switchable hard magnet in the center and the coil around it (right). Additionally, in Figure 10.16, right, the magnetic flux lines generated by the two magnetic field sources are depicted.

Figure 10.17 shows the simulated magnetic flux density along the line of the MR fluid gap, in analogy to Figure 10.9 for the first MR damper type. Due to the more compact design of the damper piston with only one single magnetic circuit, its length without the piston rod is only 40 mm. Also shown in Figure 10.17 is the magnetic flux density distribution received from the FEM simulation for the operational state in which the maximum magnetization of the AlNiCo hard magnet takes place. The highest achievable magnetic flux density in the active MR fluid gap zones was calculated to be *ca.* 600 mT (Figure 10.17). However, the flux density can be continuously diminished down to zero by a stepwise decrease of the magnetization or even a complete demagnetization of the AlNiCo magnet by the electromagnet.

Figure 10.18 exhibits the results of measurements of the magnetic flux density in the magnetic circuit at different magnetization states of the AlNiCo magnet.

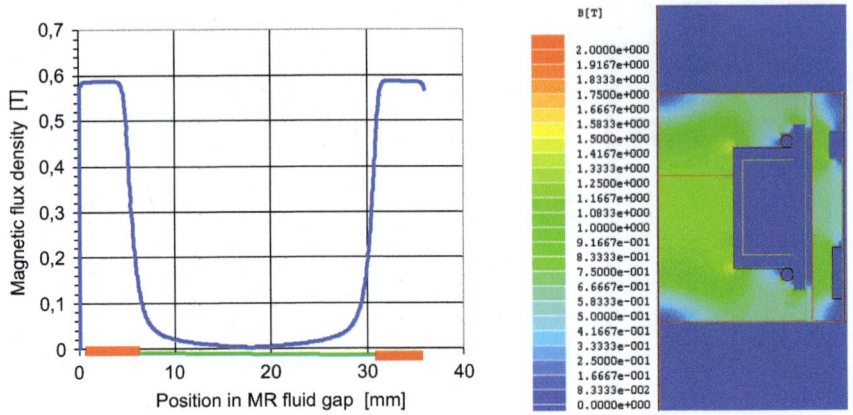

Figure 10.17   Simulated magnetic flux density along a line through the gap in the state of full magnetization of the switchable hard magnet (left) and magnetic flux density distribution (right).

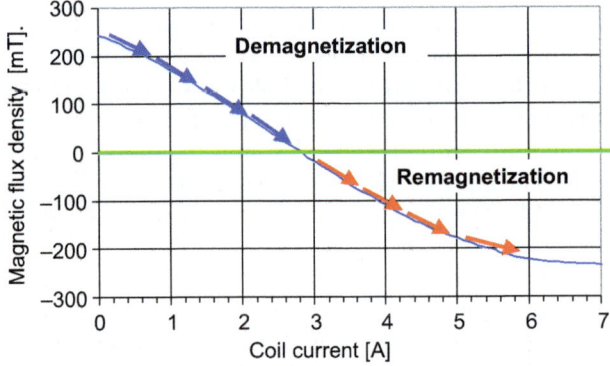

Figure 10.18   Magnetic flux density due to demagnetization and remagnetization of the AlNiCo magnet in the piston with air, measured in the MR valve.

On the left side the curve starts at a magnetic flux density of 250 mT. With rising coil current, the magnetic flux density decreases according to the demagnetization of the AlNiCo magnet. At a coil current of 2.8 A the AlNiCo magnet is completely demagnetized and with further increasing coil current a remagnetization in the reverse direction takes place.

As for the first type of MR damper, a demonstration model for this MR damper with switchable hard magnet and electromagnet was also constructed and tested. Just like the former MR damper with permanent magnets, this damper has also an air cushion in order to balance the volume variation due to the piston rod. Figure 10.19 shows the damper piston (left) as well as the complete damper (right).

The experimental investigations on this damper were performed with the same mechanical test machine as before. In these measurements, the

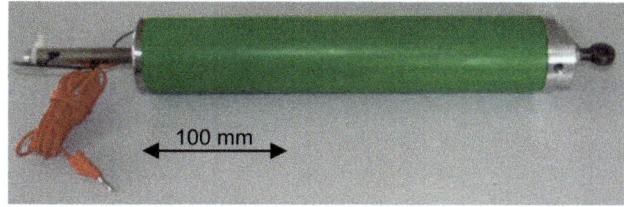

Figure 10.19 Demonstrator model of the MR damper with electromagnet and switchable hard magnet: piston (left) and complete damper (right).

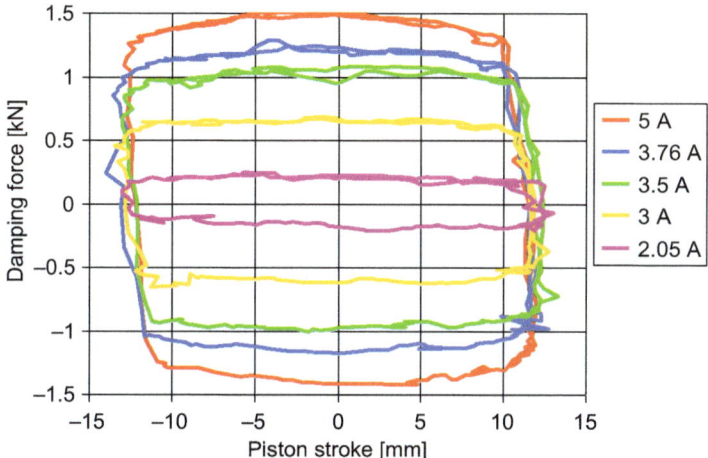

Figure 10.20 Experimental results of force measurements on the MR damper with electromagnet and switchable hard magnet. The MR fluid contains 25 vol% iron particles.

magnetization of the AlNiCo magnet was varied by short pulses of the coil current up to 5 A. The required energy for the change of the magnetization and the corresponding alteration of the damping force was estimated to be less than 1 J, demonstrating the energy-efficient operation of this novel type of MR damper. The results of the measurements are depicted in Figure 10.20. The investigations were carried out with a MR fluid containing 25 vol% iron particles. As expected, upon increasing the coil current, the magnetization of the AlNiCo magnet and the corresponding damping force are enhanced. These damping forces can even be increased with the use of a MR fluid with higher concentration of iron particles.

The temporal course of the damping force in reaction to sudden changes of the magnetization of the AlNiCo magnet is depicted in Figure 10.21. In the left diagram, a stepwise increase of the damping force followed by a corresponding decrease at a vibration frequency of 1 Hz is shown. The right diagram exhibits the reverse changes at a vibration frequency of 0.5 Hz. For the modification of

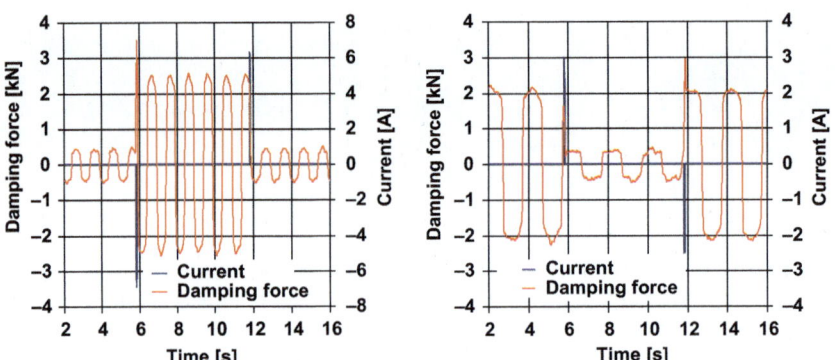

Figure 10.21   Temporal course of the damping force and the coil current of the MR damper with electromagnet and switchable hard magnet: excitation frequency of 1 Hz (left) and of 0.5 Hz (right).

the damping force, only a short pulse of the coil current is necessary, emphasizing the low energy demand of this MR damper.

Beside the described hybrid magnetic circuits with combinations of permanent magnets and electromagnets on one side and switchable hard magnets and electromagnets on the other side, it is even possible to combine all three magnetic sources in a magnetic circuit.[23] This third concept bears the advantage that through a positive superposition of the magnetic fields from the permanent and switchable hard magnets, a higher magnetic flux density in the active MR fluid gap zones can be achieved in the powerless state. However, the required effort for such magnetic circuits with three magnetic sources is higher than for the hybrid circuits discussed before.

10.4.2   Rotational MR Devices

In rotational magnetorheological devices the MR fluid is usually used for the controlled transmission of torques from an input to an output shaft. On the one hand, this mechanism is applied for clutches which have to transmit only a limited torque. On the other hand, it can also be exploited for brakes which control a rotational motion by means of a defined braking torque. However, such rotational devices can additionally be used for damping purposes to eliminate rotational vibrations. Despite this versatility, the notation clutch will be used for all such rotational devices in the following.

Rotational MR devices are usually operated in the shear mode, where the MR fluid in a thin layer mechanically connects or disconnects the input and output shaft of the clutch. Two main designs of MR clutches are known, and are designated disk-type and bell-type.[13] In the disk-type clutch, the shear gaps filled with the MR fluid are oriented perpendicular to the rotational axis of the clutch (Figure 10.22, left). In contrast to this configuration, in the bell-type MR clutch the MR fluid gaps are oriented parallel to the rotational axis (Figure 10.22, right). Figure 10.22 also shows the magnetic field lines generated

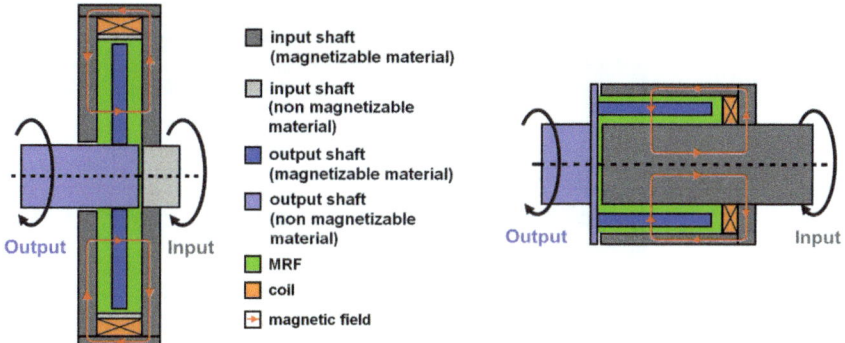

Figure 10.22 Disk-type MR clutch (left) and bell-type MR clutch (right), each with input shaft and output shaft, the magnetic circuit made from magnetisable material and the magnetic field lines penetrating the MR fluid.

by the coil which penetrate the MR fluid in the shear gaps and attach the output to the input shaft *via* the stiffening of the MR fluid.

Disk-type and bell-type clutches have different advantages and disadvantages. The disk-type clutch can be made more compact with a large number of parallel disks and corresponding shear gaps, which leads to a large transmittable torque. However, at high rotational speeds without applied magnetic field, the torque is more prone to particle separation due to the centrifugal forces acting on the heavy iron particles in the MR fluid. In contrast, in bell-type MR clutches, the particles can move only over the small distance of the thin MR fluid gap width.

In the following, three MR clutches, each equipped with a hybrid magnetic circuit containing a permanent magnet and an electromagnet, are introduced. All three clutches were designed in the bell-type configuration. Their designs differ especially in their number of mechanical parts which can rotate with respect to each other. In the most simple design, the clutch consists only of the input and the output part. In this case, the electromagnet is not stationary in the clutch operation with rotating input and output shaft. Hence, the coil has to be electrically connected by sliding contacts. Other differences between the three presented MR clutches concern their size and weight as well as their maximum transmittable torque.

10.4.2.1 MR Clutch with Simple Design

The first MR clutch with permanent magnet and electromagnet consists of two parts which are mechanically connected to the input shaft and the output shaft, respectively. Figure 10.23 exhibits the schematic design as well as the outer dimensions of the clutch. The coil is connected to the input shaft, whereas the other magnetic source, the permanent magnet, is attached to the output shaft. The torque is transmitted by the MR fluid in the two concentric gaps which are oriented parallel to the rotational axis according to the selected bell-type

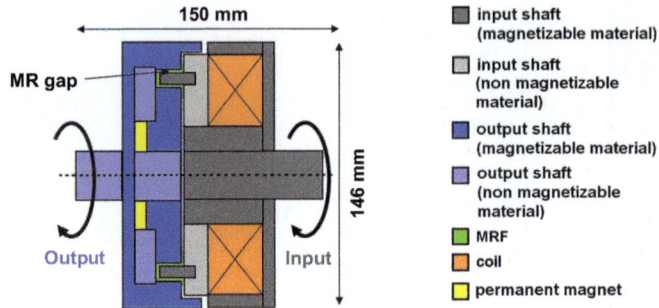

Figure 10.23   Scheme of the two-part MR clutch with permanent magnet and electromagnet.

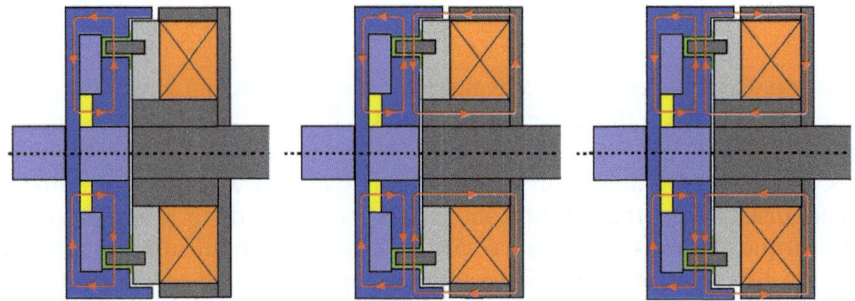

Figure 10.24   Magnetic field lines in the two-part MR clutch in three different operational modes: Magnetic field generated only by the permanent magnet (left), negative superposition or weakening of the fields (middle) and positive superposition or strengthening of the fields (right).

design. In order to distinguish between the two parts of the clutch in Figure 10.23, the input part is shown in gray and the output part in blue. The magnetic circuit which guides the magnetic flux through the MR fluid gaps is indicated by darker gray and blue colors.

It becomes apparent from Figure 10.23 that the two magnetic sources generate magnetic fields in two different magnetic sub-circuits. However, the magnetic fields superpose in the MR fluid gaps and can weaken and strengthen each other, like in case of the MR damper with permanent magnet and electromagnet in Section 10.4.1.1. Without any current in the coil, only the permanent magnet generates a magnetic field in the MR fluid.[24]

This operational mode is schematically shown in Figure 10.24, left. Here, the MR fluid in the gaps stiffens without any supply of electric energy and a corresponding torque can be transmitted by the clutch. Figure 10.24, middle, visualizes the operational state, in which a current applied to the coil causes a magnetic field which counteracts the field of the permanent magnet. As a consequence, the two fields weaken each other and the MR fluid re-liquefies. This means that the engaged and disengaged states of the clutch are reversed with respect to the known MR clutches with an electromagnet only. This is a big advantage in applications, in which the clutch normally works in the engaged state only interrupted by short disengagements. For such kind of operation, only very little electric energy has to be supplied to the clutch.

Finally, in the third operational mode in Figure 10.24, right, the magnetic field of the electromagnet strengthens that of the permanent magnet. Such an enhanced magnetic field causes an even higher yield stress in the MR fluid than in the operational mode with the permanent magnet only. This should lead to a higher torque transmission of the clutch.

In order to quantify the magnetic field, FEM simulations of the magnetic flux density distribution in the MR fluid gap were performed, in analogy to the MR damper. The results of these simulations are shown in Figure 10.25, where the position in the MR fluid gap is explained in the right sketch. In the operational mode with the permanent magnet only, the magnetic flux density in the MR fluid gap amounts to 450 mT. Decreasing the magnetic field strength with a counter-field caused by a coil current of −7.5 A results in a magnetic flux density of less than 50 mT. When the polarity of the coil current is reversed to +7.5 A, the two magnetic fields strengthen each other and the total magnetic flux density is enhanced up to 830 mT. This result demonstrates that the magnetic flux density can be varied in a very broad range and a medium flux density is generated by the permanent magnet without electric energy supply.

The described MR clutch with the hybrid magnetic circuit containing a permanent magnet and an electromagnet was designed in more detail, manufactured and tested. Figure 10.26 exhibits the design model of the MR clutch and the assembled demonstrator. In both pictures, the electromagnet is integrated in the bottom part and the permanent magnet is located in the top part of the clutch. The mass of this MR clutch is 11 kg and it can be operated at rotational speeds up to 2000 rpm.

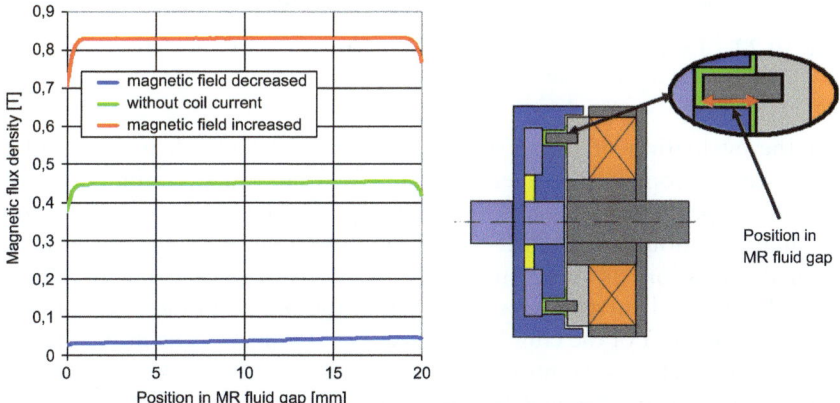

Figure 10.25 Magnetic flux density in the MR fluid gap for the three operational modes corresponding to coil currents of −7.5 A, 0 A and +7.5 A, received from FEM simulations. The position in the MR fluid gap refers to a line along the gap shown in the right figure.

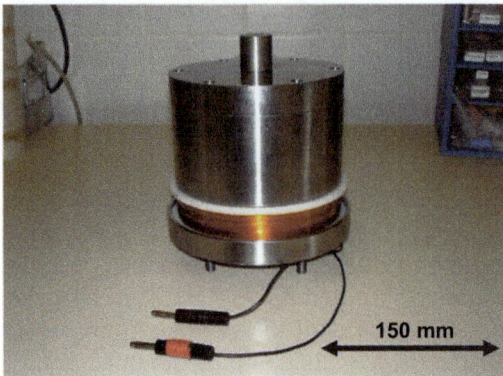

Figure 10.26 Design model of the two-part MR clutch (left) and demonstrator model (right).

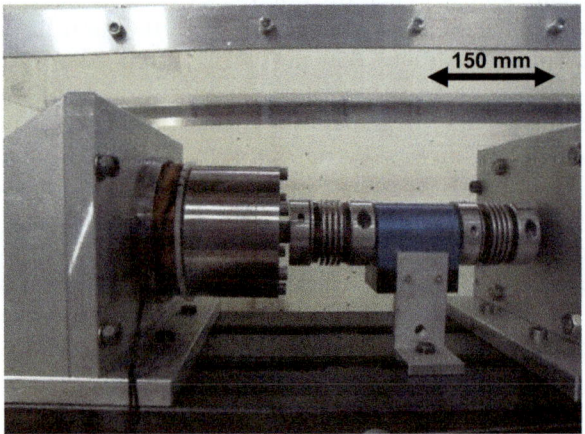

Figure 10.27 Experimental test setup for torque measurements with the two-part MR clutch integrated.

For the evaluation of the performance of the MR clutch in terms of the torque transmission, it was filled with a MR fluid which contains 35 vol% carbonyl iron particles in silicone oil and was developed in the same laboratory. The clutch was mounted in a self-constructed mechanical testing machine with rotational excitation, as shown in Figure 10.27. In these experiments, the torque at different coil currents was measured in the braking mode, which means that the output shaft of the clutch was fixed like in a brake. The input shaft was driven by an electric servomotor at low speed and the braking torque was measured with a torque sensor between the motor and the clutch.

Figure 10.28 shows the results of the torque measurements in comparison with calculated data. For the calculations, a simple model for the resulting shear force was used. The model includes the shear stress of the MR fluid depending on the magnetic flux density in the MR fluid gap. The measurements

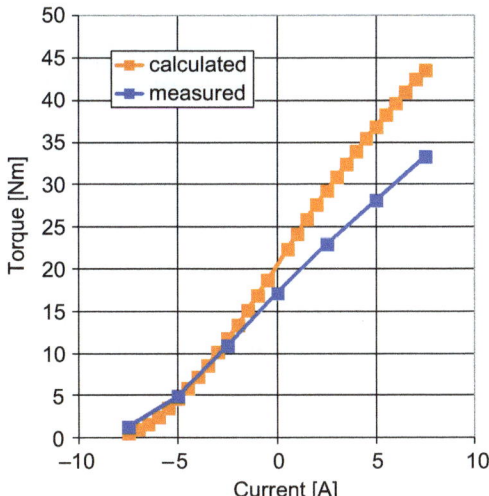

Figure 10.28 Measured and calculated torque of the two-part MR clutch in dependence on the coil current.

and the calculations were conducted at various coil currents between −7.5 A and +7.5 A. The comparison in Figure 10.28 shows that the experimental torques are lower than the calculated data. However, the general dependence of the torque on the coil current is in rough agreement with the predicted behavior. Without any coil current (0 A), a medium torque of 17 Nm can be transmitted by the clutch without electric energy supply. When a counter-field is generated in the MR fluid gap by a coil current of −7.5 A, the transmitted torque is diminished to 1.5 Nm. With a reversed coil current of +7.5 A, an enhanced transmitted torque of 34 Nm could be achieved.

The results demonstrate that the broad variability of the magnetic flux density in the MR fluid gap by the variation of the coil current corresponds to a wide range of torque transmission. The accessible range of torque transmission between *ca.* 1.5 and 34 Nm offers a high flexibility of operation of the clutch. Without any coil current and electric energy supply, a medium torque is transmitted by the clutch. As mentioned before, the operational states of engagement and disengagement can be reversed with this configuration.

10.4.2.2 MR Clutch with Versatile Design

The second MR clutch with permanent magnet and electromagnet consists in total of four parts which can rotate with respect to each other.[25] Figure 10.29 exhibits the schematic design of this more versatile clutch. The input shaft (blue in Figure 10.29) is coupled to the part with the permanent magnet. The torque is transmitted *via* the MR fluid in the axially oriented concentric gaps to the output shaft (red in Figure 10.29). In contrast to the more simple two-part MR

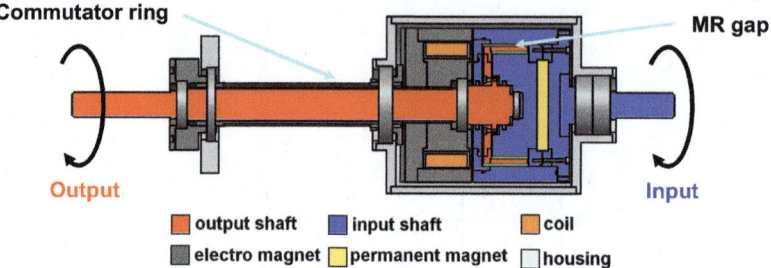

Figure 10.29 Scheme of the MR clutch with stationary electromagnet consisting of four separate parts: Input shaft, output shaft, coil part and housing.

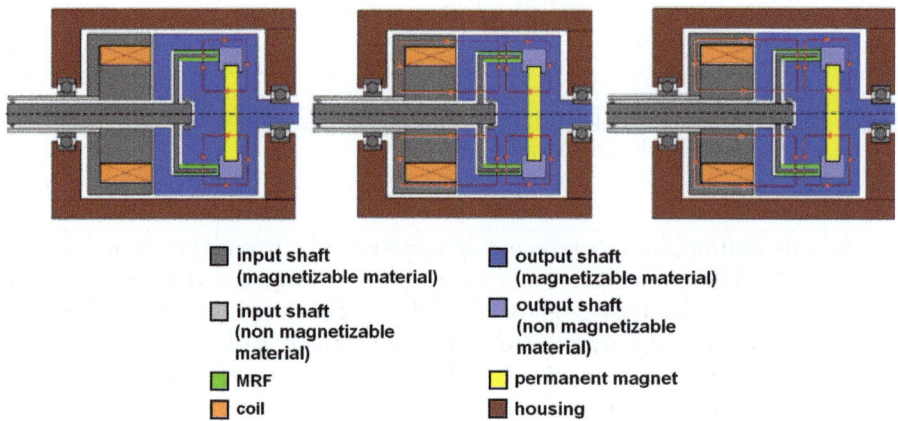

Figure 10.30 Magnetic field lines in the MR clutch with stationary electromagnet in three different operational modes: Magnetic field generated only by the permanent magnet (left), negative superposition or weakening of the fields (middle) and positive superposition or strengthening of the fields (right).

clutch explained before, the housing of this clutch is stationary and allows a tight attachment to the environment of the clutch. Finally, the fourth part of the clutch containing the coil is apart from the other three parts. It can be mechanically connected to the housing, which avoids the necessity of sliding electrical contacts. However, the relative motion of the coil with respect to the magnetic circuit in the output part of the clutch could eventually cause eddy currents. In order to eliminate such disturbing eddy currents, the coil part can alternatively be coupled to the output shaft. For this purpose, a commutator ring is attached to the output shaft (Figure 10.29).

Figure 10.30 exhibits the three working modes of the MR clutch corresponding to those of the two-part clutch. Again, without any coil current, a magnetic field is generated by the permanent magnet, which causes a torque

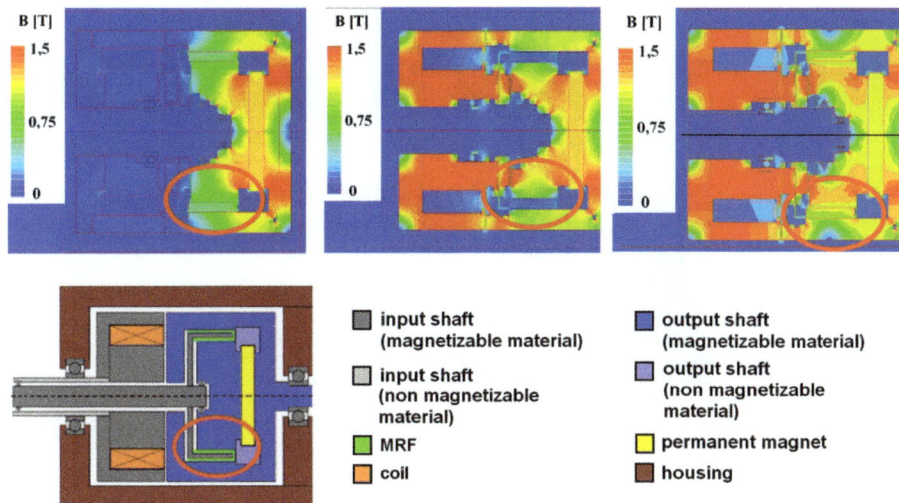

Figure 10.31 Top: Maps of the magnetic flux density in the MR clutch with stationary electromagnet in three different operational modes : magnetic field generated only by the permanent magnet (left), negative superposition or weakening of the fields (middle) and positive superposition or strengthening of the fields (right). Bottom: Corresponding scheme of the MR clutch. The red ellipses indicate the region of the MR fluid gaps.

transmission without energy supply. The magnetic field of the permanent magnet in the MR fluid gaps can be decreased or increased by the electromagnet, depending on the polarity of the coil current.

In Figure 10.31, maps of the calculated magnetic flux density in the MR fluid clutch received from FEM simulations are exhibited. The red ellipses indicate the region of the MR fluid gaps. When only the permanent magnet generates the magnetic field in the gaps, a medium flux density is achieved (Figure 10.31, left). With the counter-field generated by the electromagnet, the magnetic flux density in the MR fluid gaps is nearly canceled (Figure 10.30, middle). Finally, a positive superposition of both magnetic sources leads to an enhanced magnetic flux density in the MR fluid gaps (Figure 10.31, right).

According to the simulation results, a MR clutch bearing a stationary electromagnet was designed, constructed and tested. Figure 10.32 depicts the demonstrator model of the clutch. For the investigations on the same mechanical testing machine as before, a MR fluid with 35 vol% iron particles in silicone oil was prepared and filled into the MR clutch. The measurements were conducted similarly to those on the two-part clutch. The results in terms of the transmitted torque in dependence on the coil current are depicted in Figure 10.33. Without coil current, a medium torque of 30 Nm is transmitted. Depending on the polarity of the coil current, the torque can be decreased or increased up to 55 Nm.

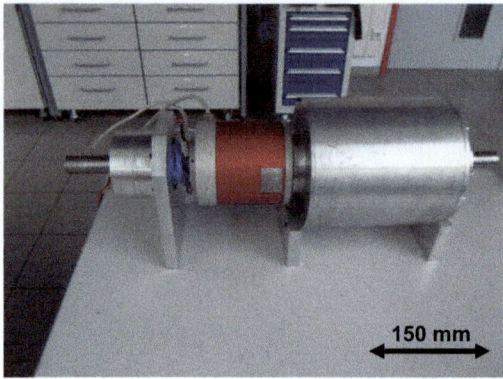

Figure 10.32   Demonstrator model of the MR clutch with stationary electromagnet.

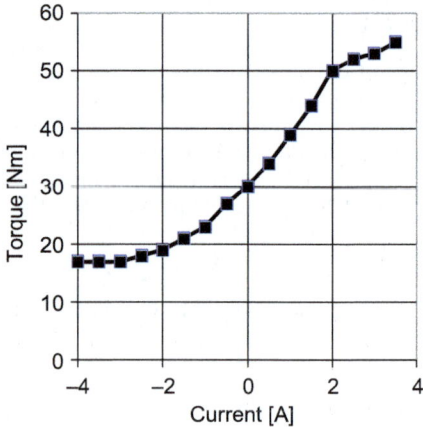

Figure 10.33   Measured torque of the MR clutch with stationary electromagnet depending on the coil current.

10.4.2.3   *MR Clutch with High Torque Transmission*

The third MR clutch with permanent magnet and electromagnet was designed to transmit very high torques. The motivation for this clutch has been created by the development of a hybrid vehicle for public transport, whose combustion engine should be connected by this MR clutch to the electric generator in the vehicle and disconnected on demand. Figure 10.34 exhibits the schematic design as well as the outer dimensions of the high torque MR clutch. The permanent magnet with a diameter of about 200 mm was assembled from a large number of small NdFeB magnets in order to avoid the effort of manufacturing a permanent magnet of this size in one piece.

The high-torque clutch was designed in more detail, manufactured and assembled. In order to achieve exceptionally high torques, a MR fluid with 50 vol% iron particles in silicone oil, which reaches a shear stress of 70 kPa at

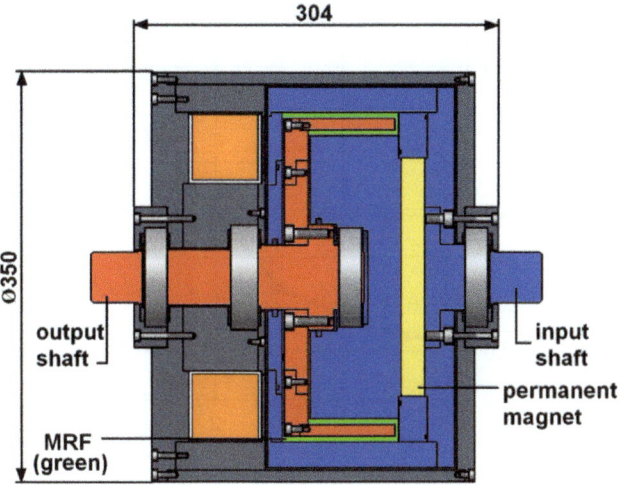

Figure 10.34 Scheme of the MR clutch with high torque transmission and outer dimensions in mm.

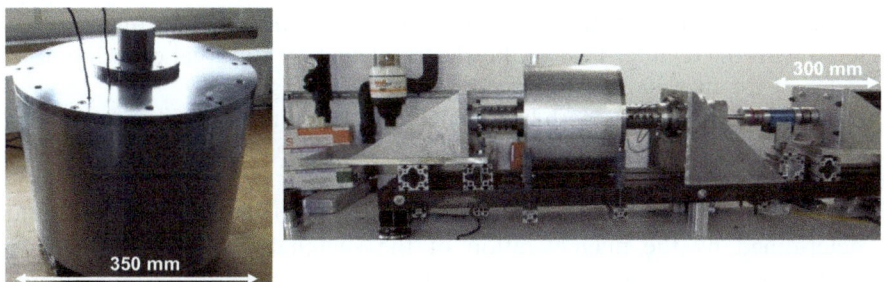

Figure 10.35 Demonstrator model of the MR clutch with high torque transmission (left) and integrated in the experimental test setup for torque measurements (right).

700 mT magnetic flux density, was selected for the clutch and manufactured. The demonstrator model of the clutch is represented in Figure 10.35. It was integrated in the mechanical testing machine for torque measurements and its performance was evaluated.

The results of the torque measurements upon variation of the coil current are shown in Figure 10.36. Starting from the operational mode without electric energy supply, the torque of 180 Nm can be enhanced by the positive superposition of the magnetic fields of permanent magnet and electromagnet up to nearly 800 Nm.

This high torque would be sufficient for the application in the hybrid vehicle, but it would be even better to accomplish this torque with the use of the permanent magnet only, *i.e.* without the supply of electric energy. Further work on this high-torque MR clutch shall be focussed on the design of a magnetic circuit,

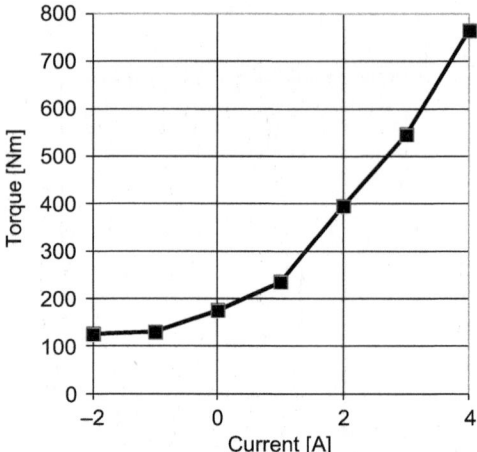

Figure 10.36 Measured torque of the MR clutch with high torque transmission depending on the coil current.

which should generate a stronger magnetic field by the permanent magnet only. Another future goal is the decrease of the minimum transmitted torque which is achieved with the counter-field of the electromagnet by a mutual cancelation of the fields of the two magnetic sources in the MR fluid gaps.

Besides the MR clutches with hybrid magnetic circuits containing permanent magnets and electromagnets, the combination of a switchable hard magnet with an electromagnet is possible, too, in analogy to the corresponding MR damper. In such a clutch the transmittable torque in the powerless state would be determined by the magnetization of the switchable hard magnet. This concept could be interesting for applications, in which the transmittable torque has to be limited.

10.5   Conclusion

In this chapter, novel concepts for magnetic circuits in MR devices with translational and rotational motion were introduced. Hybrid magnetic circuits containing permanent or switchable hard magnets in addition to the electromagnet offer the possibility to pre-set a magnetic flux density in the MR fluid gap without any supply of electric energy. The additional electromagnet is able to cancel or to enhance the flux density, depending on the current in the coil. The pre-set flux density corresponds to a basic damping force in a MR damper or to a basic torque in a MR clutch, which can be decreased or increased by the coil current.

These concepts yield various advantages for different MR devices. The failsafe behaviour of MR dampers can be improved, because a basic damping force is assured without the supply of electric energy by the magnetic field of the permanent magnet. With the integration of a switchable hard magnet, a tunable basic damping force is feasible by only a short current pulse in the coil to magnetize the hard magnet, which improves the energy-efficiency of the device. Finally, the operational states of engagement and disengagement in a MR clutch can be reversed with the combination of a permanent magnet and an electromagnet in the magnetic circuit. Also, this concept saves electric energy, if the clutch usually operates in the engaged state.

It is important to consider MR devices as comprehensive mechatronic systems, where the overall performance of the system depends on the mutual adaptation of its components or sub-systems. In order to receive an optimal performance of the MR device, the MR fluid properties have to be fitted in accordance with the system demands. Together with an optimized magnetic circuit design, such MR devices offer far-reaching perspectives for a multitude of applications in the automotive sector and in mechanical engineering.

Acknowledgements

Financial support for this work from the European Community, from the German Ministry for Education and Research as well as from the Bavarian State Ministry for Economy, Infrastructure, Traffic and Technology is gratefully acknowledged.

References

1. *Proceedings of the 12th International Conference on Electro-Rheological Fluids and Magneto-Rheological Suspensions 2010*, ed. R. Tao, World Scientific, Singapore, 2011.
2. J. D. Carlson, F. Goncalves, Controllable Fluids Come of Age, *Proceedings of Actuator 2008 – 11th International Conference on New Actuators*, 2008, pp. 477–480.
3. A. L. Smith, J. C. Ulicny, L. C. Kennedy, Magnetorheological fluid fan drive for trucks, *Proceedings of the 10th International Conference on Electrorheological Fluids and Magnetorheological Suspensions*, World Scientific, Singapore, 2007, pp. 479–486.
4. S. J. Dyke, B. F. Spencer, Jr., M. K. Sain and J. D. Carlson, An Experimental Study of MR Dampers for Seismic Protection, *Smart Mater. Struct.*, 1998, **7**, 693–703.
5. U. Lange, S. Vassileva and L. Zipser, Controllable magnetorheological dampers for shock and vibration, *Proceedings of Actuator 2002 – 8th International Conference on New Actuators*, 2002, 339–342.
6. M. Mao, W. Hu, Y.-T. Choi, N. M. Wereley, A magnetorheological damper with bifold valves for shock and vibration mitigations, *Proceedings of the 10th International Conference on Electrorheological Fluids and Magnetorheological Suspensions*, World Scientific, Singapore, 2007, 563–569.
7. H. S. Lee and S.-B. Choi, Control and response characteristics of a magneto-rheological fluid damper for passenger vehicles, *J. Intell. Mater. Syst. Struct.*, 2000, **11**, 80–87.
8. Q. H. Nguyen and S.-B. Choi, Optimal design of a vehicle magnetorheological damper considering the damping force and dynamic range, *Smart Mater. Struct.*, 2009, **18**, 015013.

9. Y.-J. Nam and M.-K. Park, Electromagnetic Design of a Magnetorheological Damper, *J. Intell. Mater. Syst. Struct.*, 2009, **20**, 181–191.

10. B. Yang, J. Luo and L. Dong, Magnetic circuit FEM analysis and optimum design for MR damper, *Int. J. Appl. Electromagn.Mech.*, 2010, **33**, 207–216.

11. J. Luo, B. Yang and L. Dong, Magnetic circuit design for magnetorheological damper, *Int. J. Appl. Electromagn. Mech.*, 2010, **33**, 815–822.

12. J. Rabinow, The magnetic fluid clutch, *AIEE Trans.*, 1948, **67**, 1308–1315.

13. D. Lampe, A. Thess and C. Dotzauer, MRF-clutch-design considerations and performance, *Proc. of Actuator 1998 – 6th International Conference on New Actuators*, 1998, 449–453.

14. D. Lampe and R. Grundmann, Transitional and sold state behaviour of a magnetorheological clutch, *Proceedings of Actuator 2000 – 7th International Conference on New Actuators*, 2000.

15. C. Kieburg, G. Oetter, R. Lochtman, C. Gabriel, H. M. Laun, P. Pfister, G. Schober and H. Stein-wender: High performance magnetorheological fluids tailored for a 700 Nm automotive 4-wheel-drive clutch, *Proceedings of the 10th International. Confereence on Electrorheological Fluids and Magnetorheological Suspensions*, World Scientific, Singapore, 2007, 101–107.

16. F. Gratzer, H. Steinwender and A. Kusej, Magnetorheological AWD clutches, *ATZ*, 2008, **110**, 26–31.

17. T. Kikuchi, K. Ikeda, K. Otsuki, T. Kakehashi and J. Furusho, Compact MR fluid clutch device for human-friendly actuator, *J. Phys.: Conf. Series*, 2009, **149**, 012059.

18. J. Ehrlich and H. Böse: Novel magnetorheological damper with outstanding fail-safe characteristics, *Proceedings of Actuator 2008 – 11th International Conference on New Actuators*, 2008, 495–498.

19. H. Böse and J. Ehrlich, Novel Adaptive Damping Systems Based on Magnetorheological Fluids, *Adv. Sci. Technol.*, 2012, **77**, 86–95.

20. H. Böse and J. Ehrlich, High-Performance Magnetorheological Damper with Hybrid Magnetic Circuit, *Proceedings of Actuator 2012 – 13th International Conference on New Actuators*, 2012, 109–112.

21. H. Böse and J. Ehrlich, Performance of magnetorheological fluids in a novel damper with excellent fail-safe behavior, *J. Intell. Mater. Syst. Struct.*, 2010, **21**, 1537–1542.

22. J. Ehrlich, J. Vrbata and H. Böse, Novel magnetorheological damper with improved energy efficiency, *Proceedings of Actuator 2010 – 12th International Conference on New Actuators*, 2010, 545–548.

23. H. Böse and J. Ehrlich, Magnetorheological dampers with various designs of hybrid magnetic circuits, *J. Intell. Mater. Syst. Struct.*, 2011, **23**, 979–987.

24. H. Böse, J. Ehrlich, T. Gerlach, C. Deak, C. Janssen and A.-M. Trendler, Magnetorheological clutch with reversed torque transmission behavior, *Proceedings of Actuator 2008 – 11th International Conference on New Actuators*, 2008, 847–851.

25. H. Böse, T. Gerlach and J. Ehrlich, Magnetorheological Torque Transmission Devices with Permanent Magnets, *Journal of Physics: Conference Series*, 2013, **412**, 012050.

CHAPTER 11

Magnetorheological Fluid-Based High Precision Finishing Technology

W. I. KORDONSKI

QED Technologies International, 1040 University Avenue, Rochester NY 14607, USA
Email: kordonski@qedmrf.com

11.1   Introduction

Projection lenses for advanced lithography used in manufacturing of integrated circuits with nanometre features as well as optics for lasers, airborne surveillance, weapon systems, medical devices, digital photography and mirrors for space telescopes are examples of modern optical applications that rely on leading-edge production technologies, especially those delivering high precision aspherical and free form surfaces. The most challenging step in the fabrication of such complex surfaces is polishing, particularly, so-called sub-aperture polishing based on zonal material removal. This process requires precision control of position and velocity of the polishing zone. Currently, it is provided by sophisticated contour-controlled precision CNC machines that execute finishing algorithms according to the prescription. Full advantage of the deterministic nature of CNC machining can only be taken if a sub-aperture polishing tool instantly adapts (conforms) to the local surface and its removal function is well characterized and stable. Commonly used mechanical tools with air pressure or an elastic cushion behind the polishing pad do not provide the required level of adaptability and stability.[1,2]

RSC Smart Materials No. 6
Magnetorheology: Advances and Applications
Edited by Norman Wereley
© The Royal Society of Chemistry 2014
Published by the Royal Society of Chemistry, www.rsc.org

Liquid substances by their nature can easily conform to any surface and attempts were made to utilize this unique property in controlled material removal including polishing.[3–5] In doing so, one or other type of flow of a low viscous fluid, commonly water, supplies energy to abrasive particles to cause surface zonal erosion and material removal. Depending on process parameters such as fluid velocity and particle size, the regime of material removal can extend from cutting to gentle polishing. For example, previous work has shown that water jets can be used to polish materials such as glass, diamond, ceramics, stainless steel and alloys.[3] The surface quality strongly depends on the size and impact angle of the abrasive grains. Surface roughness of $R_a \sim 130$ nm on glass has been achieved after processing. An appropriate adjustment of process parameters such as jet velocity, abrasive size and concentration makes reduction of surface roughness on glass to $R_a \sim 1$ nm possible.[4] Another example is so called "Elastic Emission Machining" where a hydrodynamic principle is used to provide high precision polishing.[5] In this technique, a loaded elastic polyurethane ball polishes the workpiece as it scans over the part surface. The ball is rotated rapidly in a polishing fluid and, due to hydrodynamic forces, floats above the workpiece surface. The floating gap, which is created by an elasto-hydrodynamic lubrication state, is much larger than the diameter of the abrasive particles but is still very small. The process delivers atomically smooth surfaces. The mechanism proposed for this process is an elastic bombardment of the surface by the polishing particles.

The use of a magnetically sensitive liquid media, particularly magnetorheological (MR) fluids, as a tool for precision material removal is the basis for magnetorheological finishing technology where the surface zonal erosion results from the shear flow of MR fluid containing abrasive particles. The technology is currently known under the trade mark MRF®.[6,7] In the following, this abbreviation will be used when different aspects of magnetorheological finishing are discussed. In spite of the fact that MRF is widely utilized in the optical industry, scientific fundamentals of this magnetically assisted finishing technology are scarcely covered.[8] Some attempts were made to build empirical models that correlate the removal rate in MRF with glass properties, experimentally measured surface pressure and drag force,[9] or a combination of the above.[10–14] The objective of this chapter is to discuss different scientific aspects of MRF technology and, in particular, consider a new concept of material removal in MRF. The concept is based on principles of the mechanics of suspensions. It is shown that results of modeling are sufficiently well supported by experiments. Also, some specific polishing results are presented to demonstrate effectiveness of the technology.

11.2   Magnetorheological Finishing (MRF)

11.2.1   General Consideration

The key element of MRF is a MR polishing fluid. In general, a MR fluid is a liquid composition that undergoes a change in mechanical properties and converts into a plastic material in the presence of a magnetic field. Normally, MR fluids consist of ferromagnetic particles, typically greater than 0.1 micrometres in diameter, dispersed within a carrier fluid. In the presence of a magnetic field, the particles become magnetized and are thereby organized into chains within the fluid. The chains of particles form a spatial structure, which is responsible for the change in mechanical properties, such as the increase of the yield stress. In the absence of a magnetic field, the particles return to a disorganized or free state and the initial condition of the overall material is correspondingly restored. It is commonly accepted to model MR fluid as a Bingham plastic material with the yield stress controlled by a magnetic field.[6] The model suggests that material under deformation behaves as a solid body while stress is below the yield point, and flows like a Newtonian fluid when the stress is higher than the yield. MR polishing fluid comprises such main constituents as water, magnetic particles, an abrasive and chemical additive that retard particles' corrosion and agglomeration. The use of water is determined by its unique chemical properties, which are beneficial in polishing glasses and silicon substrates. The MRF technological process includes four main constituents: a MR polishing fluid management system, CNC platform, control software and a means for providing system input (metrology). The schematic shown in Figure 11.1 gives an idea of the structure of a MR polishing fluid management system, which is the basis of MRF technology. A centrifugal pump provides circulation and agitation of water based MR polishing fluid.

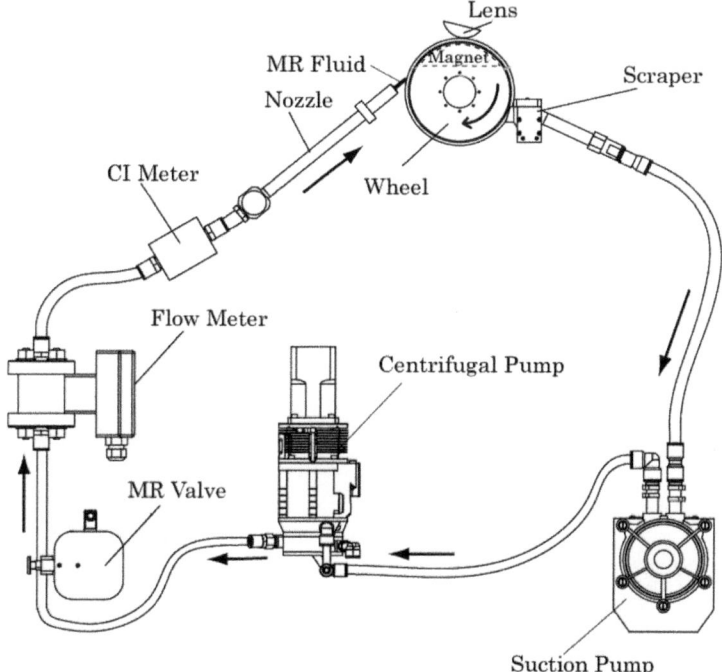

Figure 11.1　Schematic of MRF polishing fluid management system.

The fluid flow rate is measured with a flow meter that controls hydraulic resistance of a MR valve through feed-back loop thus maintaining required stable flow. The MR valve is a section of pipe snapped between pole pieces of an electromagnet thus forming a magnetically controlled hydraulic resistance when MR fluid flows through.

To compensate for water evaporation during system operation, the concentration of the magnetic component in the fluid is measured by an in-line inductive sensor (CI meter).[15] The CI meter provides an appropriate fluid moisture adjustment *via* a feed-back loop. A nozzle deposits MR fluid on the rim of a rotating wheel. The fluid is then collected from the wheel by a magnetically sealed scraper[16] and returned to the centrifugal pump with a peristaltic suction pump. The MRF polishing interface is schematically shown in Figure 11.2a. A convex lens is installed at some fixed distance from a moving wall, so that the lens surface and the wall form *a converging gap*. An electromagnet, placed below the moving wall, generates a non-uniform magnetic field in the vicinity of the gap. The magnetic field gradient is normal to the wall. The MR polishing fluid is delivered to the moving wall just above the electromagnet pole pieces to form a polishing ribbon. As the ribbon moves in the field, it acquires plastic Bingham properties. The top layer of the ribbon is saturated with abrasive due to levitation of non-magnetic abrasive particles in response to the magnetic field gradient.

Thereafter, the ribbon, which is pulled against the moving wall by the magnetic field gradient, is dragged through the gap resulting in material removal over the lens contact zone. This area is designated as the "polishing spot". Two images of the polishing zone are shown in Figure 11.2. The first (taken using high speed photography) shown in Figure 11.2b, visualizes the contact zone between a thin stationary meniscus lens and the moving rigid wall. An interoferogram of the lens surface after spot polishing under identical conditions is shown in Figure 11.2c. A comparison of Figures 11.2b and 11.2c shows that material removal does occur within the boundaries of the contact zone. The rate of material removal can be preset by the magnetic field strength, geometrical parameters of the interface, such as thickness of the gap, and

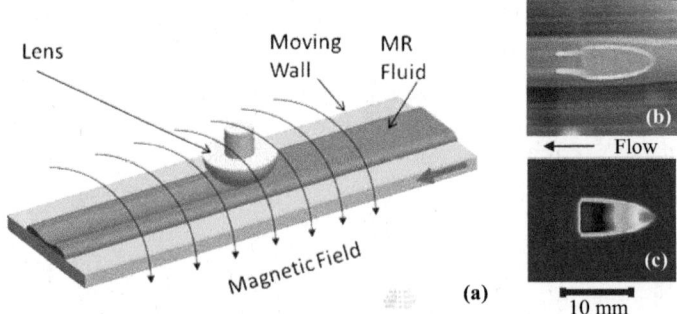

Figure 11.2   Schematic of MRF: (a) MRF interface; (b) an image of the contact zone; (c) polishing spot interferogram.

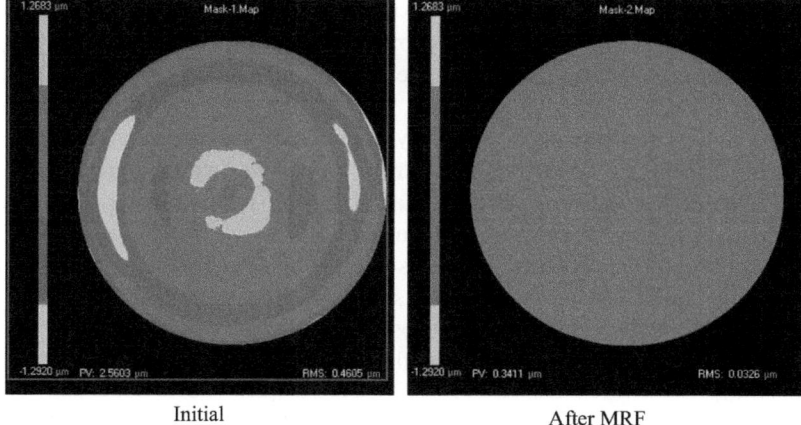

Figure 11.3   Aspherical surface finished with MRF.

moving wall velocity. Inputs to the polishing program are the material removal function or the polishing spot and the initial surface figure. The PC based software uses a series of complex algorithms and convolves the removal function with the error function to derive an operating program for the CNC platform. The CNC platform executes the velocity (dwell time) schedule and constantly maintains the position of the rotating workpiece through the polishing spot. Because of its conformability and subaperture nature, the MRF polishing tool may finish complex surface shapes like aspheres that have constantly changing local curvature. A fundamental advantage of MRF over competing technologies is that the polishing tool does not wear since the recirculating fluid is continuously monitored and maintained. Polishing debris and heat are continuously removed. The technique requires no dedicated tooling or special setup.

Results of finishing a 7″ parabolic surface are shown in Figure 11.3. This is an example of MRF exceptional performance on aspheres. Initial figure error of 2.6 μm P-V and 0.46 μm rms was reduced to 0.34 μm P-V and 0.03 μm rms in 4 iterations (15× convergence).

11.3   Modeling of Material Removal in MRF

11.3.1   Concept of an Elastic Pad

Abrasive particle load is a key problem in consideration of material removal with abrasive slurries in polishing. Polishing is most commonly carried out by pressing an elastic pad with embedded abrasive particles against the moving surface to be polished. According to Preston,[17] the removal rate in this case is proportional to the applied pressure and relative pad velocity. Removal rate also depends on the properties of the polishing interface such as the mechanical properties (like elasticity) of the polishing pad which transmits indentation load to the abrasive particle. Taking into account that MR fluid in a magnetic field stiffens and acquires essential elastic properties, it is not unreasonable to suggest that such magnetized material can be considered as a moving polishing pad similar to conventional polishing tools. Along this line, an assumption also should be made that stresses caused by such 'pad' deformation in the converging gap are lower than the yield stress across the whole 'pad' body. To evaluate credibility of this hypothesis, appropriate mechanical properties of typical MR polishing fluid were measured with an Anton Paar MCR 301 magneto-rheometer at a magnetic field strength and field-to-shear orientation corresponding to MRF.[18] Measurements were made for sample internal magnetic field strength of 150 kA m^{-1} and oscillation frequency of 1.592 Hz.

Results of measurements of the MR fluid storage modulus G' are shown in Figure 11.4a.

At low strains (<10%) magnetized MR fluid does exhibit essential elastic properties ($G' \sim 0.5$ MPa), that sharply diminish after some yield point. This can be associated with the yield strength of a structure formed by magnetic particles in the magnetic field. It means that at high strains (>10%) or in the developed shear flow, when shear stress is higher than the yield stress and the structure is destroyed, no essential elastic properties of MR fluid are expected. Young's modulus of ~ 1 MPa ($E \sim 2G'$) that was obtained at low strain for MR fluid is significantly lower than the Young's modulus of conventional pads (~ 50–100 MPa).[19] Even assuming that there is no shear flow in the MRF polishing interface, it is reasonable to suggest that particle load, which would be sufficient to support removal rates demonstrated by MRF (3 microns/min and higher), cannot be generated by deformation of a much softer analog of a conventional pad. Another possible source of abrasive particle load for surface indentation can be the MR fluid normal stress associated with the change of the fluid structure morphology. This change occurs due to squeezing of magnetic particles into chains as a result of strong dipole–dipole interaction.[20] Evaluation of this stress can be drawn from the measurements with an Anton Paar magneto-rheometer of the MR fluid 1st normal stress difference taken at the same conditions as above. Results of measurements are shown in Figure 11.4b.

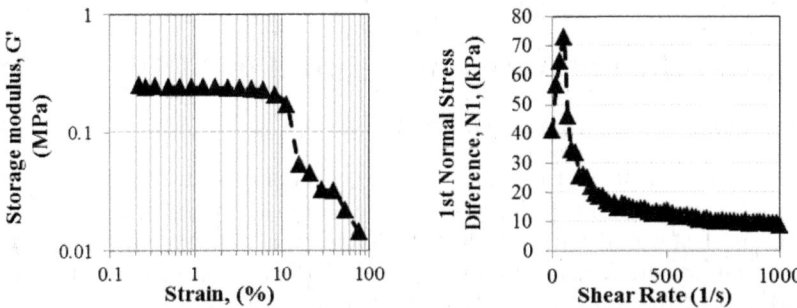

Figure 11.4 Mechanical properties of MR polishing fluid induced by magnetic field: (a) storage modulus; (b) normal stress.

The actual surface normal stress of "N1"/4 ~ 5kPa is much lower than the normal stress generated with conventional pads (80–140 kPa) at a typical pad pressure of 40–70kPa and asperities density of ~ 0.5.[19] As it follows, neither the MR pad elasticity, normal stress nor the hydrodynamic pressure generated by MR fluid viscous flow in the converging gap (as it was shown earlier[8,10]) can be considered as a load for abrasive particles.

11.3.2 Concept of the Shear Flow

Alternatively, the abrasive particle's load can be provided by a fluid flow, such as in conditions of the shear flow of a concentrated mixture of solid particles. At sufficiently high shear rates, such flow is characterized by particles intensive interaction and collision between each other and the surface. In the case of a binary (bimodal) mixture, and according to the principle of conservation of momentum, larger particles may supply considerable load for smaller particles. When such an event takes place near the surface, it may result in effective surface indentation by the smaller particle, especially if the particle possesses appropriate mechanical properties. As applied to polishing, this conceptual model suggests that larger or basic particles energized by shear flow, provide an indentation load for smaller abrasive particles to penetrate the surface and remove material. Such a mechanism of material removal is shown below to analyze MR fluid-based polishing processes. In doing so, the assumption is made that some form of shear flow of a highly concentrated suspension (~ 50 vol%) of relatively large magnetic particles (microns) and much smaller abrasive particles (tens of nanometres) occurs in the polishing interface.

As the starting point for the problem modeling and particle force evaluation, an assumption is made that the particles' dynamics in the considered case are similar to the general features of granular shear flow described elsewhere.[21–23] In general, granular flow encompasses the motion of discrete particles or grains. The particles are macroscopic (>1 micron) and there is no Brownian motion. When the concentration of particles is relatively low and particles supported by a carrier fluid do not collide, it is deemed that multiphase flow occurs. Such flow can be thought of as a disperse phase interacting only with a fluid phase. As concentration increases, interaction of particles takes effect in the form of instantaneous collisions resulting in particles oscillation and elevated dissipation of energy. In this case granular flow takes place.

The granular flow approach allows evaluation of the surface stress and particle load using constitutive relations accepted for the granular flow, namely, the dependence of the wall normal stress on the shear rate. It was found that as the solid concentration increases up to 0.7, keeping all other parameters constant, the stress is nearly proportional to the square of the shear rate, then goes down through a sharp transition and finally becomes independent of the shear rate at high solid concentration. For relatively moderate concentrations, the wall normal stress takes the form of

$$\tau_{22} = K\rho_p d_p^2 \dot{\gamma}^2 \qquad (11.1)$$

and consequently particle force takes the form of

$$G_p = K \frac{\pi}{4} \rho_p d_p^4 \dot{\gamma}^2 \qquad (11.2)$$

where ρ_p is the density of particle, d_p is the diameter of particle and $\dot{\gamma}$ is the shear rate. The dimensionless coefficient K takes into account other flow parameters such as concentration, mechanical properties of particles, carrier fluid damping properties, flow geometry, *etc.*

According to eqn (11.2), the problem of evaluation of the particle force is mainly reduced to determining the flow shear rate at the surface of interest. In the following analysis the shear rate is obtained by numerical modeling of the particular shear flow, taking into account rheological properties of the media. The shear rate is then used for calculating the force of the basic particle assuming that this force is a load for the abrasive particle.

The following analysis is restricted to the qualitative comparison of experimental removal rate profiles in the polishing spot with calculated profiles of surface loading by particles in the contact zone.

The effective shear rate was determined by modeling (using a commercially available computational fluid dynamics (CFD) package[24]) of the Bingham flow through a converging gap in the geometry similar to the one depicted in Figure 11.2. The model gap was formed on a cylinder rather than a spherical surface used in MRF in order to simplify the task and avoid some software limitations. Other parameters were the same as in experiments described below: surface radius of curvature of 75 mm; wall velocity of 3 m s^{-1}; gap thickness of 2 mm; plunging depth of 0.5 mm; fluid rheological properties. The three-dimensional solution was found using the free surface volume of fluid (VOF) method and Perzyna hypothesis for effective viscosity of Bingham plastic[25]

$$\mu = Min \begin{cases} A\mu_\infty \\ \mu_\infty + \frac{\tau_0}{\dot{\gamma}} \end{cases} \qquad (11.3)$$

Here A is an arbitrary, dimensionless multiplier supplied by the user (typically, $A = 10^3$–10^5), $\mu\infty$ is the viscosity in the limit of very large strain (the fully plastic limit), τ_0 is the yield stress (in the case under consideration depends on magnetic field strength and magnetic particles concentration) and $\dot{\gamma}$ is the shear rate. Rheological parameters required by eqn (11.3) were obtained with an Anton Paar Magneto Rheometer MSR 301 and are shown in Figure 11.5.

In addition, some considerations were given to the boundary conditions, the time step and mesh size so that an accurate and stable solution could be achieved in a reasonable amount of time. The evaluation of accuracy was based on the magnitude of the pressure at the lens apex, where it should be equal to zero in the case of a simple Newtonian fluid. An error of less than 1% was achieved.

The snapshot of the computer simulation given in Figure 11.6 shows both the map of velocity profiles (vectors) and the map of shear stress distribution in the center plane along the flow.

For the purpose of illustration, a threshold is set on the value of the shear stress higher than the yield stress, in this case equal to 4.7 kPa. As it would be

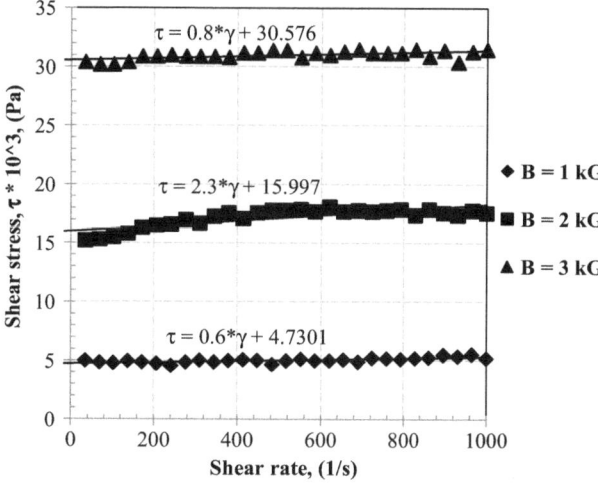

Figure 11.5   Rheological properties of MR fluid induced by magnetic field.

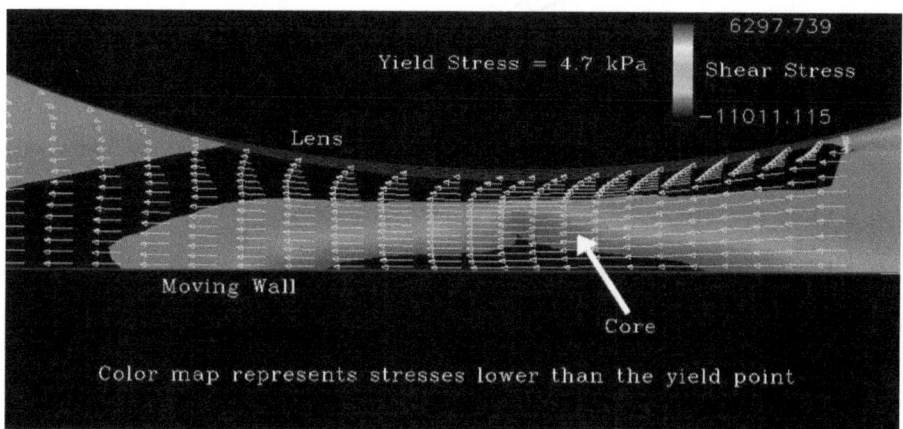

Figure 11.6   Result of the computer simulation of MR fluid flow in the converging gap.

expected, [8,26,27] modeling reveals the formation of a thin layer of sheared fluid sandwiched between the lens surface and a core of un-sheared material attached to the moving wall. The shear stress is lower than the yield stress in the core domain and exceeds the yield point of 4.7 kPa in the thin zone near the surface of the lens. This fact is also illustrated by the shear stress distribution and the velocity profile across the gap shown in Figure 11.7 for the MR fluid with the yield stress of 20kPa.

The shear stress is lower than the yield stress of 20 kPa in the core domain and exceeds the yield point in the thin zone near the surface. The velocity profile is essentially flat in the core region. Thus, in this case, the core moves with the velocity of the wall and initial interface with a large gap of 2 mm is effectively

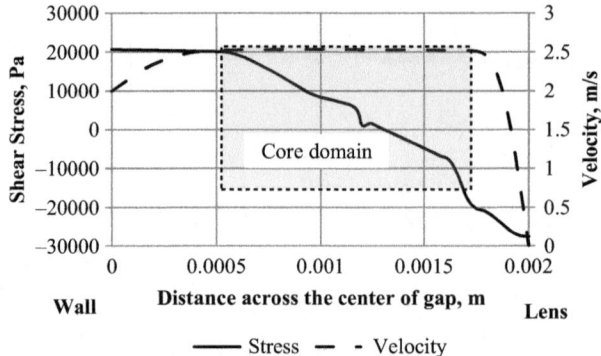

Figure 11.7 Calculated shear stress and fluid velocity distribution across the gap formed by the moving wall and the surface of lens.

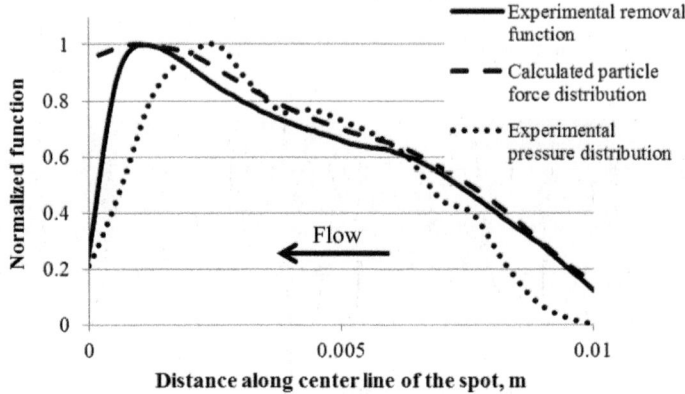

Figure 11.8 Comparison of calculated material removal rate profile with both experimental removal rate profile and pressure distribution.

transformed into the new one with a much smaller gap of ~ 0.2 mm resulting in associated significant increase in the effective shear rate.

The shear rate in the sheared zone was determined and used to calculate the particle force distribution along the center line of flow using (11.2). Thereafter, normalized values were plotted together with normalized experimental removal rate profiles. Experimental removal functions (spots) were taken on flat parts with a wheel radius equal to the radius of the lens used in the modeling. As an example, results for fluid with the yield stress of 16 kPa are shown in Figure 11.8. One can see that the correlation between the experimental and calculated profile is reasonably good and counts in favor of hydrodynamic (shear flow) mode of material removal.

Experimental pressure distribution was obtained with an ultra-thin, tactile pressure sensor Tekscan attached to the lens surface. The sensor comprises numerous individual sensing elements, or sensels in the form of matrix allowing mapping of pressure distribution.[28] As shown in Figure 11.8 (dotted line) the

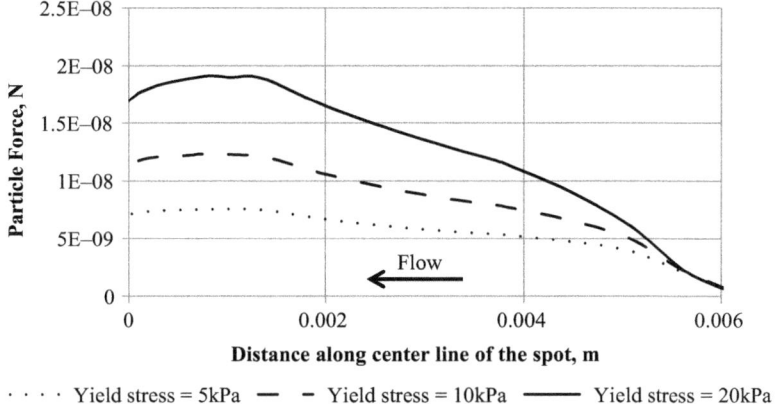

Figure 11.9 Dependence of particle force (removal rate) on MR fluid yield stress.

pressure distribution is shifted towards flow and does not correlate with the distribution of the removal rate. This points to the fact that, generated by plastic flow, hydrodynamic pressure (fluid normal stress) does not contribute to material removal in the contact zone.

The model adequately validates MRF regularity as an increase of removal rate with the magnetic field and concentration of magnetic particles due to appropriate increase in the fluid yield stress. The change in the yield stress results in the change of the thickness of the core (the sheared zone) with appropriate change in the shear rate, which in turn, results in a change in the particle force. This is illustrated by Figure 11.9 where distribution of calculated particle force is shown for fluids with different yield stress (5, 10 and 20 kPa). As one can see, the particle force increases with the fluid yield stress. This increase in the particle force with the yield stress is in a reasonable accordance with the dependence of the removal rate on magnetic field strength, as shown in Figure 11.10. Here, both normalized peak of the particle force and the peak of removal rate are plotted against magnetic field strength. In doing so, the yield stresses for particle force calculation as well as the experimental removal rate were determined at the same magnetic field strength.

The model also revealed that the removal rate depends on the geometry of the converging gap. As it follows from the results of calculations shown in Figure 11.11a, the particle force for the gap geometry formed with the surface of 50 mm in radius, is higher compared to the force corresponding to the gap formed with the surface of 75 mm in radius. This prediction was confirmed experimentally. Two spots were taken with the 150 mm diameter wheel on the fused silica (FS) glass: one on a convex sphere with a radius of 35 mm (Figure 11.11b) and another one on a flat surface (Figure 11.11c). The peak removal rate of 5.73 μm min^{-1} was obtained on the sphere and lower peak of 3.84 μm min^{-1} was obtained on the flat surface when all conditions were equal.

According to eqn (11.2), the abrasive particle load is very sensitive to the size of the basic particle. This should result in an increase of removal rate as the size of basic particle increases. As experimental results show, this prediction was

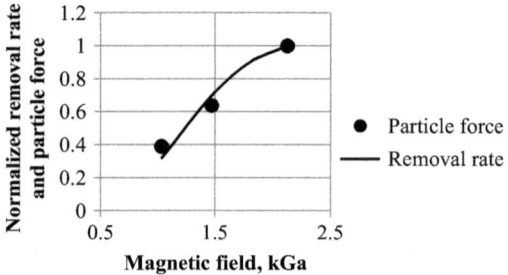

Figure 11.10   Effect of magnetic field on the calculated particle force and the experimental removal rate.

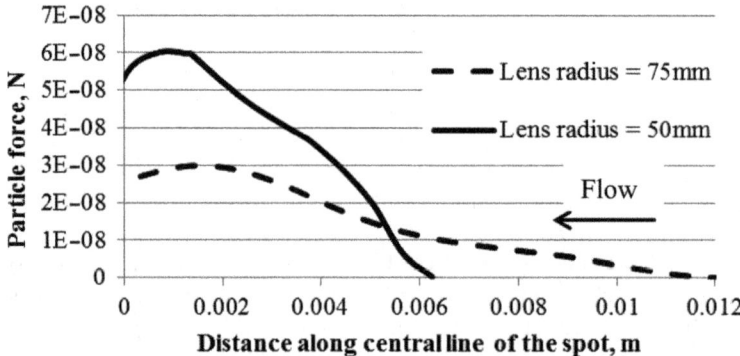

Figure 11.11   Effect of gap geometry on the particle force and the removal rate: (a) calculated removal rate profiles; (b) experimental polishing spot on a convex surface; (c) experimental polishing spot on a flat surface.

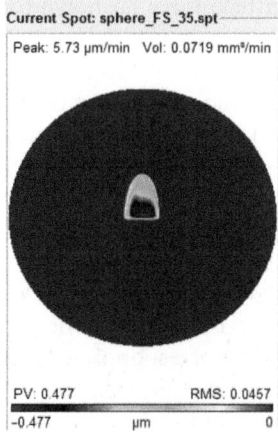
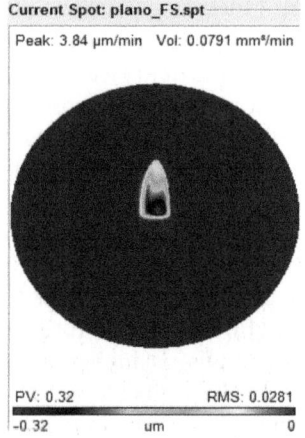

also born out. MRF spots shown in Figure 11.12a,b were taken on FS glass with two fluids composed of magnetic particles of sizes 1um and 4 μm but at different magnetic fields to equalize the fluids' yield stress.

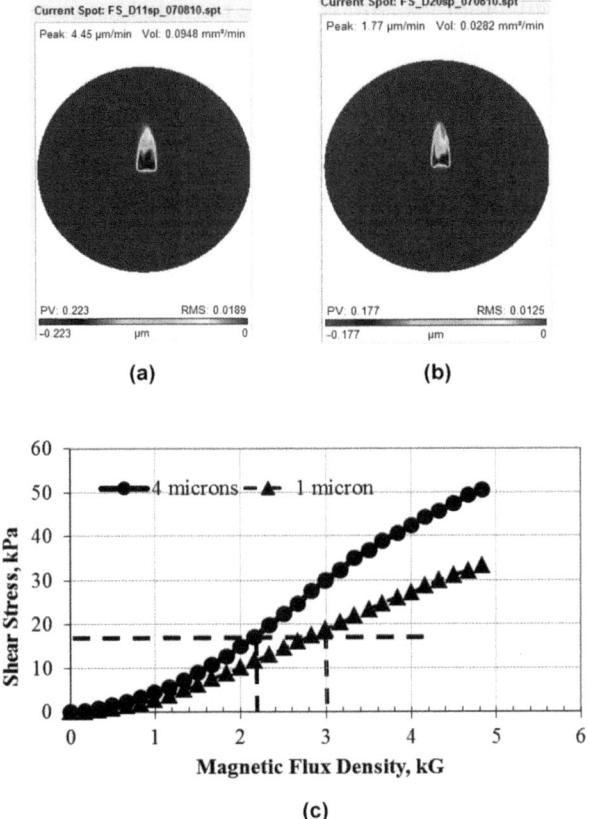

Figure 11.12 Effect of particle size on the removal rate: (a) polishing spot taken with particle size of 1 micron; (b) polishing spot taken with particle size of 4 micron; (c) rheological measurements allow equalization of the yield stress.

Removal rate of 4.45 μm min^{-1} which corresponds to the fluid with larger magnetic particles (Figure 11.12a), is higher compared to the one (1.77 μm min^{-1}) obtained with fluid composed of 1 micron particles (Figure 11.12b) with other conditions being equal. Appropriate field strength was determined with magneto rheological measurements made with an Anton Parr magneto-rheometer at low shear rate as shown in Figure 11.12c.

Some quantitative model evaluation can be made using the Hertzian theory of surface penetration[29], which is the generally accepted approach in modeling of material removal on glass.[30] In the case of a spherical indenter, the tensile stress generated over a contact area is given by:

$$\sigma_p = \frac{(1 - 2 \cdot \vartheta_M)G_p}{2\pi r_c^2} \qquad (11.4)$$

Here r_c is the contact radius, and G_p is the particle load (contact force). The contact radius is given by:

$$r_c = \left[\left(\frac{3}{4} G_p r_a \right) k_E \right]^{\frac{1}{3}}$$ (11.5)

and

$$k_E = \left(\frac{1 - \vartheta_M^2}{E_M} + \frac{1 - \vartheta_a^2}{E_a} \right)$$ (11.6)

where r_a is the radius of the abrasive particle (indenter) ϑ_M and E_M are the Poisson's ratio and Young's modulus for the material (glass) and ϑ_a and E_a are Poisson's ratio and Young's modulus for the abrasive particle.

The Hertzian theory also predicts the depth of penetration in the form of[27]

$$h_t = \left(\frac{9}{16} \right)^{\frac{1}{3}} \left(\frac{G_p}{E_r} \right)^{\frac{2}{3}} \left(\frac{1}{r_a} \right)^{\frac{1}{3}}$$ (11.7)

where $E_r = \frac{1}{k_E}$ is the reduced elastic modulus.

In order to evaluate the magnitude of particle load and corresponding contact stress generated by abrasive particles in addition to the shear rate, it is necessary to have a grasp of the size of the basic particles which are most likely aggregates of the original magnetic particles. The size of this fluid sub-structure depends on a ratio between restoring (magnetic) and destroying (hydro-dynamic) forces acting on the aggregate. The ratio is known as the Mason number[31]

$$M = \frac{\mu_0 \kappa_a H^2}{\eta_0 \dot{\gamma}}$$ (11.8)

where μ_0 is the magnetic permeability of vacuum, κ_a is the aggregate susceptibility, H is magnetic field strength, η_0 is fluid dynamic viscosity and $\dot{\gamma}$ is the shear rate.

As it was shown,[29] the aggregate size, particularly its aspect ratio (or length of particles chain), decreases as the Mason number decreases. At relatively high shear rates of $\sim 10^4$ s^{-1}, $H = 150$ kA m^{-1} and $\kappa_a = 5$, which are characteristic for the case under consideration, the Mason number of 14 predicts the aspect ratio of ~ 1–2 suggesting that the size of the aggregate is small. To evaluate the surface tensile stress and depth of penetration, an assumption was made that the aggregate consists of 4 spherical particles of 4 microns in diameter. This aggregate may be considered as an ellipsoid with an aspect ratio of 1.5. Calculations of tensile stress with (11.2) and (11.4) were performed for cerium oxide abrasive particles 100 nm in size and fused silica glass. In doing so, the coefficient K in (11.2) was taken as 1.[19,20] Corresponding results are shown in Figure 11.13 (dashed line). It is worth noticing that the calculated tensile stress in the range of hundreds of MPa is comparable to the ultimate tensile strength for glass (33 MPa) and even some harder materials. Taking into account that

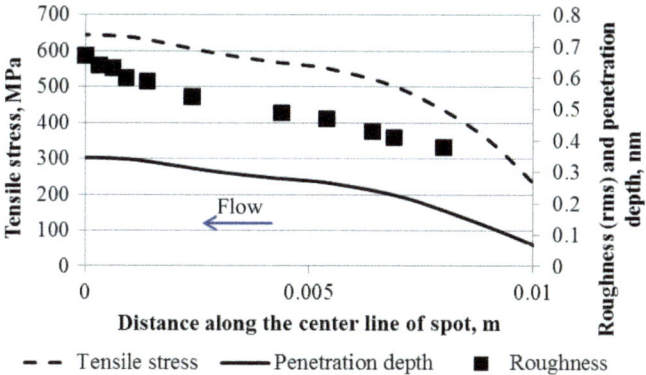

Figure 11.13 Results of calculations of surface indentation by abrasive particles.

ultimate tensile strength is a limit state of tensile stress that leads to tensile failure in the manner of ductile failure or in the manner of brittle failure, the predicted values of stress are quite sufficient to result in observed material removal, giving some quantitative support to the model.

Another approach in model verification can be the comparison of values of experimental surface roughness with the penetration depth calculated with eqn (11.7). This comparison is shown in Figure 11.13 (right axis) where actual roughness distribution along the center line of a polishing spot taken on FS glass (square solid markers) is in good qualitative and reasonable quantitative agreement with the corresponding calculated penetration depth (solid line). In general, the penetration depth of a few Angstroms is close to actual experimental results for surface roughness observed in MRF on glasses.[32]

11.4 Conclusion

Magnetorheological finishing (MRF®), as a leading-edge production technology, is widely used in the optics industry for manufacturing modern optical elements including those with high precision aspherical and free form surfaces. The objective of this chapter has been to discuss scientific aspects of MRF technology, and particularly the modeling of the mechanism of material removal in MRF. The model of elastic pad formed from MR fluid in magnetic field appeared to be unsound in view of the measurements and analysis of mechanical properties of a magnetized MR fluid. In this regard, it was shown that an analog of conventional pad formed with such material cannot support abrasive particle load, which would be appropriate to provide removal rates typical to MRF. An alternative approach is based on the principle of conservation of particle momentum and suggests that a load for surface nano-indentation by abrasive particles is provided at their interaction near the wall with larger and heavier basic magnetic particles fluctuating due to collision in the shear flow of concentrated binary suspension. Regularity of the granular shear flow and numerical simulation are used in modeling. The model is in good qualitative and reasonable quantitative agreement with experimental results for MRF. Presented results of finishing of some challenging surfaces demonstrate efficiency of the technology.

Acknowledgements

The author appreciates valuable contribution from Sergei Gorodkin, Bob James, Arpad Sekeres and Robin Townell.

References

1. I. Marinescu, E. Uhlmann, and T. Doi, *Handbook of Lapping and Polishing*, CRC Press, Taylor & Francis Group, 2006.
2. D. D. Walker, A. T. H. Beaucamp, D. Brooks, R. Freeman, A. King, G. McCavana, R. Morton, D. Riley and J. Simms, *Proc. SPIE 47th Annual Mtg, Seattle*, 2002, **4451**, 267.
3. A. Momber and R. Kovacevic, *Principles of Abrasive Water Jet Machining*, Springer, New York, NY, 1998.
4. S. M. Booij, Fluid Jet Polishing, *PhD Thesis*, Technische Universiteit Delft, printed in the Netherlands by PrintPartners IP Skamp B.V., Enschede, ISBN 90-9017012-X, 2003.
5. Y. Mori, K. Yamauchi and K. Endo, *J. Jpn Soc. Precision Eng.*, 1988, **10**(1), 24.
6. W. Kordonski, *J. Intell. Mater. Syst. Struct.*, 1993, **4**(1), 65.
7. W. Kordonsky, I. Prokhorov, S. Gorodkin, G. Gorodkin, L. Gleb and B. Kashevsky, US Patent 5 449 313, 1995.
8. W. Kordonski and S. Jacobs, *Int. J. Modern Phys. B*, 1996, **10**(23&24), 2837.
9. J. Lambropoulos, C. Miao and S. Jacobs, *Opt. Express*, 2010, **18**(19), 19713.
10. A. B. Shorey, S. D. Jacobs, W. I. Kordonski and R. F. Gans, *Appl. Opt.*, 2001, **40**, 20.
11. J. E. DeGroote, A. E. Marino, J. P. Wilson, A. L. Bishop, J. C. Lambropoulos and S. D. Jacobs, *Appl. Opt.*, 2007, **46**, 7927.
12. C. Miao, S. N. Shafrir, J. C. Lambropoulos and S. D. Jacobs, *SPIE Conference 7426: Optical Manufacturing and Testing VIII*, San Diego, CA, Aug. 4–5, 2009, (CD) 7426-11.
13. C. Miao, S. N. Shafrir, J. C. Lambropoulos, J. Mici and Stephen D. Jacobs, *Appl. Opt.*, 2009, **48**, 2585.
14. Y.-F. Dai, C. Song, X.-Q. Peng and F. Shi, *Appl. Opt.*, 2010, **49**, 298.
15. W. Kordonski, A. Sekeres and R. James, US Patent 7 557 566, 2009.
16. W. Kordonski, A. Price, J. Carapella and A. Sekeres, US Patent 7 156 724, 2007.
17. F. Preston, *J. Soc. Glass. Tech.*, 1927, **11**, 214.
18. W. Kordonski and S. Gorodkin, *J. Phys.: Conf. Ser.*, 2009, **149**, 012064.

19. G. B. Basim, I. U. Vakarelski and B. M. Moudgil, *J. Colloid Interface Sci.*, 2003, **263**, 506.
20. H. M. Laun, C. Gabriel and G. Schmidt, *J. Non-Newtonian Fluid Mech.*, 2008, **148**, 47.
21. H. H. Shen, *15th ASCE Eng. Mech. Conf.*, Columbia University, New York, June 2–5, 2002.
22. A. Karion and M. Hunt, *Powder Technol.*, 2000, **109**(1–3), 145.
23. W. Losert, L. Bocquet, T. C. Lubensky and J. P. Gollub, *Physical Review Lett.*, 2000, **85**, 1428.
24. http://www.adaptive-research.com (last accessed June 2011).
25. P. Perzyna, *Adv. Appl. Mech.*, 1966, **9**, 343.
26. J. A. Tichy, *J. Rheol.*, 1991, **35**(4), 477.
27. K. P. Gertzos, P. G. Nikolakopoulos and C. A. Papadopoulos, *Tribol. Int.*, 2008, **41**, 1190.
28. http://www.tekscan.com (last visited March 2011).
29. S. F. Ang, T. Scholz, A. Klocke and G. A. Schneider, *Dent. Mater.*, 2009, **15**, 1403.
30. L. M. Cook, *J. Non-Crystalline Solids*, 1990, **120**, 152.
31. Z. P. Shulman, V. I. Kordonski, E. A. Zaltsgendler, I. V. Prokhorov, B. M. Khusid and S. A. Demchuk, *Int. J. Multiphase Flow*, 1986, **12**, 935.
32. A. Shorey, S. Gorodkin and W. Kordonski, *Tech. Digest SPIE*, 2003, **TD02**, 69.

CHAPTER 12

Adaptive Magnetorheological Energy Absorbing Mounts for Shock Mitigation

NORMAN M. WERELEY,* HARINDER J. SINGH AND YOUNG-TAI CHOI

Smart Structures Laboratory, Department of Aerospace Engineering, University of Maryland, College Park, MD 20742, U.S.A.
*Email: wereley@umd.edu

12.1 Introduction

Intense shock loads resulting from harsh operating environments in vehicles, or high sink rate landings or crashes in helicopters, have the potential to cause severe injuries to seated operators as well as crew members.[1,2] Such intense impacts can be significantly attenuated if the seat suspension is outfitted with a simple passive energy absorber (EA) with a prescribed stroking load. However, a passive EA with a fixed stroking load cannot mitigate the variety of shock pulses, sink rates, and seated occupant weights that would be encountered. Thus, passive EAs, also known as fixed load energy absorbers (FLEAs), cannot optimally protect occupants under varying impact conditions. To provide adequate protection for the expected variation in impact events, a variable load energy absorber (VLEA) is needed.

Magnetorheological energy absorbers (MREAs) are a type of VLEA that can provide adaptive stroking load capabilities to achieve shock mitigation and crashworthiness for vehicles, high-speed boats, and helicopters. MREAs

RSC Smart Materials No. 6
Magnetorheology: Advances and Applications
Edited by Norman Wereley

have attractive features, such as rapidly adjustable stroking load in response to an applied current input. An MREA is similar to a conventional hydraulic shock absorber in that the fluid is pushed through an orifice by the motion of piston inside the hydraulic cylinder.[3,4] However, the orifice is typically integrated with an electromagnet housed in the piston. MREAs employ magnetorheological (MR) fluids, which are typically composed of 0.3–10 micron diameter carbonyl iron particles suspended in a hydrocarbon-based fluid.[5,6] The magnetic field generated by feeding current into the electro-magnetic coil induces magnetic induction between the carbonyl particles and thereby changes the apparent viscosity of the MR fluid, further, enabling adjustment of the MREA stroking load. Another major advantage of MREAs is low power consumption. Unlike systems that use force generators or actuators in conjunction with active feedback control, instabilities such as control spillover can be avoided or eliminated because MREAs are inherently dissipative devices.

Under consideration is the optimal control of a single degree of freedom system representing a rigid payload descending at a prescribed drop velocity. The MREA isolates the payload from the shock, and the energy dissipated is related to the area under the load-displacement curve. A key goal is to exploit the entire EA stroke during the shock event such that the payload energy is dissipated over the entire stroke, and payload deceleration is minimized and the potential for damage to the payload is minimized. If the MREA stroking load is too large, then the payload would come to rest before utilizing the available EA stroke and payload decelerations would be larger than necessary. On the other hand, if the MREA stroking load is too small, then the MREA will bottom out, thus, producing an undesirable severe end-stop impact. However, the MREA stroking load can be optimally selected for a given payload mass and impact (or drop) velocity or sink rate, such that the suspension payload comes to rest after fully utilizing the available MREA stroke; that is, a soft landing. The optimal stroking load of the MREA, characterized by a unique optimal Bingham number, enables the optimal control of the terminal trajectory of the payload mass. This chapter describes the procedure by which such an optimal Bingham number, which depends on payload mass, drop velocity and EA stroke, can be selected to optimally control payload to achieve a soft landing; that is, the payload comes to rest after fully utilizing the available stroke of the MREA.

12.2 Magnetorheological Energy Absorbers (MREAs)

The configuration of a single degree of freedom system employing an MREA for drop-induced shock mitigation is shown in Figure 12.1 with the payload mass, m, subjected to initial drop velocity, v_0. The available EA stroke before the impact is S.

The governing equation of motion is

$$m\ddot{z}(t) = -f_d - mg \qquad (12.1)$$

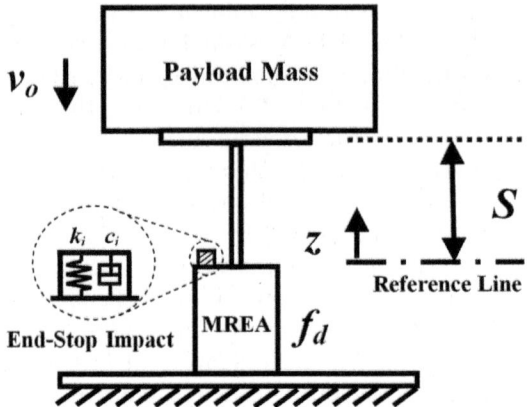

Figure 12.1   Configuration of magnetorheological energy absorbing mounts for drop-induced shock mitigation.

The MREA damping force is the sum of viscous damping (passive) and MR yield force (controllable) and given as follows

$$f_d = c\dot{z}(t) + f_y sign\{\dot{z}(t)\} \tag{12.2}$$

with the initial conditions given by

$$z(0) = S; \quad \dot{z}(0) = -v_0 \tag{12.3}$$

Here, f_d is the total MREA force, c is the viscous damping constant of the MREA, $z(t)$ is the displacement of the MREA from the reference line or payload displacement, f_y is the MR yield force and g is the acceleration due to gravity.

The governing equation, eqn (12.1) can be rewritten in terms of the velocity, $\dot{z}(t) = v(t)$.

$$\dot{v}(t) = -\frac{c}{m}v(t) - \frac{f_y}{m}sign\{v(t)\} - g \tag{12.4}$$

Integrating eqn (12.4) and using the initial condition for velocity given by eqn (12.3) we obtain

$$\dot{z}(t) = v(t) = -v_0 \left\{ \left[1 - \frac{f_y sign\{v(t)\}}{cv_0} - \frac{mg}{cv_0} \right] e^{-\frac{ct}{m}} + \frac{f_y sign\{v(t)\}}{cv_0} + \frac{mg}{cv_0} \right\} \tag{12.5}$$

The Bingham number, *Bi*, is defined as the ratio of the MR yield force (controllable) to the viscous damping force (passive).

$$Bi = \frac{f_y}{cv_0} \tag{12.6}$$

The Bingham number can be interpreted as the nondimensional yield force, or the *control variable*. Note that $v(t)$ is negative during the shock event because the payload moves downward, therefore *Signum* function attains a value of -1.

Using the Bingham number in eqn (12.6), the payload velocity in eqn (12.5) can be rewritten as follows

$$\dot{z}(t) = v(t) = -v_0 \left\{ \left[1 + Bi - \frac{mg}{cv_0} \right] e^{-\frac{ct}{m}} - Bi + \frac{mg}{cv_0} \right\} \tag{12.7}$$

By integrating eqn (12.7) again and using the initial condition given by eqn (12.3), we obtain the displacement given as

$$z(t) = \frac{mv_0}{c} \left(1 + Bi - \frac{mg}{cv_0} \right) \left(e^{-\frac{ct}{m}} - 1 \right) + v_0 t \left(Bi - \frac{mg}{cv_0} \right) + S \tag{12.8}$$

The deceleration of the MREA is obtained by differentiating eqn (12.7).

$$\ddot{z}(t) = \frac{cv_0}{m} \left[1 + Bi - \frac{mg}{cv_0} \right] e^{-\frac{ct}{m}} \tag{12.9}$$

12.3 Terminal Trajectory Control

The terminal trajectory control seeks to maximize the shock attenuation by adopting two key goals.[7] The first goal is to utilize the entire EA stroke such that the kinetic energy of the payload is dissipated over the entire stroke. In other words, the energy dissipation per unit EA stroke is minimized. The second goal is to eliminate end-stop impact, *i.e.*, the condition when the MREA runs out of stroke. These two control objectives are the terminal conditions given as follows

$$z(t_e) = 0$$
$$\dot{z}(t_e) = 0 \tag{12.10}$$

where t_e is the time at which the payload comes to a complete halt after the shock event. The simplicity of this approach lies in the fact that a constant Bingham number for a given shock intensity achieves these terminal conditions.

To calculate the optimal Bingham number, the Bingham number satisfying the velocity terminal condition and the displacement terminal condition are evaluated separately using eqn (12.7), (12.8) and (12.10).[7] The time at which the Bingham numbers corresponding to the displacement and velocity terminal condition coincide is the stoppage time. At this coinciding point all the terminal conditions are satisfied. The optimal Bingham number, Bi_o, is given by the following equation.[7]

$$Bi_o = \frac{mg}{cv_0} - \frac{1}{1 - e^{\left\{ \frac{cS}{mv_0} - 1 - W \left[e^{\left(\frac{cS}{mv_0} - 1 \right)} \left(\frac{cS}{mv_0} - 1 \right) \right] \right\}}} \tag{12.11}$$

where $W[\cdot]$ is the Lambert W Function or product log function.[8]

12.4   Optimal Bingham Number

The optimal Bingham number, Bi_o, varies with viscous damping constant, c, and drop velocities, as shown in Figure 12.2. In this case, the EA stroke was taken as $S = 15$ cm, and payload mass as $m = 30$ kg. From this chart, the optimal Bingham number, Bi_o, decreases as viscous damping constant increases. Because the viscous forces in the MREA are directly proportional to the viscous damping constant, reducing the viscous damping implies a reduction in the Bingham number in eqn (12.6). Moreover, the total MREA stroking load is the sum of the passive viscous force and the controllable MR yield force as defined by eqn (12.2). Therefore, absorbing a given amount of kinetic energy corresponding to a particular drop velocity is a trade-off between viscous damping force and MR yield force (or choice of Bingham number as the control variable). In other words, kinetic energy is dissipated by two stroking load components of the MREA *i.e.* passive viscous force and MR yield force. If the passive force is relatively low, then a high MR yield force or optimal Bingham number is required to reach the terminal conditions in eqn (12.10). In contrast, if the passive force is relatively high, then a lower MR yield force or optimal Bingham number is required to reach the terminal conditions in eqn (12.10). Figure 12.2 depicts this design trade-off for an MREA, and a particular design can be selected based on the requirements of viscous damping constant and the optimal Bingham number.

A second key observation is that if the drop velocity increases, then so too does the required MR yield force or optimal Bingham number. If the drop velocity increases, then the kinetic energy that must be absorbed by the shock isolation mount also increases, which implies that the stroking load must also

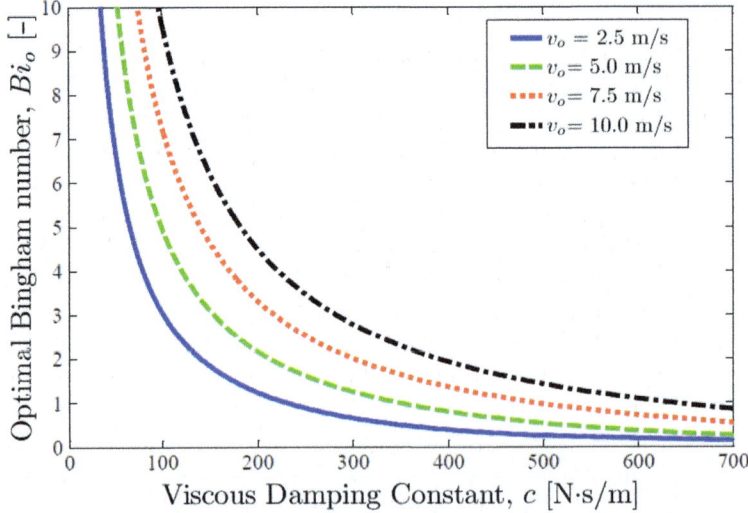

Figure 12.2   Optimal Bingham number variation with viscous damping coefficients and drop velocities.

increase for the available stroke. Because the viscous damping force is fixed for a particular MREA, the increase in kinetic energy is dissipated by an increase in the MR yield force or optimal Bingham number.

12.5   Optimal Time Response of MREA

Different MREA responses based on optimal and non-optimal Bingham numbers for drop-induced shock mitigation are shown in Figure 12.3 for a

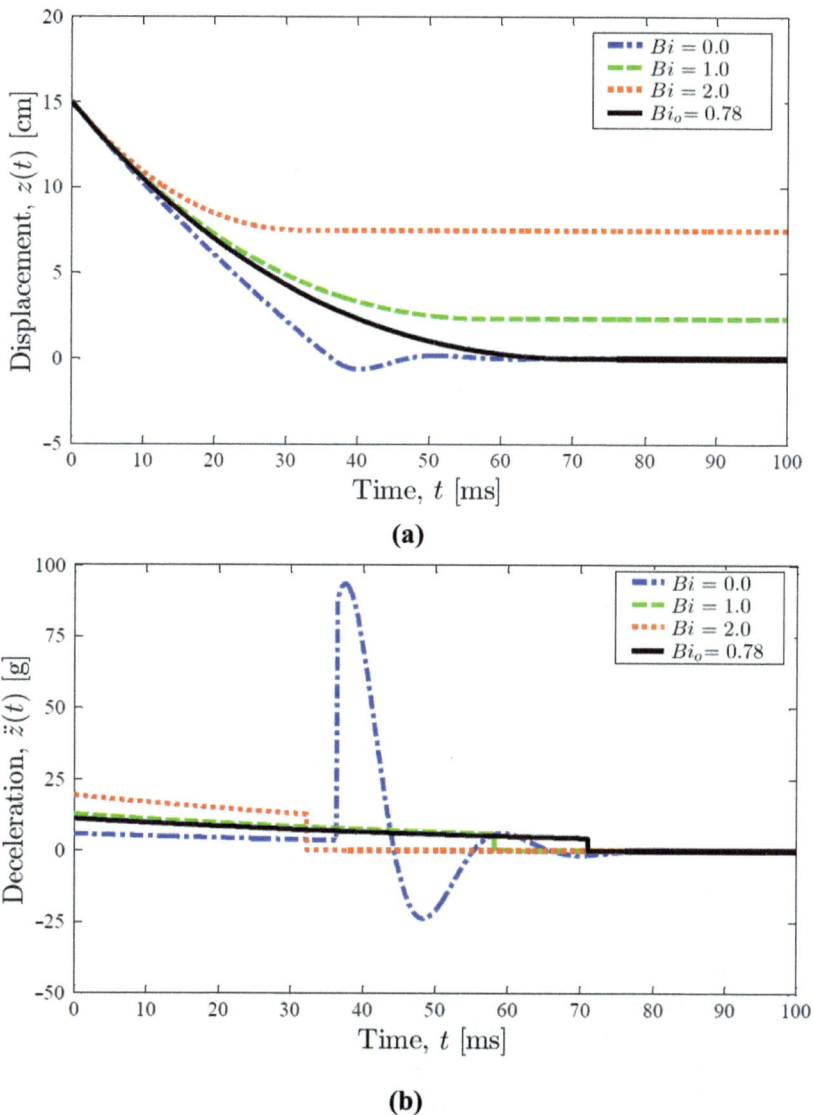

Figure 12.3   Payload (a) displacement and (b) deceleration for optimal and non-optimal Bingham numbers.

payload mass of $m = 30$ kg. The optimal Bingham number was found to be $Bi_o = 0.78$ corresponding to a viscous damping constant, $c = 400$ N s m^{-1}, and drop velocity, $v_o = 5$ m s^{-1}. The end-stop impact was modeled using a very stiff spring, $k_i = 3000$ kN m^{-1}, and damping ratio $c_i = 7500$ N s m^{-1}, as shown in Figure 12.1.[9,10]

It is clear from Figure 12.3a that the payload utilized the complete EA stroke without experiencing an end-stop impact only when the optimal Bingham number control was implemented, that is, $Bi = Bi_o$. If the Bingham number is less than the optimal Bingham number, or $Bi < Bi_o$, then the payload completes the EA stroke with a non-zero velocity and incurs an end-stop impact. On the other hand, if $Bi > Bi_o$, then the payload did not fully utilize the EA stroke because the MR yield force is too high.

Figure 12.3b presents the deceleration of the payload mass for different Bingham numbers. The payload incurred large peak deceleration due to an end-stop impact when $Bi < Bi_o$. Such excessive deceleration and corresponding loads may result in potential payload damage and are, therefore, undesirable. For cases where $Bi > Bi_o$, the maximum deceleration was much less than the peak decelerations experienced for end-stop impacts. However, the maximum payload decelerations were greater than that for optimal Bingham number control because the MREA stroke was not fully utilized, which led to excessive energy dissipation per unit stroke.

12.6 Optimal Response of MREA for Varying Shocks

This section compares the optimal responses of the payload mass incurred for sink rates of $v_o = 5$ and 10 m s^{-1}. Two different MREA designs are also compared for the same payload mass and MREA stroke, where the designs varied based on choice of viscous damping constants, either low viscous damping, $c = 100$, or high viscous damping, $c = 700$ N s m^{-1}. The optimal Bingham number for each case is tabulated in Table 12.1.

The displacement *vs.* time (Figure 12.4a) and deceleration *vs.* time (Figure 12.4b), achieved using optimal Bingham number control, are shown for the two MREA designs. For all cases utilizing optimal Bingham number control, the payload exhibited a soft landing for either drop velocity or either viscous damping constant, and satisfied the optimal terminal conditions, as shown in Figure 12.4a.

Table 12.1 Optimal Bingham Numbers, Bi_o.

Viscous damping constant, $c/N\ s\ m^{-1}$	Drop velocity, $v_o/m\ s^{-1}$	Optimal Bingham number, Bi_o
100	5	4.933
	10	9.633
700	5	0.229
	10	0.847

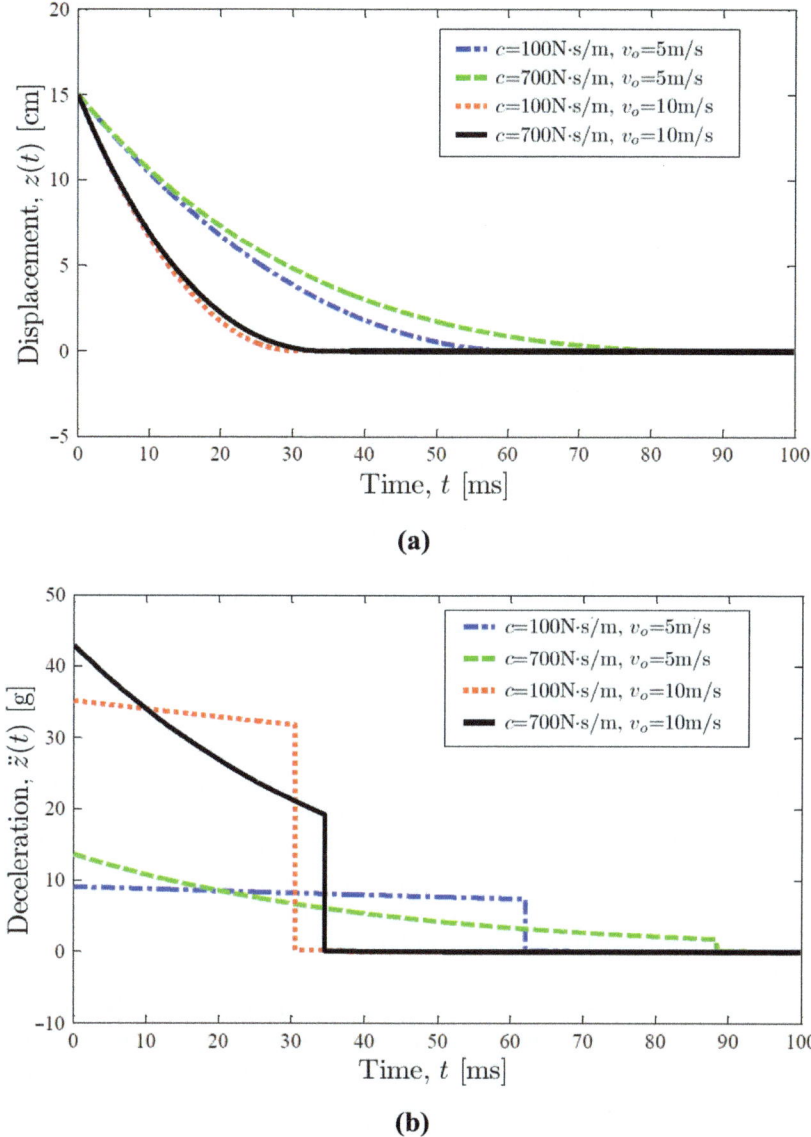

Figure 12.4 Payload (a) displacement and (b) deceleration for varying damping coefficients and drop velocities for optimal Bingham numbers.

For the high drop velocity case, $v_o = 10$ m s^{-1}, the payload for both MREA designs using optimal Bingham number control rapidly achieved the soft landing and utilized the available MREA stroke. It should be noted that the low viscous damping case achieved a soft landing faster than the high speed damping case (Figure 12.4b). However, the low viscous damping case required a much higher optimal Bingham number in the controller than was the case for the high viscous damping case. This resulted in a deceleration for the low viscous damping case that decreased more moderately than that for the high viscous damping case (Figure 12.4b). Also, the duration of the deceleration for the low viscous damping case was slightly shorter than that for the high viscous damping case (Figure 12.4b). This indicates that it is more efficient to rely on MR yield force over viscous force when seeking to minimize the stroke required for a soft landing. This is demonstrated in Figure 12.4b where the deceleration is slowly ramping down for the low viscous damping case. In contrast, the high damping case relies much more on the viscous force to absorb energy, which is less efficient because as the payload slows and the stroking load reduces, the energy absorbed per unit of payload displacement is reduced.

For the low speed drop case, $v_o = 5$ m s^{-1}, and for both MREA cases, the payload achieved the soft landing much more slowly than for the high speed case and, again, utilized the available MREA stroke (Figure 12.4a). However, to achieve a soft landing for the low drop velocity case required a smaller optimal Bingham number than for the high drop velocity case. This is because the stroking load, required to absorb the reduced kinetic energy, is also reduced for the low drop velocity case. For the low drop velocity case, the deceleration for the low viscous damping case occurred over a much shorter interval than the deceleration for the high viscous damping case. Alternatively, the stroking load for the high viscous damping case is dominated by the viscous force, so that the magnitude of the stroking load becomes small when the damper velocity becomes small, especially near the end of its stroke. As shown in Figure 12.4b, the stroking load is nearly constant for the low viscous damping case and most efficiently absorbs energy for the entire stroke; however, the high damping case relies mostly on viscous damping for energy absorption, which is less efficient because as the payload slows, so that consequently the stroking load is reduced, the energy absorbed per unit of payload displacement is falling rapidly. As before, the low viscous damping case achieved a soft landing faster than for the high speed damping case, indicating that it is more efficient to rely on MR yield force over viscous force in terms of minimizing time to achieve a soft landing.

Based on these results, there is a trade-off in MREA design between MR yield force and viscous damping. It is advantageous to implement an MREA with low viscous force and high MR yield force, because energy dissipated per unit of payload displacement can be maximized. This implies that an MREA capable of achieving the largest possible Bingham number would be best for MR shock isolation, as long as the maximum allowable deceleration of the payload is not violated.

12.7   Conclusions

The drop-induced shock mitigation of a single degree of freedom system employing an adaptive MREA was theoretically analyzed. Terminal trajectory control achieved *via* selection of an optimal Bingham number was demonstrated *via* analysis. This optimal Bingham number control algorithm avoided end-stop impact and enabled the payload to utilize the entire MREA stroke for energy absorption. Sub-optimal Bingham numbers resulted in either end-stop impact ($Bi < Bi_o$) or under-utilization of MREA stroke ($Bi > Bi_o$), in which sub-optimal solutions led to higher payload decelerations than necessary, thereby increasing probability of damage to the payload. The optimal Bingham number increased as drop velocity, v_o, increased. Therefore, if the impact becomes more intense, then higher yield force is necessary to mitigate the shock load. Also, as viscous damping, c, increased, then the optimal Bingham number decreased, which implies that an MREA design trade-off exists between viscous damping and MR yield force.

By analyzing MREAs with different viscous damping constants, it was shown that it is advantageous to implement an MREA with low viscous force and high MR yield force, because energy dissipated per unit of stroke is maximized. An MREA capable of achieving the largest possible Bingham number should be used for MR shock isolation, as long as the maximum allowable deceleration of the payload is not violated.

References

1. G. J. Hiemenz, Y.-T. Choi and N. M. Wereley, *J. Aircraft*, 2007, **44**(3), 1031–1034.
2. S. P. Desjardins, *J. Am. Heli. Soc.*, 2006, **51**(2), 150–163.
3. E. Cook, W. Hu and N. M. Wereley, *J. Intell. Mater. Syst. Struct.*, 2007, **18**(12), 1197–1203.
4. M. Mao, W. Hu, N. M. Wereley, A. L. Browne and J. Ulicny, *Proc. ASME Conf. SMASIS*, 2009, Oxnard, CA, USA.
5. J. Jeon and S. Koo, *J. Magn. Magn. Mater.*, 2012, **324**(4), 424–429.
6. G. Cha, Y. S. Ju, L. A. Ahure and N. M. Wereley, *J. Appl. Phys.*, 2010, **107**(9), 09B505.
7. N. M. Wereley, Y.-T. Choi and H. J. Singh, *J. Intell. Mater. Syst. Struct.*, 2011, **22**, 515–519.
8. R. M. Corless, G. H. Gonnet, D. E. G. Hare, D. J. Jeffrey and D. E. Knuth, *Adv. Comput. Math.*, 1996, **5**, 329–359.
9. S. M. M. Jafri, Modeling of impact dynamics of a tennis ball with a flat surface, *Masters Thesis*, Texas A & M University, 2004.
10. W. T. Thomson, *Theory of Vibration with Applications*, Prentice-Hall International, Inc., New Jersey, 1988.

CHAPTER 13

Semi-Active Isolation System Using Self-Powered Magnetorheological Dampers

YOUNG-TAI CHOI, HYUN JEONG SONG, WEI HU
AND NORMAN M. WERELEY*

Smart Structures Laboratory, Department of Aerospace Engineering,
University of Maryland, College Park, MD 20742, U.S.A.
*Email: wereley@umd.edu

13.1 Introduction

In recent years, there has been much interest in semi-active dampers because they exhibit continuously controllable damper force, have simpler designs than active hydraulic dampers, no instability problem, and low power requirements. Magnetorheological (MR) fluid-based dampers are advantageous because damper force can be continuously and rapidly controlled.[1,2] In addition, power requirements of MR dampers are typically much less than other semi-active or active damper systems. However, MR dampers require an external power source, such as a battery, in order to operate and adjust the damper force. A simple DC battery power supply can be implemented in locations where power lines are hard or impossible to access. However, batteries are heavy with a short operational life, and require extensive maintenance. Emerging power harvesting technology[3–7] is making the goal of a practical self-powered MR damper feasible, that is, an MR damper that can be operated using power harvested from its environment without recourse to an external power source.

RSC Smart Materials No. 6
Magnetorheology: Advances and Applications
Edited by Norman Wereley

In our prior work,[8] we theoretically established the feasibility of a semi-active vibration isolation system using a vibration based self-powered MR damper. In addition, we showed that the self-powered MR damper, with the power harvesting dynamic vibration absorber (DVA), could achieve good vibration isolation performance using neither a sensor to measure motion as a feedback signal nor a control algorithm. Therefore, the key objective of this study was to experimentally evaluate a semi-active vibration isolation system using a vibration based self-powered MR damper. For doing so, a power harvesting DVA was designed and fabricated that can convert mechanical energy from vibration or shock into electrical energy by means of electromagnetic induction. The power harvesting DVA consists of an electromagnetic coil winding, moving masses, coil springs, guided poles, and four permanent magnets. By electronically connecting the power harvesting DVA with an MR damper for a seat suspension of the amphibious Expeditionary Fighting Vehicle (EFV), a self-powered MR damper was constructed in this study. The energy harvesting characteristics of the power harvesting DVA in the frequency domain were experimentally measured under different acceleration excitations of 0.3–1.2 g (where $g = 9.81$ ms^{-2}). In addition, the damper force of the self-powered MR damper (*i.e.*, in this study, the MR damper used in the EFV seat suspension with the added power harvesting DVA) was experimentally measured in time and frequency domains. To evaluate the vibration isolation performance of a semi-active system using the self-powered MR damper, an EFV seat suspension mockup using the self-powered MR damper was constructed. The vibration isolation performance of the EFV seat suspension mockup using the self-powered MR damper was experimentally evaluated under eight different representative random excitation accelerations.

13.2 Self-Powered MR Damper

13.2.1 Power Harvesting Dynamic Vibration Absorber (DVA)

Figure 13.1 presents the configuration of the power harvesting DVA. The key components of the power harvesting DVA are permanent magnets, an electromagnetic coil winding, and coil springs. In the gap between the four magnets, the moving masses with the electromagnetic coil winding were supported by coil springs coaxially located outside linear guided poles. If ambient vibration is transmitted to the power harvesting DVA, the moving masses with the electromagnetic coil winding slide up and down. Within the magnetic field, the motion of the electromagnetic coil winding generates an induced voltage and its magnitude is proportional to the rate of change of the magnetic field. To maximize the power harvested from ambient vibration, the moving mass and the coil springs were chosen so that the resonance frequency of the power harvesting DVA matched the target system for which the vibration was to be suppressed.

Figure 13.2 presents a photograph of the fabricated power harvesting DVA. The target system to be isolated using the power harvesting MR damper was an

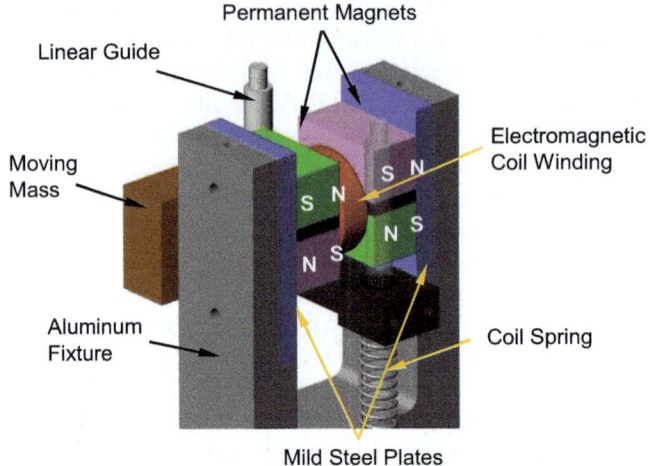

Figure 13.1   Configuration of the power harvesting dynamic vibration absorber (DVA).

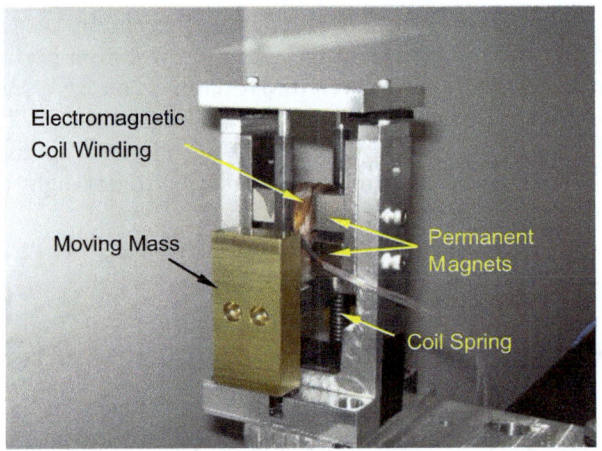

Figure 13.2   Photograph of the power harvesting DVA fabricated.

EFV seat suspension mockup[9,10] with a resonance frequency of around 4.5 Hz. Therefore, the resonance frequency of the power harvesting DVA was also chosen to be close to 4.5 Hz. Dimensions of the power harvesting DVA are as follows: 2 in ×2.5 in ×4.5 in. The moving mass was made of brass to prevent any interaction with the permanent magnets, and the electromagnetic coil winding was mounted on the moving masses. The 25 AWG wire was used for the electromagnetic coil winding, and the total size of the electromagnetic coil winding was 1.25 inch in diameter and 0.5 inch in width. Four NdFeB permanent magnets were used and each magnet size had 1 in ×0.5 in ×0.5 in. To minimize the weight of the power harvesting DVA, the main fixture was made of aluminum. Each air gap between the electromagnetic coil winding and the permanent magnet was about 0.01 inches. The effective magnetic density in the gap center between the permanent magnets was about 0.4 T. Two moving masses were used, each with a mass of 132 grams. The mass of the electromagnetic coil winding was 54 grams and the total mass of the power harvesting DVA was 854 grams.

13.2.2   Harvesting Characteristics of the Self-Powered MR Damper

Figure 13.3 is a photograph of the experimental setup for measuring induced voltage and damper force of the self-powered MR damper. In this case, the prototype MR damper for the EFV seat suspension was used. A hydraulic shaker excited the damper cylinder, and, in turn, the power harvesting DVA. Displacement was monitored by an LVDT (linear variable differential transformer) sensor. The output lines of the power harvesting DVA were electronically connected to the MR damper. The induced voltage generated by the power harvesting DVA was measured under different acceleration levels.

Figure 13.4 presents the measured maximum voltage generated from the power harvesting DVA with open-circuit (*i.e.*, the power harvesting DVA was electronically disconnected from the MR damper) under different excitation accelerations. In this case, acceleration levels of 0.3, 0.6, 0.9, and 1.2 *g* were used. As seen in this figure, the maximum voltage was generated around the target resonance frequency (*i.e.*, around 4.5 Hz) of the power harvesting DVA. However, as acceleration levels increased, the frequency at the peak of maximum voltage also increased. End-stop rubber bushings was implemented, on the bottom and top sides inside the power harvesting DVA, to minimize

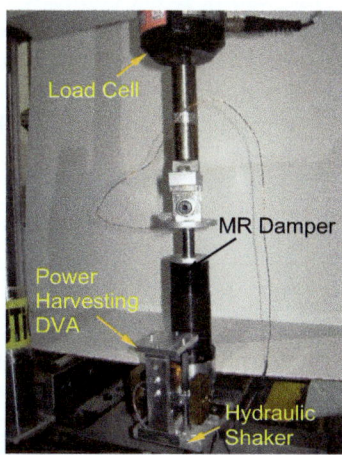

Figure 13.3   Photograph of the experimental testing setup of the self-powered MR damper for measuring induced voltage and damper force of the self-powered MR damper.

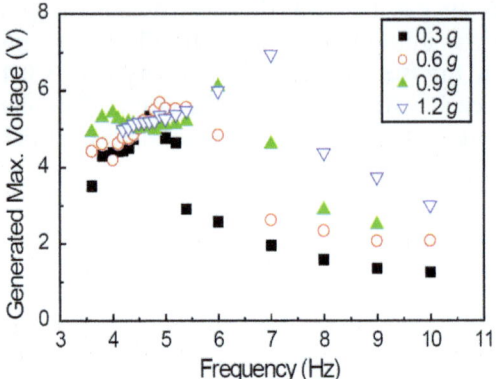

Figure 13.4   Measured maximum voltage generated from the power harvesting DVA
with open-circuit (*i.e.*, the power harvesting DVA was electrically disconnected from the MR damper) under different excitation accelerations.

damage to the electromagnetic coil winding as a result of end-stop impact.
When acceleration levels increased, the electromagnetic coil winding contacted
the end-stop bushings, thus producing impact dynamics. This phenomenon
may increase the effective stiffness of the power harvesting DVA.

Figures 13.5 and 13.6 present the measured maximum and RMS voltage,
current, and power generated from the power harvesting DVA under different
excitation accelerations. In this case, the power harvesting DVA was electronically connected to the MR damper. Comparing to the maximum voltage
of the power harvesting DVA with open-circuit as shown in Figure 13.4, the
maximum voltage levels of the power harvesting DVA with the MR damper
became smaller. In particular, at 0.3 *g* excitation acceleration, the maximum
voltage level of the power harvesting DVA connected to the MR damper became significantly smaller than the maximum voltage of the open-circuit case.
The reason is that the electromagnetic coil winding of the MR damper worked
as shunt damping[3,4,11] and thus dampened the motion of the electromagnetic
coil winding of the power harvesting DVA. On the other hand, the generated
current, I_s of the power harvesting DVA with the MR damper was calculated
from the relation below:

$$I_s = \frac{V_s}{\sqrt{(R_d + R_c)^2 + (L_d + L_c)^2 \omega^2}} \qquad (13.1)$$

Here, V_s is the induced voltage of the power harvesting DVA with the MR
damper and ω is the frequency of the excitation acceleration. R_d and L_d are the
electrical resistance and inductance of the MR damper for seat suspensions of
the EFV, respectively. R_c and L_c are the electrical resistance and inductance of
the electromagnetic coil winding, respectively. In this study, the measured
resistance and inductance of the MR damper at room temperature were $R_d =$
6.8 Ω and $L_d = 45$ mH, respectively. In addition, the measured resistance and

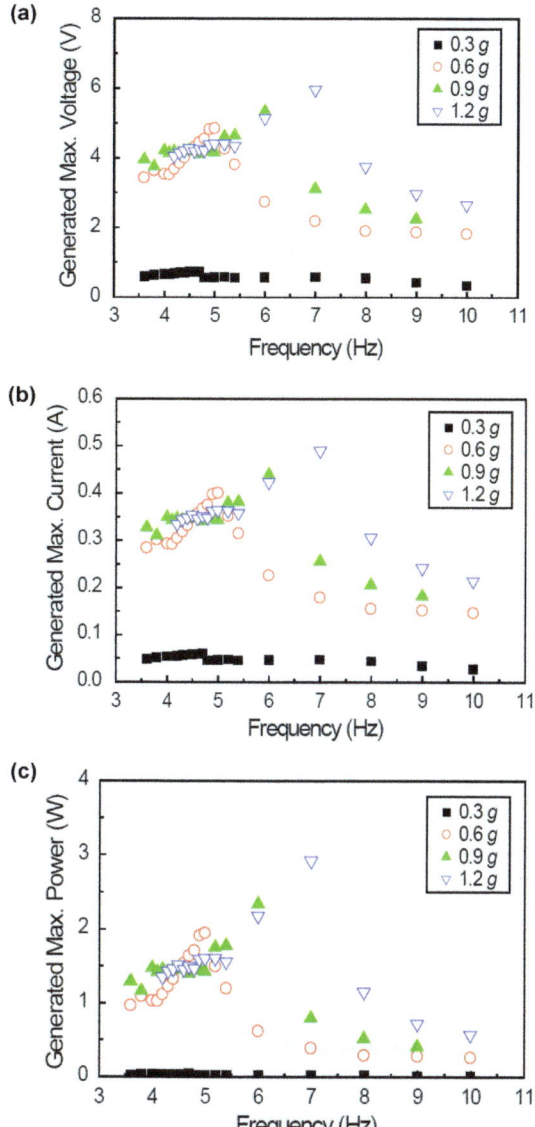

Figure 13.5   Measured maximum voltage, current, and power generated from the power harvesting DVA under different excitation accelerations. Note that the power harvesting DVA was electronically connected to the MR damper; (a) maximum induced voltage, (b) maximum current generated, (c) maximum power harvested.

inductance of the electromagnetic coil winding were $R_c = 5.2$ Ω and $L_c = 3.5$ mH, respectively. Therefore, the total resistance and inductance of the self-powered MR damper were 12 Ω and 48.5 mH, respectively. The peaks of the maximum voltage and current generated from the power harvesting DVA were

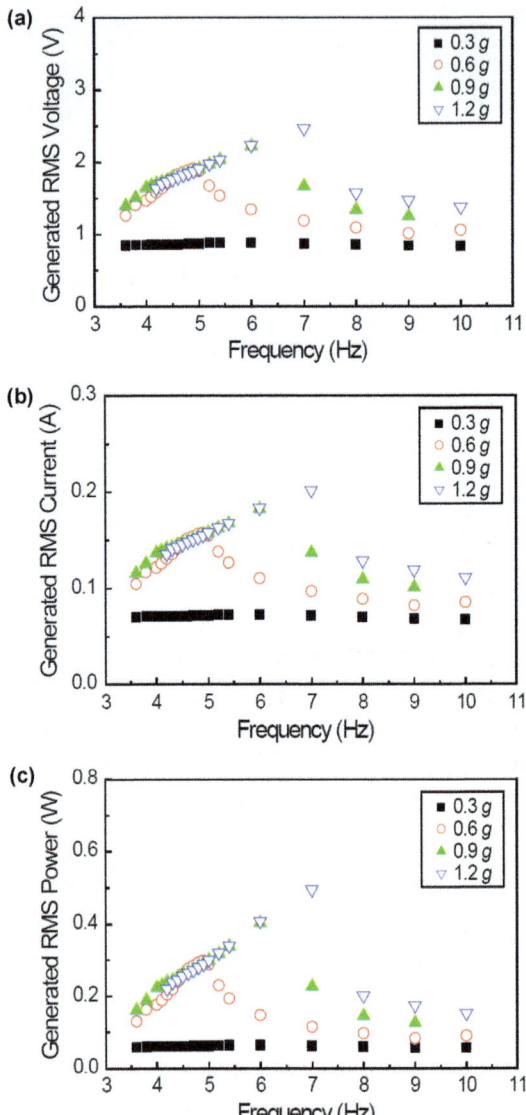

Figure 13.6 Measured RMS voltage, current, and power generated from the power harvesting DVA under different excitation accelerations. Note that the power harvesting DVA was electronically connected to the MR damper; (a) RMS induced voltage, (b) RMS current generated, (c) RMS power harvested.

6.0 V and 0.49 A @ 1.2 *g*. Corresponding peak maximum harvested power was 2.92 W @ 1.2 *g*. Note that because the output power lines of the power harvesting DVA were directly connected to the MR damper without passing through any rectifier, the induced voltage, current, and power in this study were in AC type. The peak RMS voltage, current, and power of the power harvesting DVA were 2.45 V_{rms}, 0.2 A_{rms}, and 0.49 W_{rms}, respectively. Also, as seen in these figures, the induced voltage or current generated by the power harvesting DVA grew larger near resonance, but reduced substantially above resonance. Thus, the power harvesting DVA achieves vibration isolation using neither a sensor nor control algorithm because high damping is required near the resonance frequency for good vibration isolation performance and low damping above the resonance frequency.[12]

13.2.3   Damper Force Performance of the Self-Powered MR Damper

Figure 13.7 presents the measured RMS damper force of the self-powered MR damper (*i.e.*, the MR damper with the power harvesting DVA) under different excitation accelerations. In this case, the damper force of the self-powered MR damper was measured in the frequency domain under constant amplitude acceleration excitation over the tested frequency range. Therefore, the damper force of the self-powered MR damper decreased as the excitation frequency increased. The yield force of the self-powered MR damper (*i.e.*, the damper force of the self-powered MR damper minus the damper force of the

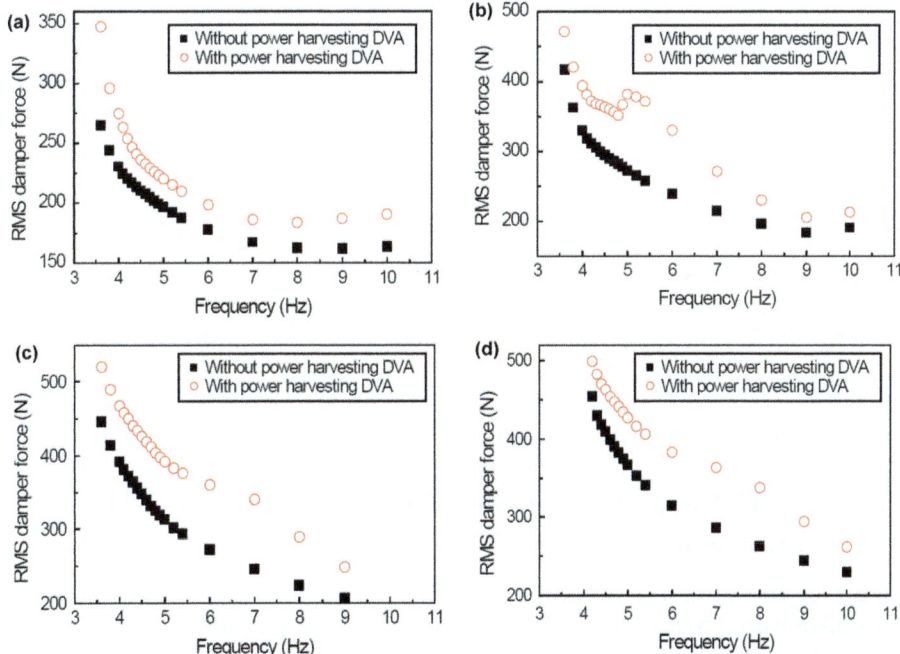

Figure 13.7   Measured RMS damper force of the self-powered MR damper under different excitation accelerations; (a) at ± 0.3 *g* acceleration, (b) at ± 0.6 *g* acceleration, (c) at ± 0.9 *g* acceleration, (d) at ± 1.2 *g* acceleration.

self-powered MR damper without the power harvesting DVA), became relatively large near the resonance frequency. However, above the resonance frequency of 4.5 Hz, the yield force of the self-powered MR damper decreased.

Figure 13.8 presents measured damper force characteristics of the self-powered MR damper under 0.6 *g* excitation acceleration with 5 Hz. Figure 13.8(a) shows the damper force in the time domain. As seen in this figure, the maximum damper force (658 N @ 0.6 *g*) of the self-powered MR damper with the power harvesting DVA was larger than the maximum damper force (340 N @ 0.6 *g*) without the power harvesting DVA. At this condition, the maximum yield force of the self-powered MR damper was 339 N at 0.6 *g* excitation acceleration. Figure 13.8(b) shows the induced voltage of the power harvesting DVA. As seen in this figure, the time history of the induced voltage is not sinusoidal because the electromagnetic coil winding slid up and down vigorously and experienced direction change of the magnetic flux from the four permanent magnets. In addition, the induced voltage (–5.0 V and 4.8 V @ 0.6 *g*) was slightly biased from zero voltage because the initial position of the electromagnetic coil winding was below the center of the coil winding stroke. As a result, the electromagnetic coil winding had longer upstroke (corresponding to negative voltage in this figure) than downstroke (corresponding to positive voltage in this figure). Figure 13.8(c) shows the damper force *versus* piston displacement, which verifies that the damper force of the self-powered MR damper greatly increased. At 0.6 *g* excitation, the damper force of the self-powered MR damper was augmented at high speeds (low displacement), much like viscous damping.

Figure 13.9 presents the measured damper force characteristics of the self-powered MR damper under 1.2 *g* excitation acceleration with 7 Hz. As observed in Figure 13.5 and 13.6, the maximum induced voltage at 1.2 *g* excitation acceleration occurred at the frequency of 7 Hz. In Figures 13.9(a) and 13.9(b), the damper force and induced voltage of the self-powered MR damper at 1.2 *g* excitation acceleration were biased more than those at 0.6 *g* excitation acceleration. At this condition, the maximum yield force of the self-powered MR damper was 335 N @ 1.2 *g* excitation acceleration. The maximum induced voltage of the self-powered MR damper was –6.0 V and 4.9 V @ 1.2 *g* excitation acceleration. Figure 13.9(c) shows the damper force *versus* piston displacement. In contrast to the damper force behavior at 0.6 *g* excitation, the damper force of the self-powered MR damper at 1.2 *g* excitation was augmented more at low speed or peak displacement, much as would be the case for Coulomb friction.

13.3 Seat Suspension Using the Self-Powered MR Damper

Figure 13.10 shows a photograph of the EFV seat suspension mockup using the self-powered MR damper. In this figure, the bottom frame was excited by a hydraulic shaker that produces typical excitation acceleration, and the seat suspension was installed on the bottom frame. In addition, the power harvesting DVA was mounted on the bottom frame and electronically connected to the MR damper. The seat suspension mockup used in this study was

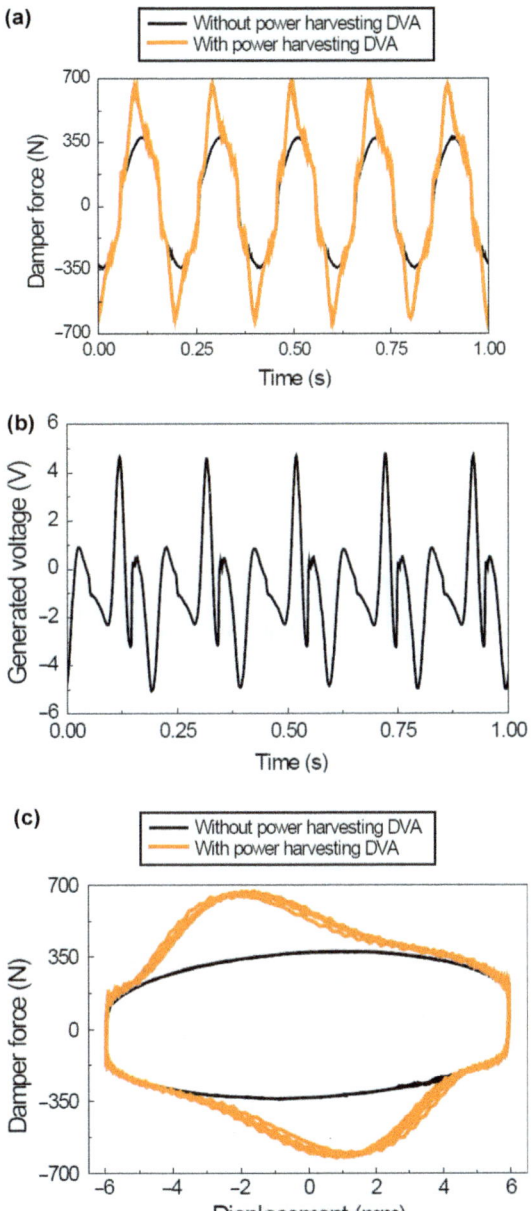

Figure 13.8 Measured damper force characteristics of the self-powered MR damper under ± 0.6 *g* excitation acceleration with 5 Hz; (a) damper force in time domain, (b) generated voltage in time domain, (c) damper force versus piston displacement.

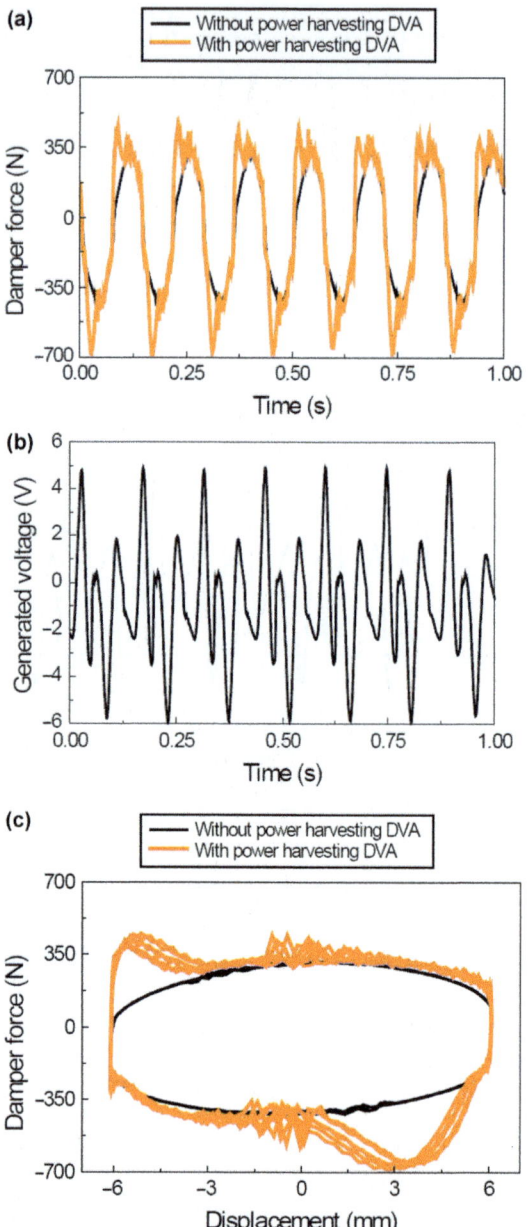

Figure 13.9 Measured damper force characteristics of the self-powered MR damper under ± 1.2 *g* excitation acceleration with 7 Hz; (a) damper force in time domain, (b) generated voltage in time domain, (c) damper force versus piston displacement.

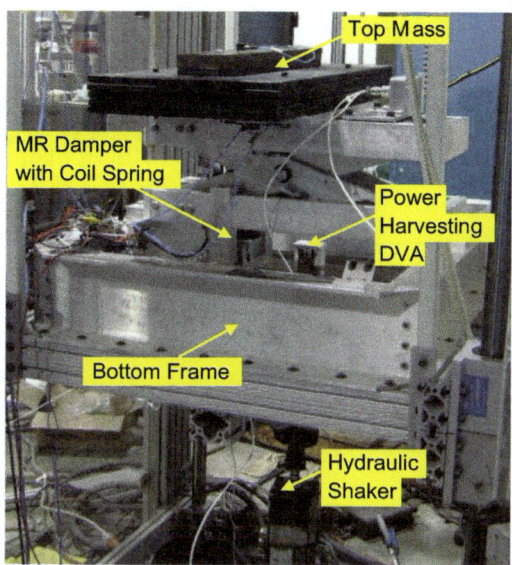

Figure 13.10   Photograph of the EFV seat suspension mockup using the self-powered MR damper. Note that, in this figure, the power harvesting DVA was mounted on the bottom frame.

a single-degree-of-freedom system having a top mass, coil spring, and self-powered MR damper. The resonance frequency of this mockup system was around 4.5 Hz. The vibration isolation performance of the seat suspension mockup using the self-powered MR damper was evaluated using eight different representative random excitation accelerations (in this study, denoted as V1–V8 acceleration levels).[9]

Figure 13.11 presents the time responses of the EFV seat suspension mockup using the self-powered MR damper under the V1 acceleration level. In the V1 case, the power harvesting DVA was mounted on the bottom frame. In addition, the passive hydraulic damper implies that the MR damper is unpowered (*i.e.*, no power harvesting DVA was connected to the damper). As seen in this figure, the self-powered MR damper can reduce the vibration acceleration of the top mass and the peak acceleration level as well. Figure 13.11(c) shows the voltage generated from the self-powered MR damper. In this case, the maximum generated voltage level was 9.6 V.

Figure 13.12 presents vibration isolation performance of the EFV seat suspension mockup using the self-powered MR damper in terms of maximum acceleration reduction. In this case, eight different representative random excitation accelerations were used. In addition, two different installations of the power harvesting DVA were considered. In the first configuration, the power harvesting DVA was mounted on the bottom frame or raft. In the second configuration, the power harvesting DVA was mounted on the top or seat mass. As seen in these figures, all three damper configurations (*i.e.*, the passive hydraulic damper, the self-powered MR damper with the DVA mounted on the top mass, and the self-powered MR damper with the DVA mounted on the

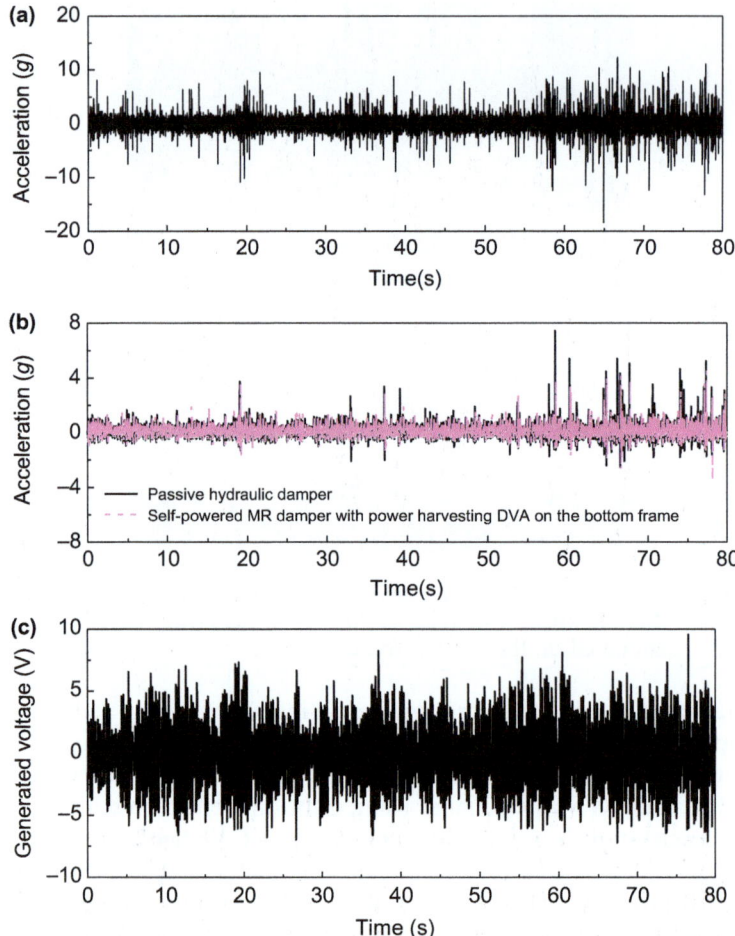

Figure 13.11   Time responses of the EFV seat suspension mockup using the self-powered MR damper under the V1 acceleration level; (a) V1 excitation acceleration, (b) top mass acceleration, (c) generated voltage.

bottom frame) reduced peak acceleration of the top mass relative to the input acceleration levels. However, for V3, V5, V6, V7, and V8 excitation cases, the self-powered MR damper with the DVA mounted on the top mass did no better than the passive hydraulic damper in reducing top mass acceleration levels. However, the self-powered MR damper with the DVA on the bottom frame performed better than the passive hydraulic damper for most acceleration levels. If the power harvesting DVA was mounted on the top mass and the top mass acceleration was isolated, then induced voltage also subsequently decreased reducing performance. This fact can be verified using data plotted in Figure 13.13. Figure 13.13 presents the voltage generated from the power harvesting DVA on the top mass under the V1 acceleration level. Comparing to the voltage level (*i.e.*, 9.6 V) of the power harvesting DVA on the bottom frame

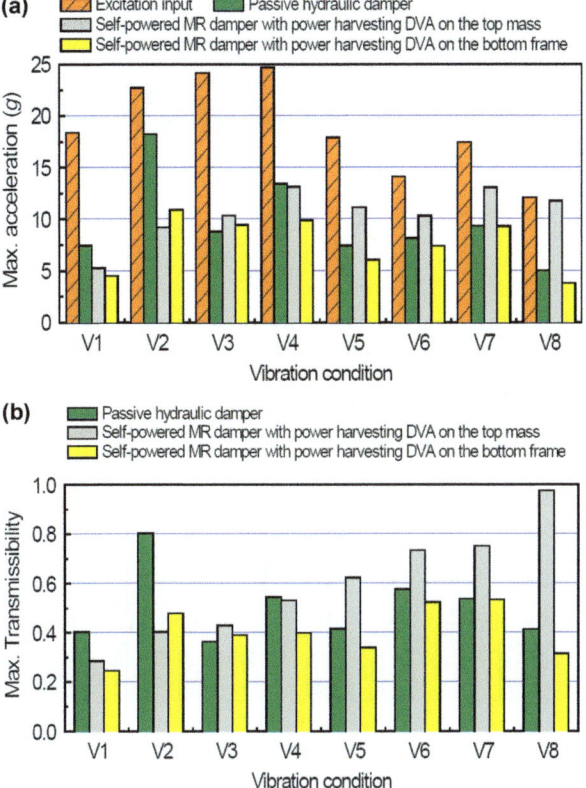

Figure 13.12 Vibration isolation performance of the EFV seat suspension mockup using the self-powered MR damper in terms of maximum acceleration reduction; (a) maximum acceleration of the top mass, (b) maximum transmissibility.

as shown in Figure 13.11(c), the voltage level (*i.e.*, 5.8 V) generated from the power harvesting DVA on the top mass was much smaller. To evaluate how well each damper reduced top mass acceleration, we calculated the maximum transmissibility as follows:

$$\text{Max. transmissibility} = \frac{\text{Max. top mass acceleration}}{\text{Max. excitation acceleration}} \quad (13.2)$$

The best and worst maximum transmissibility values of the passive hydraulic damper in this study were 0.36 and 0.80, respectively. Note that lower transmissibility value implies lower top mass acceleration level and subsequently better vibration isolation performance. The best and worst maximum transmissibility values of the self-powered MR damper with the power harvesting DVA mounted on the top mass were 0.29 and 0.98, respectively. For the self-powered MR damper with the power harvesting DVA mounted on the bottom frame, the best and worst maximum transmissibility values were 0.25

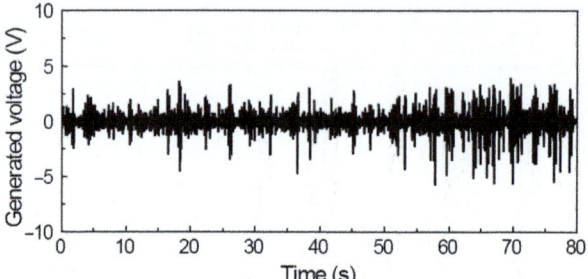

Figure 13.13   The voltage generated from the power harvesting DVA on the top mass under the V1 acceleration level.

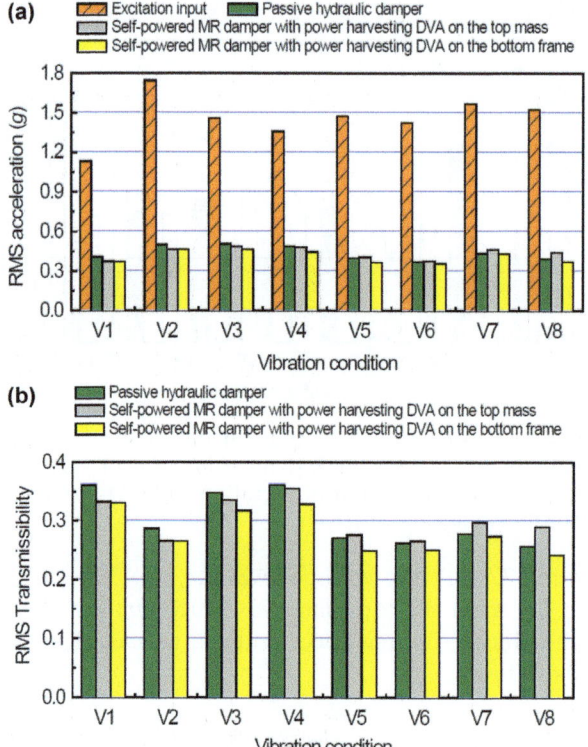

Figure 13.14   Vibration isolation performance of the EFV seat suspension mockup using the self-powered MR damper in terms of RMS acceleration reduction; (a) RMS acceleration of the top mass, (b) RMS transmissibility.

and 0.53, respectively. Among them, the self-powered MR damper with the power harvesting DVA on the bottom frame provided consistently good maximum acceleration reduction performance.

Figure 13.14 presents vibration isolation performance of the EFV seat suspension mockup using the self-powered MR damper in terms of RMS acceleration reduction. As seen in these figures, all three damper configurations

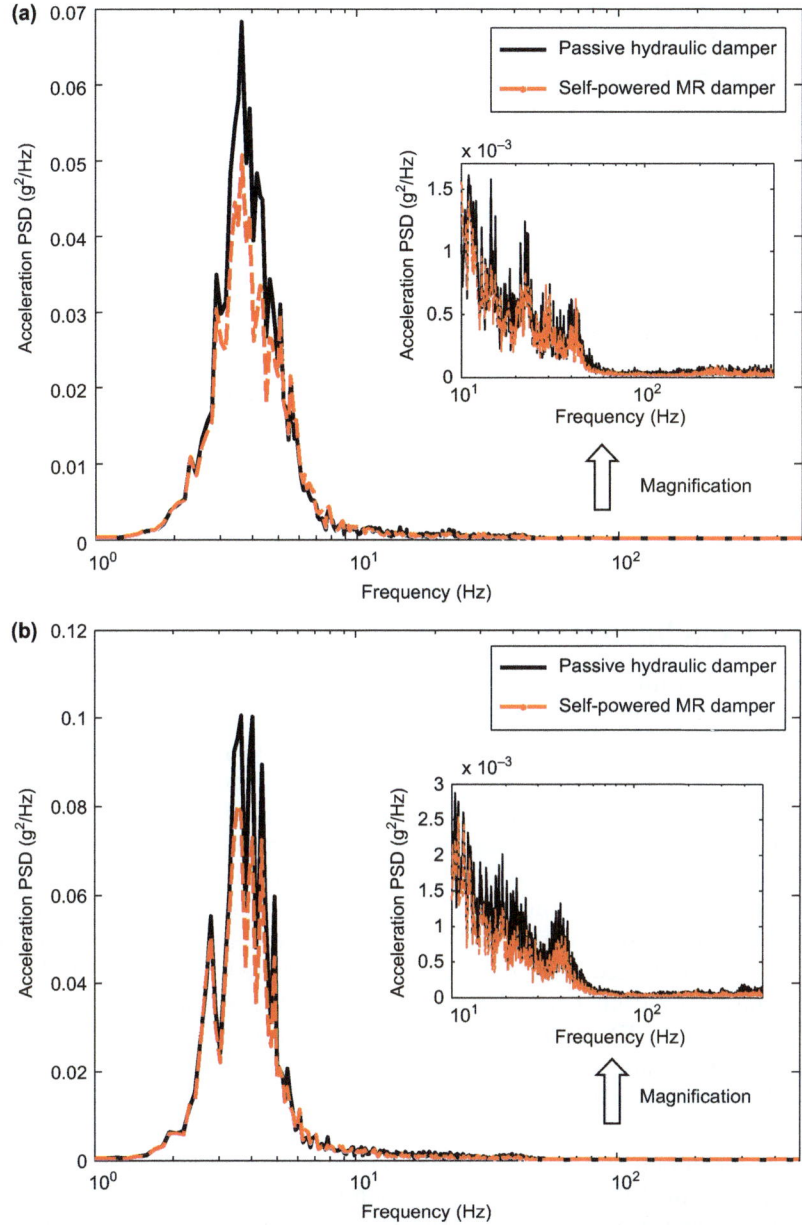

Figure 13.15   Power spectral density (PSD) of the acceleration of the top mass on the EFV seat suspension mockup using the self-powered MR damper. Here, the self-powered MR damper has the power harvesting DVA on the bottom frame; (a) under the V1 acceleration level, (b) under the V4 acceleration level.

reduced the RMS acceleration of the top mass relative to the RMS input excitation acceleration. The self-powered MR damper with the power harvesting
DVA on the bottom frame showed best performance among the three damper
configurations tested in this study. For these cases, we also calculated the RMS
transmissibility using similar relationship to the one given in eqn (13.2). The
best and worst RMS transmissibility values of the passive hydraulic damper
were 0.26 and 0.36, respectively. The best and worst RMS transmissibility
values of the self-powered MR damper with the power harvesting DVA on the
top mass were 0.27 and 0.36, respectively. For the self-powered MR damper
with the power harvesting DVA on the bottom frame, the best and worst RMS
transmissibility values were 0.24 and 0.33, respectively.

Figure 13.15 presents the power spectral density (PSD) of the acceleration of
the top mass on the EFV seat suspension mockup using the self-powered MR
damper. In this case, the V1 and V4 excitation cases were used. As seen in these
figure, the self-powered MR damper with the power harvesting DVA on the
bottom frame showed better vibration isolation performance than the passive
hydraulic damper in not only the lower frequency range, but also in the higher
frequency range.

13.4 Conclusions

Experimental evaluation of a semi-active seat suspension using a self-powered
MR damper that consisted of an MR damper with a power harvesting dynamic
vibration absorber (DVA) was conducted in this study. To this end, a power
harvesting DVA was fabricated and electronically connected to a prototype
MR damper for seat suspensions of the Expeditionary Fighting Vehicle (EFV).
The power harvesting characteristics of the self-powered MR damper were
experimentally investigated and its performance was also evaluated *via* damper
force testing. To evaluate the vibration isolation performance of a semi-active
system using the self-powered MR damper, an EFV seat suspension mockup
using the self-powered MR damper was constructed. Using eight different
representative random excitation accelerations, the vibration isolation performance of the EFV seat suspension mockup using the self-powered MR
damper was experimentally evaluated. From the results presented in this study,
we reached the following conclusions.

1. The maximum voltage levels of the power harvesting DVA electrically connected to the MR damper were less than the maximum voltage of the power
   harvesting DVA with open-circuit (*i.e.*, electrically disconnected from the
   MR damper). In particular, at 0.3 *g* excitation acceleration, the maximum
   voltage level of the power harvesting DVA connected to the MR damper
   became significantly smaller. The reason is that the electromagnetic coil
   winding of the MR damper worked as shunt damping and thus dampened
   the motion of the electromagnetic coil winding of the power harvesting DVA.
2. The measured voltage, current, and power generated from the power
   harvesting DVA were maximized near the resonance frequency, and

quickly reduced above the resonance frequency. The peak values of the maximum voltage, current, and power generated from the power harvesting DVA were 6.0 V, 0.49 A, and 2.92 W @ 1.2 g excitation acceleration. In addition, the peak values of the RMS voltage, current, and power generated from the power harvesting DVA were 2.45 V_{rms}, 0.2 A_{rms}, and 0.49 W_{rms} @ 1.2 g excitation acceleration.

3. Under constant amplitude acceleration excitation, the damper force of the self-powered MR damper decreased as the excitation frequency increased. The yield force (*i.e.*, the damper force of the self-powered MR damper minus the field-off damper force) was typically a maximum near resonance of the DVA and quickly decreased above the resonance frequency.

4. Comparing to the seat suspension using the passive hydraulic damper (*i.e.*, the MR damper with no field input), the seat suspension using the self-powered MR damper showed better reductions in both peak acceleration as well as RMS acceleration. However, the largest benefit was in the reduction of peak acceleration, as expected.

5. Two different installations of the power harvesting DVA were considered. The seat suspension using the self-powered MR damper on the bottom frame showed better vibration isolation performance than the seat suspension using the self-powered MR damper on the top mass. This was because when the power harvesting DVA was mounted on the top mass, if the top mass acceleration was isolated, then the induced voltage also decreased.

6. It was experimentally verified that the seat suspension using the self-powered MR damper could achieve good vibration isolation performance using neither a sensor to measure motion as a feedback signal nor a control algorithm.

References

1. Y.-T. Choi and N. M. Wereley, *J. Intell. Mater. Syst. and Struct.*, 2002, **13**(7/8), 443–451.
2. J. D. Carlson, D. M. Catanzarite and K. A. Clair, *Int. J. Mod. Phys. B*, 1996, **10**, 2857–2865.
3. H. J. Song, Y.-T. Choi, G. Wang and N. M. Wereley, 2009, *ASME J. Mech. Design*, **131**(9), 091008.
4. H. J. Song, Y.-T. Choi, A. S. Purekar and N. M. Wereley, *J. Intell. Mater. Syst. Struct.*, 2009, **20**, 2077–2088.
5. S.-W. Cho, H.-J. Jung and I.-W. Lee, *Smart Mater. Struct.*, 2005, **14**, 707–714.
6. K.-M. Choi, H.-J. Jung, H.-J. Lee and S.-W. Cho, *Smart Mater. Struct.*, 2007, **16**, 2323–2329.
7. B. Sapinski, *J. Theor. Appl. Mech.*, 2008, **46**(4), 933–947.
8. Y.-T. Choi and N. M. Wereley, *ASME J. Vib. Acoustics*, 2009, **131**, 1.

9. G. J. Hiemenz, W. Hu and N. M. Wereley, *J. Phys.: Conf. Ser.*, 2009, **149**(1), 012054.

10. W. Hu, N. Wilson, G. J. Hiemenz and N. M. Wereley, *Proc. ASME Conf. Smart Mater., Adapt. Struct. Intell. Syst. 2008*, Oct. 28–30, 2008, Ellicott City, Maryland, USA, SMASIS2008-542, 831–838.

11. G. A. Lesieutre, G. K. Ottman and H. F. Hofmann, *J. Sound Vib.*, 2004, **264**, 991–1001.

12. Y.-T. Choi, N. M. Wereley and Y.-S. Jeon, *AIAA J. Aircraft*, 2005, **42**(5), 1244–1251.

CHAPTER 14

Controllable Magnetorheological Damping in Advanced Helicopter Rotors

GRUM T. NGATU,[a] WEI HU,[a] NORMAN M. WERELEY,*[a]
CURT S. KOTHERA[b] AND GANG WANG[c]

[a] Dept. of Aerospace Engineering, University of Maryland, College Park,
MD, 20742, USA; [b] InnoVital Systems, Inc., Beltsville, MD, 20705, USA;
[c] Dept. of Mechanical and Aerospace Engineering, University of Alabama in
Huntsville, Huntsville, AL, 35758, USA
*Email: wereley@umd.edu

14.1 Introduction

Modern soft-in-plane helicopter main rotors are equipped with lead-lag dampers to alleviate mechanical instabilities such as ground resonance resulting from the interaction of the lightly damped regressing lag modes of rotor blades with the body support modes, as well as aeromechanical instabilities such as air resonance.[1] Soft-in-plane rotors are often fitted with lag dampers fabricated from energy dissipating materials such as elastomers.[2–5] In comparison to conventional hydraulic lag dampers, elastomeric lag dampers have a reduced parts count, are lighter in weight, are easier to maintain, and are more reliable. Unlike hydraulic dampers, elastomeric dampers do not produce extremely high damping forces at high lead-lag velocities. To provide sufficient damping, highly hysteretic elastomeric materials, such as a filled elastomer, are often utilized. Under dynamic lag motions, the elastomer is sheared, so that energy

RSC Smart Materials No. 6
Magnetorheology: Advances and Applications
Edited by Norman Wereley
Published by the Royal Society of Chemistry, www.rsc.org

dissipation is achieved by converting mechanical energy into heat. In addition, elastomer stiffness can be used to modify the natural frequency of the lag mode to avoid resonances.[6,7] However, highly damped elastomers exhibit non-linear hysteretic response to dynamic loading. Furthermore, damping and stiffness properties of elastomeric dampers are non-linear functions of lag frequency, dynamic lag amplitude, and operating temperature. Elastomeric damping and stiffness levels diminish markedly as the amplitude of damper motion increases.[2-7] In addition, there is a reduction in damping as the excitation frequency is increased. At small lead-lag displacements, elastomeric dampers have exhibited low loss factors and high stiffness, resulting in unfavorable limit cycle instabilities.[4] In forward flight conditions, the blade lead-lag motion in helicopters occurs at two frequencies, the lead-lag frequency and 1/rev frequency, and as the 1/rev amplitude is increased, it substantially reduces damping at lower lag/rev amplitudes, which may also cause undesirable limit cycle oscillations.[4,6]

To address these undesirable effects of elastomeric dampers, a damper that combines hydraulic and elastomeric damping mechanisms, referred to here as a Fluid-Elastomeric (FE) lag damper, was developed by Lord Corporation.[2,4,8,9] FE lag dampers alleviate the undesirable effects of non-linear elastomeric damper behavior by minimizing the dependence of the lag mode damping and stiffness on amplitude and frequency to achieve nearly linear performance. Because damping is supplied primarily *via* viscous damping in the FE damper, the elastomeric material is selected based on stiffness and shear fatigue properties, as opposed to damping properties.[8]

A second approach to address the undesirable effects of elastomeric dampers is the introduction of adaptive or smart lag damping employing magnetorheological (MR) fluids. Even though FE lag dampers provide substantially improved performance over elastomeric dampers, they are limited to providing fixed or passive damping. Since damping augmentation is only required over certain flight regimes where there is a potential for instabilities to occur,[10] a passive damper providing a fixed damping could produce large periodic loads in the rotor hub. Furthermore, passive dampers tend to present damping loss as temperature increases either due to in-service self-heating or hot operating conditions. Under these circumstances, elastomer softening and/or fluid thinning occurs which adversely affects damper performance.[2-4,11,12] Thus, an adaptive damper, which can produce the desired amount of damping without a corresponding increase in periodic loads and can be adjusted to compensate for performance losses at extreme environmental conditions, would be of considerable value. Magnetorheological (MR) fluids typically consist of spherical micron-sized magnetic particles (microspheres) suspended in a liquid medium such as silicone or hydraulic oil.[13,14] Their rheological properties, that is their yield stress and viscosity, can be rapidly and continuously controlled by varying the applied magnetic field. Conventional MR fluids utilize spherical 7–10 micron diameter iron particles at high weight fractions (65–80 wt%) and have been shown to exhibit yield stress as high as 100 kPa.[15,16] The fast response and controllability of MR fluids are advantageous for implementation in semi-active smart vibration-absorption systems,[17] primary vehicle suspension systems,[18,19] landing gear for aircraft,[20-22] adaptive crew seats for vibration,[23-25] and shock isolation.[26,27] Because the yield stress of the MR fluid exhibits a substantial controllable range when a magnetic field is applied, many MR fluid based devices were designed such that the damping level can be controlled in feedback by applying a magnetic field.[28,29] MR fluids have also been considered for application to helicopter rotor lag dampers.[7,10,30,31] Much of this work has been done to evaluate the feasibility and capabilities of employing MR dampers in conjunction with elastomeric materials in helicopter lead-lag damping applications.[10,31] A pair of 1/6th Froude-scale Comanche helicopter fluid-elastomeric lag dampers were modified to be hybrid elasto-meric-MR dampers and their capabilities were demonstrated for lag mode damping applications. Recently, a full-scale linear stroke tubular magnetorheological fluid-elastomeric (MRFE) lag damper, which can be fully integrated into an actual helicopter rotor system, was developed.[30] The linear stroke MRFE lag damper was developed as a retrofit to an existing elastomeric lag damper, configured to shear elastomeric material sandwiched between two concentric tubes that move relative to each other. In the inner metallic tube of the elastomeric damper, a piston, an MR valve and MR fluid were introduced. In this arrangement, the elastomeric and MR damping effects are decoupled systems, such that their individual contribution can be nearly superposed to yield the total MRFE damper performance.

In this chapter, full-scale FE and MR fluid-elastomeric (MRFE) lag dampers are studied. These devices were experimentally evaluated for their controllable damping capability and dynamic range under sinusoidal excitation, consistent with helicopter lead-lag damper loading conditions. Due to kinematical complexity in modern hingeless helicopter main rotors, a snubber type FE lag damper is usually made from a laminated stack of alternating elastomeric-metallic rings, and damping fluids are included in the flexible body to increase dynamic range of the snubber damper.[32] In the present FE lag damper, a laminated stack of elastomeric and metallic rings forms the cylindrical damper body having two distinct internal chambers. The flexing of the elastomeric container forces fluid to flow from one chamber into the other through flow ports, thus providing the required damping. This FE damper has an advantage over conventional linear stroke hydraulic dampers because it has no moving parts, thus avoiding the issue of wear and leakage. Although the elastomer introduces a significant stiffness to the system, its damping contribution is minimal compared to the viscous damping generated. The MRFE damper utilizes the existing flexible damper body and the flexible center wall of the FE damper, with the two port holes in the center wall retrofitted with one MR valve each. The MRFE damper provides controllable damping that can augment damping under key operating conditions, and increase stability of a helicopter rotor. Further, the MRFE damping components provide fail-safe passive damping in the event of reduced or no-power operation, albeit at reduced effectiveness. The MRFE damper is similar to the FE damper in that neither damper has moving parts.

The complex modulus method is employed to characterize and compare the FE damper to the MRFE damper. It is shown that the field-off damping of the MRFE damper is less than the FE damper, and a maximum controllable damping level, which is greater than that of the FE damping level, is provided by the MRFE damper when magnetic field is applied. However, the complex modulus linearization technique does not describe the non-linear behavior of the MRFE damper under sinusoidal excitation resulting from the magnetorheological effect. Various models based on lumped parameter approaches have been proposed to model the non-linear hysteretic behavior of elastomeric lag dampers,[3,5,33–35] MR dampers,[10,31,36–40] and hybrid elastomeric-MR and MRFE lag dampers[7,10,30,31] under harmonic excitations. In an elastomeric material, where non-linear behavior arises from excitation amplitude, frequency and temperature, the complex modulus can be computed at each test condition to reconstruct the hysteretic characteristics with an acceptable accuracy. However, most elastomeric lag dampers are filled elastomers and exhibit hysteretic behavior that is also displacement and velocity dependent, resulting in a non-elliptical shear strain *vs.* shear stress curve when undergoing harmonic excitation. In this regard, Snyder[3] proposed a non-linear model that combines a Kelvin chain in parallel with a cubic spring and a linear elasto-slide element. Hu[5] developed a model based on the triboelastic theory that combined rate-dependent elasto-slide elements with a yield-distribution function to account for the yield force. Classic Bingham–Plastic and Herschel–Bulkley based models have been utilized to model ER and MR devices.[36–38] Further, mechanisms-based and/or physically motivated models have been developed to characterize MR dampers. Kamath[10,31] presented a hysteretic model combining a stiffness plus viscoelastic-plastic model with a non-linear weighting function. Hu *et al.* formulated the rate-dependent elasto-slide model, which uses a rate-dependent slide and a parallel viscous damper to emulate the yield behavior of the MR fluid, and uses a stiff spring in series to represent the pre-yield behavior.[40] These models were able to capture and reconstruct the non-linear dynamic hysteresis behavior of elastomeric and MR dampers for the tested amplitude and frequency spectrum. All these parametric models are based on the physical phenomena of the system whereby a series of mechanisms are lumped to relate the input to the output. However, these models do not describe the actual physical flow phenomena or flow dynamics in fluid based MR dampers. They do not show the interaction between mechanical and hydraulic systems, which is an inherent feature of hydraulic based MRFE dampers.

Thus, a hydro-mechanical analysis, which can delineate the physical flow motion of the system and accurately describe the non-linear hysteretic behavior of the MRFE damper is proposed. Although hydro-mechanical analysis has been applied in passive systems[41,42] and active ER and MR damper systems,[43,44] there only have been limited studies in their application in helicopter damping systems and specifically in snubber type MRFE lag dampers, which are coupled systems of MR fluid and elastomeric material. The hydro-mechanical analysis explored in this chapter describes the FE damper using a series of lumped hydraulic and mechanical parameters representing various components of the damper. The model employs such physical parameters as inertia, damping, yield force and compliance. These physical parameters are dependent on damper geometry, as well as the material properties of components that can be determined *a priori*. To introduce the magnetorheological effects in the MRFE damper, prior efforts have resorted to a decoupled system, which modeled the FE and MR components of the dampers as a parallel configuration of elastomeric and MR damper elements.[7,30] In the MRFE lag damper presented here, the hydro-mechanical model can account for coupling – between the elastomeric material and the MR fluid flow – that is manifested through chamber compliance. Using a parameter identification technique, the model is shown to accurately simulate the measured force *versus* displacement time history under single frequency excitations.

14.2   Fluid-Elastomeric (FE) Lag Damper

An isometric view of the FE (damper for a Bell 430 helicopter) damper is shown in Figure 14.1. The flexible body of the damper is made of metallic rings interspersed with elastomeric layers, or a multiple lamination of metallic and elastomeric ring layers. The laminate stack is bonded to metallic plates at both ends that form the top and bottom of the container. A flexible center wall is placed in the middle of the cavity formed by the damper body. This flexible center wall, made of rubber, runs along the diameter of the flexible cylinder, completely molded at both ends to the elastomeric damper body. This creates two distinct chambers into which a viscous fluid can be poured and completely

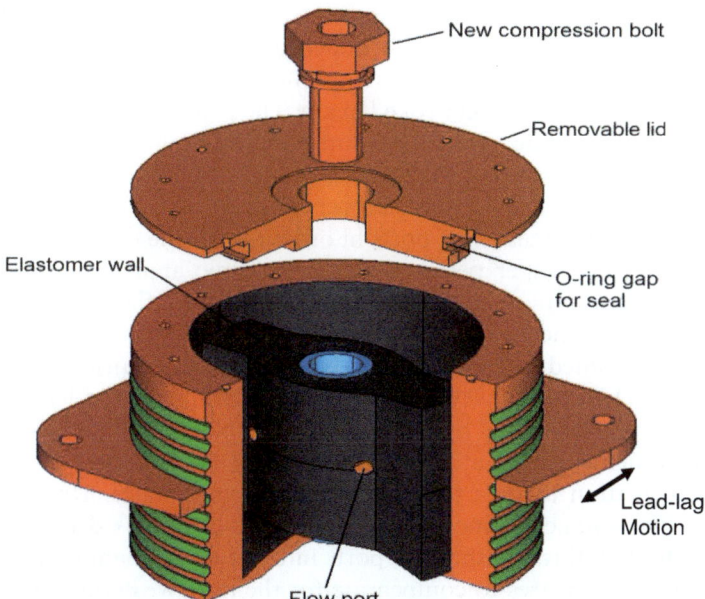

Figure 14.1   Isometric view of the FE lag damper.

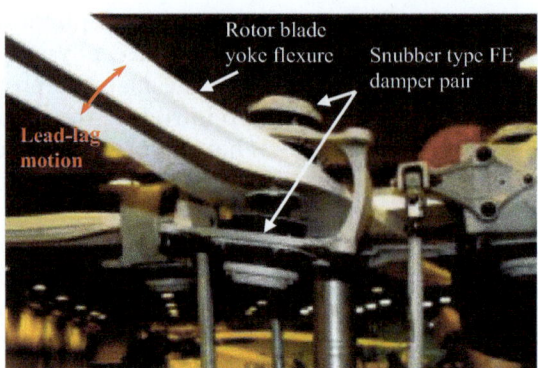

Figure 14.2   Schematic of main rotor blade with paired FE lag dampers (Courtesy of
Bell Helicopter).

sealed. Fluid can be exchanged from one reservoir to the other only by passing
through the two flow ports located in the center wall (Figure 14.1). There are
four threaded holes for filling and draining fluid, two on each side of the
damper. Once fluid is poured into the chambers, the holes are sealed with bolts
and a compression bolt is mounted through the center hole of the damper
assembly, which creates a positive internal pressure. This damper comprises a
compact volume with a height of about 107 mm (4.20 in) and diameter of
118 mm (4.65 in). In the helicopter rotor system shown in Figure 14.2, each blade
is equipped with a pair of FE lag dampers. The lag dampers, which are housed in
the pitch horn adapter assembly, are attached to the inboard side of the blade
cuff or torque tube, which in turn is attached to the blade and flexure at its
outboard end. The paired lag dampers are interconnected by a shear restraint
that is installed in the composite yoke flexure. The relative motion of the flexure
yoke and the torque tube, which constitutes the lead-lag motion, deforms the
paired lag dampers. The flow ports are located in the middle of the flexible center
wall, and the deformation of the snubber damper body forces fluid to flow
between chambers via these flow ports in order to provide damping force.

To evaluate FE damper performance, a dynamic analysis is needed. For
small displacement applied to the center of the damper, the FE damper is assumed to deform as shown in Figure 14.3. In the actual rotor blade setting, the
lag damper is intended to accommodate both radial and translational lead-lag
motions of the blades. However, for small displacement, the lead-lag blade
motion can be approximated by the translation depicted in Figure 14.3(b). The
forcing function is applied at the center plate of the damper, which creates
damper deformation relative to the upper and lower plates. Under such deformation, the volume of one of the elastomeric chambers decreases, which
forces fluid to flow through the flow ports into the other chamber as the other
chamber volume increases to compensate for the in-flow. Because the damper
geometry is complex, the input displacement excitation varies along the axis of
the compression bolt. It varies from zero at the top and bottom caps to a

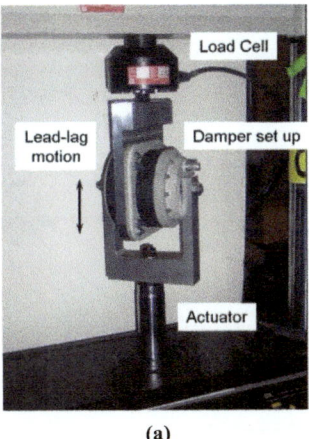

(a)

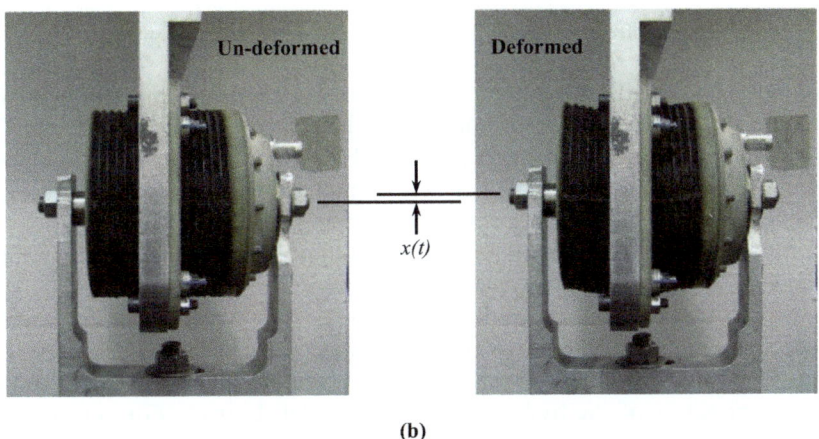

(b)

Figure 14.3 FE lag damper test set-up on MTS testing machine. (a) FE lag damper test set-up on MTS® testing machine. (b) Deformation of FE lag damper on MTS® testing machine.

maximum at the center plate of the damper. However, all damper analysis is evaluated based on the input excitation at the damper center plate.

To mimic the impact of lead-lag motion of helicopter blades as applied to the FE lag damper, a test fixture was fabricated to hold the dampers in a 24.5 kN MTS servo-hydraulic testing machine. Figure 14.3(a) shows the FE lag damper test set-up where the center plate is connected to the load cell *via* an extension rod. The lower mounting bracket is attached to the lag damper through a mounting rod that is slotted in at the mounting hole, located at the lag damper center, passing through the center wall. The lower mounting bracket is attached to the MTS actuator through an extension rod. The center wall is oriented horizontally such that the motion of the MTS actuator connected to the lower bracket forces fluid to flow from one chamber to the other, by deforming the

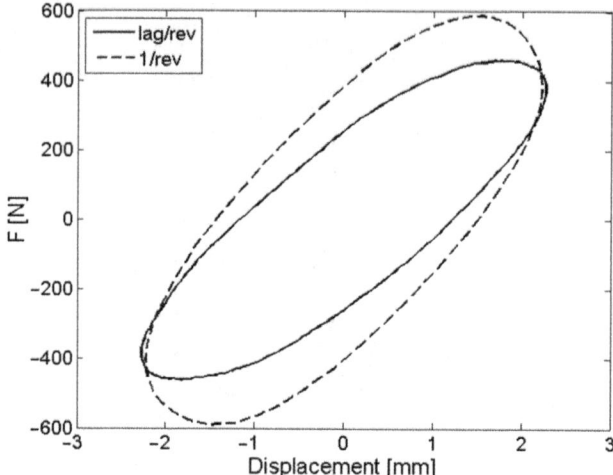

Figure 14.4 Typical force *vs.* displacement hysteresis results at single frequency.

elastomeric outer wall, and emulating the lead-lag motion. Displacement was measured using an LVDT sensor and force was measured using a load cell. The excitation frequencies were chosen to be 3.8 Hz and 5.8 Hz, which correspond to typical rotor system lag/rev and 1/rev or main rotor RPM. Damper testing was carried out for excitation amplitudes ranging from 0.762 mm (30 mil) to 3.81 mm (150 mil) in increments of 0.50 mm (20 mil) at both lag/rev and 1/rev frequencies.

For single frequency evaluation, the damper was subjected to sinusoidal displacements with varying excitation amplitudes as above. The single frequency force *vs.* displacement and force *vs.* velocity data were utilized to identify parameters of the hydro-mechanical model. Using a Fourier analysis, the measured displacement and velocity data were filtered, so that only the first harmonic was used:

$$x(t) = X_s \sin(\omega t) + X_c \cos(\omega t)$$

$$\dot{x}(t) = \omega X_s \cos(\omega t) - \omega X_c \sin(\omega t)$$

(14.1)

For the selected amplitude of 2.286 mm, Figure 14.4 shows the force *vs.* displacement of the FE damper at lag/rev and 1/rev frequencies. In each force *vs.* displacement trace, the area enclosed by the hysteresis cycle is proportional to the amount of energy dissipated per cycle. It is observed that, at the 1/rev frequency, the dissipated energy per cycle is greater than for the lag/rev frequency.

14.2.1   FE Damper Characterization

A typical approach for characterization of damper performance is the complex modulus method. It is a linear characterization technique of damper properties which treats the complex stiffness $k*$ as a combination of the in-phase stiffness k' and the loss stiffness k'', given as:

$$k* = k' + ik'' = k'(1 + i\eta) \tag{14.2}$$

Here, the loss factor, η, is defined as the ratio of the loss (quadrature) stiffness to the in-phase stiffness. The damper force is estimated by the first Fourier sine and cosine components at the excitation frequency:

$$F(t) = F_s \sin(\omega t) + F_c \cos(\omega t) = k'x(t) + \frac{k''}{\omega}\dot{x}(t) \tag{14.3}$$

The stiffness k' and k'' are determined by substituting the displacement and velocity from eqn (14.1) into the force equation above to obtain:

$$k' = \frac{F_c X_c + F_s X_s}{X_c^2 + X_s^2} \tag{14.4}$$

$$k'' = \frac{F_c X_s - F_s X_c}{X_c^2 + X_s^2} \tag{14.5}$$

The equivalent damping C_{eq} is given by:

$$C_{eq} \cong \frac{k''}{\omega} \tag{14.6}$$

This linearization technique is an approximation because the complex stiffness assumes steady state harmonics at the excitation frequency. However, it gives an acceptable representation of the linearized in-phase stiffness and equivalent damping of the FE damper for comparing overall damping performance under different loading conditions and temperature variations. The in-phase stiffness k' and equivalent damping C_{eq} of the FE damper at lag/rev and 1/rev are shown in Figure 14.5. The equivalent damping of the device is moderately dependent on amplitude. The in-phase stiffness variation of the FE damper is fairly small at all dynamic amplitudes tested at both frequencies.

14.3 Modeling of FE Damper

The aforementioned complex modulus, or Kelvin model, is a linearized characterization of the FE damper. However, the complex modulus does not describe the flow dynamics within the FE damper, which is, essentially, a hydraulic damper system. This approach treats the damper as a 'black box,' and fits the Kelvin model to the overall damper behavior relating input to output. It does not describe the inherent hydro-mechanical coupling phenomena of the system and does not distinguish between the contributions of the elastomeric material and the hydraulic flow physics. To better understand the behavior of this FE damper, a hydro-mechanical model, based on lumped fluid and mechanical systems, is proposed.

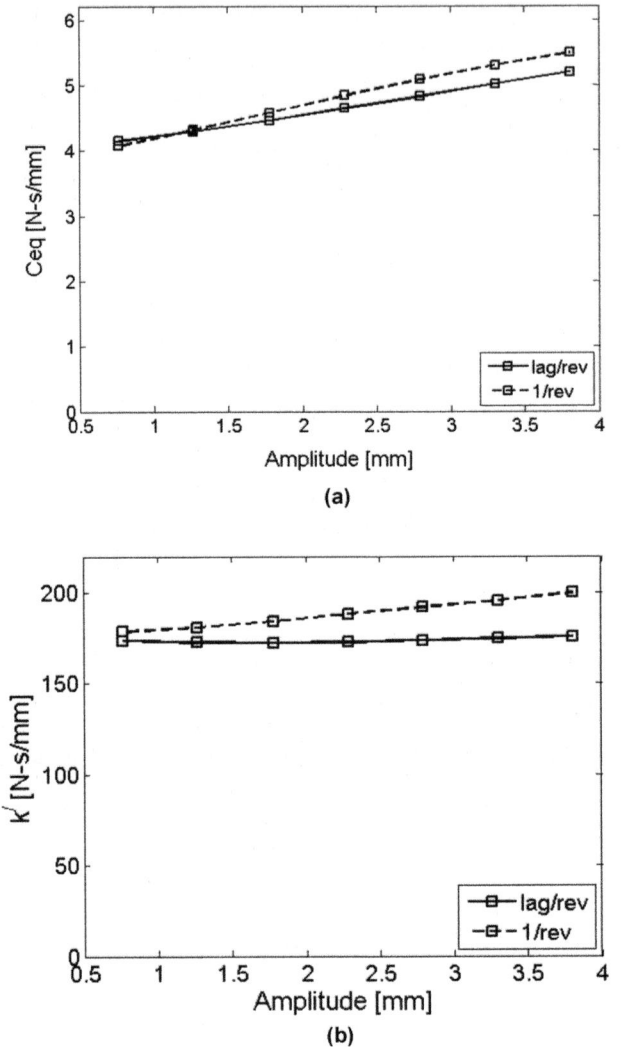

Figure 14.5   FE damper characterization. (a) Equivalent damping. (b) In-phase stiffness.

The proposed lumped parameter control volumes are shown in Figure 14.6. For clarity, the two flow ports or valves are identified with subscripts a and b. There are two volume chambers (1 and 5), which are lumped into fluid pressures P_1 and P_5 and compliances S_1 and S_5. Both S_1 and S_5 represent the compliances of the fluid and the chamber. Control volumes $a2$, $a3$, $a4$, $b2$, $b3$ and $b4$ are lumped into fluid inertances I_{a2}, I_{a3}, I_{a4}, I_{b2}, I_{b3} and I_{b4} and passive flow resistances R_{a2}, R_{a3}, R_{a4}, R_{b2}, R_{b3} and R_{b4}. One-dimensional fluid flow is assumed such that the velocity and pressure of a control volume are span-wise constant for a particular instant in time; however, these quantities do vary with time. In addition, incompressible and laminar flow is assumed.

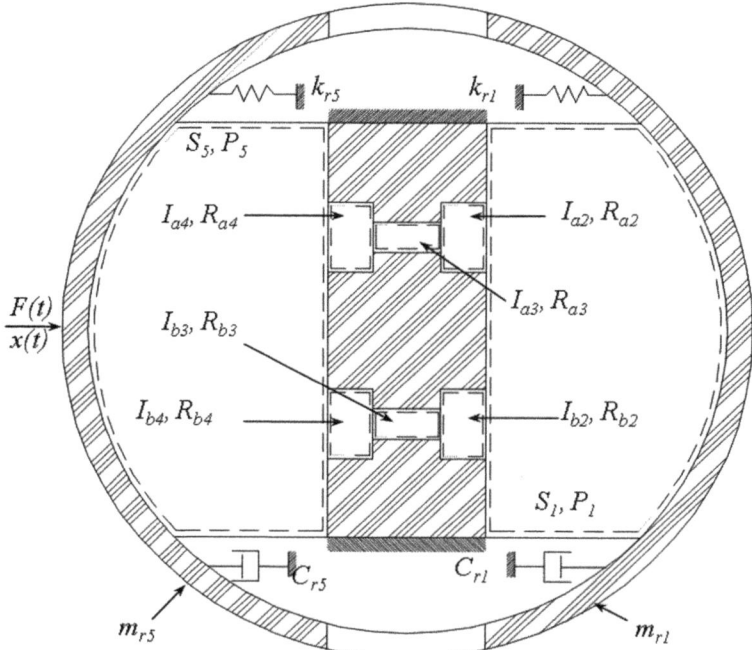

Figure 14.6 Section view of proposed hydromechanical model of FE damper.

The model is first formulated for a general case where the two flow ports and two chambers are assumed to be different. The Kelvin model is assumed for the FE damper elastomeric body in shear deformation, where it is modeled as a linear spring and linear dashpot combination. The two elastomeric chambers are broken into two semi-circular walls connected at the center, and they are assigned masses m_{r1} and m_{r5}. The shear mode in-phase stiffness and equivalent damping of the FE damper elastomeric body from the complex modulus method are utilized as the spring constants k_{r1} and k_{r5} and damping coefficients C_{r1} and C_{r5}. These values were estimated by emptying the FE damper and testing the elastomeric body alone at varying amplitudes.

The lumped parameter model includes three lumped masses (m_A, m_B and m_r) and three dynamic degrees of freedom ($x(t)$, $x_A(t)$ and $x_B(t)$). The governing equation of motion of the elastomer is given by:

$$F(t) - (k_{r1} + k_{r5})x(t) - (C_{r1} + C_{r5})\dot{x}(t)$$
$$- A_{\mathrm{p}}(P_5(t) - P_1(1)) = (m_{r1} + m_{r5})\ddot{x}(t) \tag{14.7}$$

Similar to an elastomer deformed in shear, bulging of an elastomer due to volumetric expansion produces damping as well. This bulge damping is captured by introducing a resistance parameter[45] on the net flow into the compliant chambers resulting in volumetric expansion. Thus, applying a continuity argument to volume chambers 1 and 5, yields:

$$A_p \dot{x}(t) - q_a(t) - q_b(t) = S_5(\dot{P}_5(t) - R_5(A_p \ddot{x}(t) - \dot{q}_a(t) - \dot{q}_b(t))) \qquad (14.8)$$

$$-A_p \dot{x}(t) + q_a(t) + q_b(t) = S_1(\dot{P}_1(t) - R_1(-A_p \ddot{x}(t) + \dot{q}_a(t) + \dot{q}_b(t))) \qquad (14.9)$$

where R_1 and R_5 are the elastomeric resistances. The flow rates through flow ports a, and b, in terms of the fluid inertance displacements x_a and x_b are given by:

$$q_a(t) = A_{a2} \dot{x}_{a2}(t) = A_{a3} \dot{x}_{a3}(t) = A_{a4} \dot{x}_{a4}(t) \qquad (14.10)$$

$$q_b(t) = A_{b2} \dot{x}_{b2}(t) = A_{b3} \dot{x}_{b3}(t) = A_{b4} \dot{x}_{b4}(t) \qquad (14.11)$$

where A_{ai} and x_{ai} are the cross sectional area and fluid inertance displacement of the ith lump in flow port a, and similarly for flow port b. The combined compliances of the fluid chambers for control volume 1 and 5 are given by:

$$S_1 = \frac{A_p^2}{k_{c1}}, \quad S_5 = \frac{A_p^2}{k_{c5}} \qquad (14.12)$$

Here, k_{c1} and k_{c5} represent the bulge stiffness effects of the two volume chambers, which are dominated by the elastomer bulge stiffness contribution. Due to the complexity of the damper design, deformation and the absence of a well defined piston, an equivalent piston area, A_p, is formulated to account for the total displaced fluid volume.

Also from the hydraulic model shown in Figure 14.6, the total pressure drop in the two flow ports or valves (control volumes 2 to 4) due to fluid flow is given by:

$$(P_5(t) - P_1(t)) = I_a \dot{q}_a(t) + R_a q_a(t) \qquad (14.13)$$

$$(P_5(t) - P_1(t)) = I_b \dot{q}_b(t) + R_b q_b(t) \qquad (14.14)$$

The terms I_a and R_a are given by:

$$I_a = \rho \sum_{i=2}^{4} \frac{l_{ai}}{A_{ai}} \qquad (14.15)$$

$$R_a = \frac{128\mu}{\pi} \sum_{i=2}^{4} \left(\frac{l_{ai}}{D_{ai}^4}\right) + R_{\text{minor losses}} \qquad (14.16)$$

where ρ is fluid density and l_{ai} and D_{ai} are length and diameter of the ith lump in flow port a. The equation for the resistance is derived from laminar flow analysis. The above equations hold true for the second flow port by replacing the subscript a with b. The viscosity μ was measured using a rheometer. Due to the laminar flow assumption with low Reynolds number, minor viscous losses are neglected.

By eliminating the internal variables P_1 and P_5 from eqn (14.8)–(14.11) and (14.13)–(14.14), we obtain the following two equations in terms of time varying variables:

$$(m_{r1} + m_{r5})\ddot{x}(t) + (C_{r1} + C_{r5})\dot{x}(t) + (k_{r1} + k_{r5})x(t)$$

$$+ (C_{c1} + C_{c5})(\dot{x}(t) - (1 + \psi)\dot{x}_A(t)) + (k_{c1} + k_{c5})(x(t) - (1 + \psi)x_A(t)) = F(t)$$

$$(14.17)$$

$$m_A\ddot{x}_A(t) + C_A\dot{x}_A(t) - (C_{c1} + C_{c5})(\dot{x}(t) - (1 + \psi)\dot{x}_A(t))$$

$$- (k_{c1} + k_{c5})(x(t) - (1 + \psi)x_A(t)) = 0$$

$$(14.18)$$

where the valve factor $\psi(i\omega)$ is given by:

$$\psi(i\omega) = \frac{m_B i\omega + C_B}{m_A i\omega + C_A} \tag{14.19}$$

The generalized inertance displacement x_A and mass m_A, and the viscous damping C_A for valve a and the bulge damping C_{c1} and C_{c5} are expressed as:

$$x_A = (A_{a4}/A_p)x_{a4}, \quad m_A = I_a A_p^2, \quad C_A = R_a A_p^2, \tag{14.20}$$

$$C_{c1} = R_1 A_p^2, \quad C_{c5} = R_5 A_p^2 \tag{14.21}$$

This holds true for the second flow path by replacing the subscript a with b. Eqn (14.17)–(14.18) are the basic equations of the hydro-mechanical model describing the FE damper system. The generalized inertance variables x_A and x_B are coupled through the valve factor function ψ. The valve factor ψ theoretically varies from zero (no flow in flow valve a) to infinity (no flow in flow valve b). The valve factor is a function of the head losses encountered in the flow ports as shown in eqn (14.19). C_{c1} and C_{c5} represent the damping contribution of the elastomeric body due to volumetric flexing of volume chambers.

Figure 14.7 depicts the lumped parameter hydro-mechanical model in terms of generalized inertance variable x_A. The system is composed of parallel and series combinations of linear springs and dashpots elements. The analogous mechanical system (Figures 14.6 and 14.7) and associated governing eqn (14.17)–(14.18) show that the hydro-mechanical coupling behaviors are manifested in the chamber compliances (C_{c1}, C_{c5}, k_{c1}, k_{c5}). The generalized inertance masses (m_A, m_B) and viscous damping (C_A, C_B) terms represent contributions from the hydraulic component of the device.

In the present design, the two flow ports and elastomeric chambers are identical, so that the following assumptions will hold: $k_{c1} = k_{c5} = k_c$,

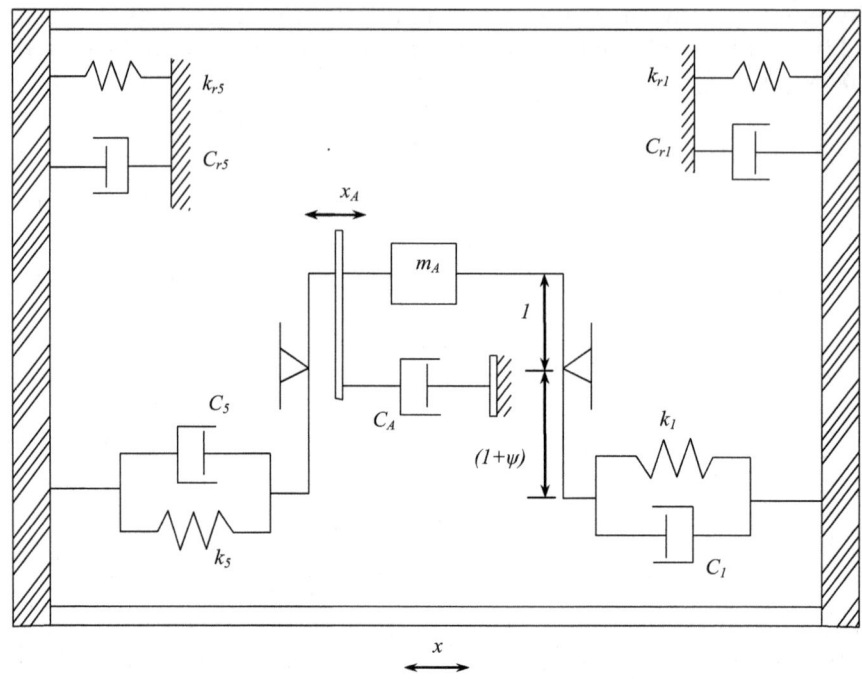

Figure 14.7   Analogous mechanical system of the hydro-mechanical model in terms of x_A (valve a).

$C_{c1} = C_{c5} = C_c$, $m_{r1} = m_{r5} = m_r$, $k_{r1} = k_{r5} = k_r$, $C_{r1} = C_{r5} = C_r$ and $\psi = 1$. Thus, the equations of motion from eqn (14.17)–(14.19) will reduce to:

$$2m_r\ddot{x}(t) + 2C_r\dot{x}(t) + 2k_r x(t) + 2C_c(\dot{x}(t) - 2\dot{x}_A(t)) + 2k_c(x(t) - 2x_A(t)) = F(t)$$

$$(14.22)$$

$$m_A\ddot{x}_A(t) + C_A\dot{x}_A(t) - 2C_c(\dot{x}(t) - 2\dot{x}_A(t)) - 2k_c(x(t) - 2x_A(t)) = 0 \quad (14.23)$$

Several parameters were determined *via* measurement or geometry in order to reduce the number of parameters that must be identified *via* analysis. Table 14.1 outlines these parameters, as well as identified and dependent parameters. The dependent parameters are functions of both the measured and estimated parameters. In order to minimize error between the measured and predicted damping forces of the FE damper at each test temperature, a constrained least-mean-squared (LMS) error minimization technique was employed to estimate and optimize the equivalent piston area A_p, the bulge stiffness k_c and the bulge damping C_c. The rubber mass is neglected in this optimization. The error function for the hydro-mechanical model is expressed as:

$$E(A_p, k_c, C_c) = \sum_{j=1}^{N} (F(t_j) - F^*(t_j))^2 \quad (14.24)$$

Table 14.1 Measured, identified and dependent model parameters.

Measured Parameters	Identified Parameters	Dependent Parameters
Viscosity, μ	Eqv. Piston area, A_p	Mass, m_A
Resistance, R	Bulge stiffness, k_c	Viscous damping, C_A
Density, ρ	Bulge damping, C_c	
Inertance, I	Yield force, F_y (MRFE only)	
Shear stiffness, k_r		
Shear damping, C_r		

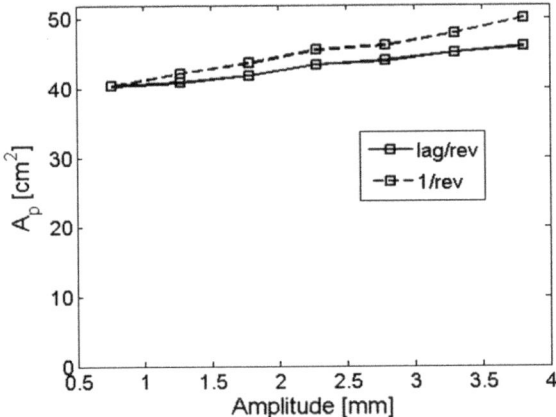

Figure 14.8 Optimized equivalent piston area A_p.

where $F(t_j)$ is the measured force, $F^*(t_j)$ is the model calculated force, t_j is the time at which the jth sample was measured and N is the number of data points per cycle.

Neglecting the density variation with temperature, the inertance was assumed to remain constant. The viscosity of the working fluid was measured using an Anton Paar Physica MCR300 parallel disc rheometer. Using the measured viscosity value, the resistance in the port holes was found to be 480 MPa s m^{-3}. Figure 14.8 shows the optimized equivalent piston areas obtained from the error minimization procedure at different amplitudes and frequencies. The equivalent area increases as the stroke amplitude is increased. The stiffness contributions from shear and volumetric deformation of the elastomeric chamber of the FE damper at lag/rev and 1/rev are given in Figure 14.9. At both frequencies, the bulge stiffness is much higher than the shear stiffness of the elastomer. Both stiffness values remain relatively constant with varying amplitudes. The damping contributions from the hydraulic and mechanical components are given in Figure 14.10, which show that the majority of the FE damping is provided through viscous flow, while elastomeric shear deformation contributes minimal damping. The viscous damping, which is proportional to the square of the equivalent piston area, increases with increasing amplitude.

The damping force *versus* velocity cycles at lag/rev and 1/rev frequencies are given in Figure 14.11. The plots show that the hydro-mechanical model closely

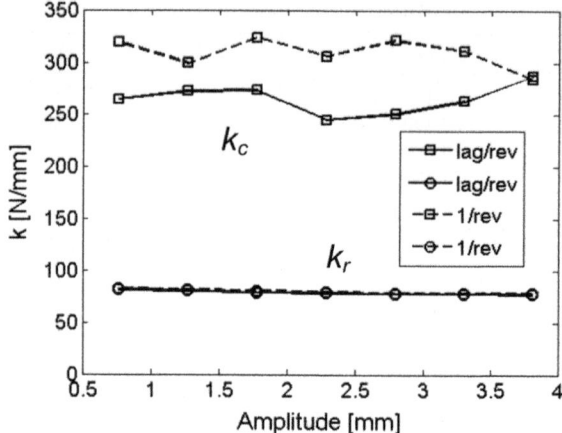

Figure 14.9   Bulge and shear stiffness of FE damper.

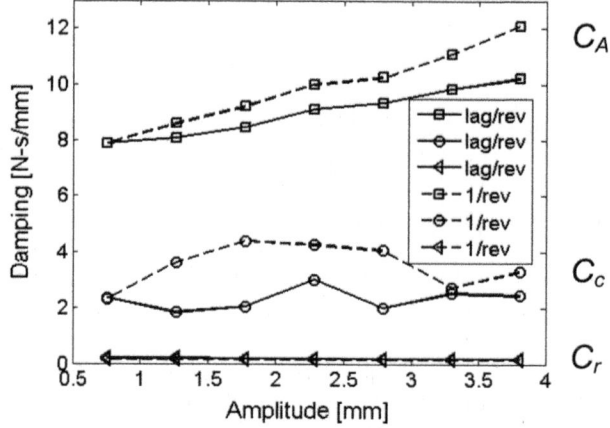

Figure 14.10   Damping of FE damper.

estimates the actual hysteretic damping force behavior of the FE damper measured in the experiments.

14.4   Magnetorheological Fluid-Elastomeric (MRFE) Lag Damper

The design of the MRFE damper requires the development and retrofitting of two MR valves inside the existing FE damper body and exchanging the existing passive, hydraulic fluid with field-controllable, MR fluid (provided by LORD Corp.). The development of the MR valves involves fitting the existing two port holes located in the center wall with two flow mode MR valves. The development of the MR valves inside the existing lag damper center wall is shown in Figure 14.12. These internal MR valves are placed at the two existing flow ports

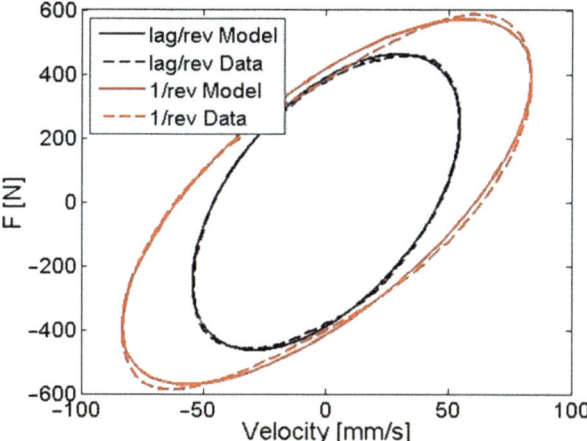

Figure 14.11   Comparison between measured and predicted damper forces.

located in the center wall, one in each hole on each side of the fluid chambers. Figure 14.12(a) shows the top view (top plate removed) of the lag damper with MR valves installed. The new MR retrofit design introduces only the MR valves. Thus, the MRFE damper still maintains the durability, and leakage free feature of the FE damper. In addition, this helps minimize the weight penalty incurred due to the addition of components. Figure 14.12(b) shows a detailed sketch of one of the incorporated MR valves with its basic components. The MR valve includes a bobbin with coil and a flux return, which in this case is designed in the shape of a stepped hollow cylinder, with the wider portion accommodating the bobbin, and the narrower one to be threaded into the flow ports. The diameter of the flow port has been oversized as much as possible to reduce associated viscous losses. A magnetic field is provided by an electromagnet enclosed in the valve such that the MR fluid flowing through the flow valve can be activated. The top plate of the original damper has also been modified to have access to the enclosed electromagnet's wires to apply current.

Figure 14.12(b) also shows the flow path of the MR fluid in the valve. The damper is assembled as a sealed unit, so that minimal possibility exists for the fluid to pass from one chamber to the other without traveling through the flow port of the MR valve. The electromagnetic coil in the MR valve is wound around the bobbin, which is inserted into the flux return. The annular gap thickness can be varied by changing the radius of the bobbin, r, changing the inner diameter of the flux return ring, or changing both at the same time. The active valve length, L_c, can be adjusted to increase or decrease the magnetic flux that activates the MR fluid, as well. Overall, the maximum size of the valve, specified by the flux return outer diameter, D_o, and the valve length, L_p, determine whether the valve fits within the constraints of the existing damper body and does not affect the required stroke of the damper to avoid contact with the outer cylindrical wall, which are the goals of the design. Figure 14.12(b) also depicts the path of the magnetic flux which is initiated

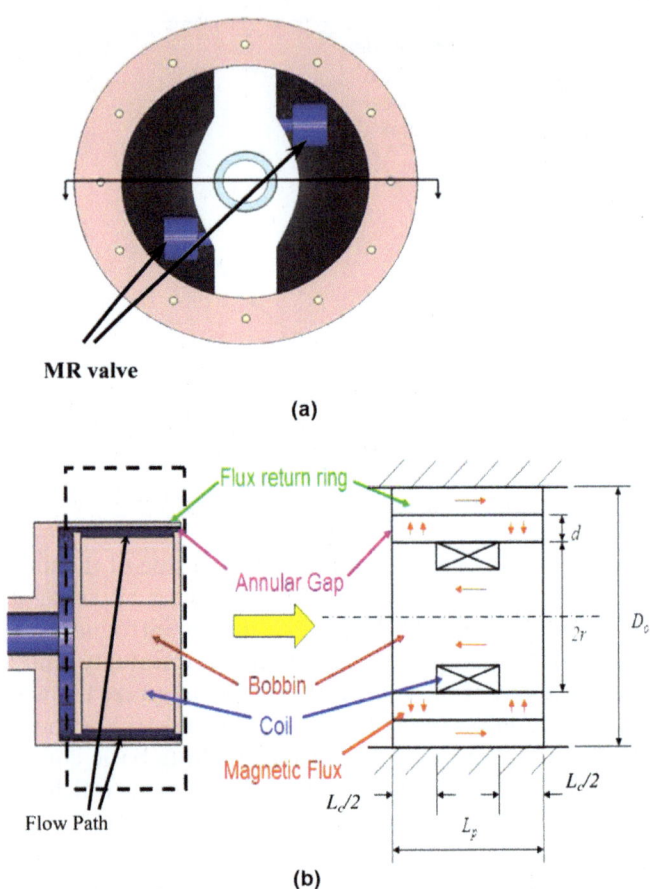

MR valve

(a)

(b)

Figure 14.12 MRFE damper design details. (a) Damper top view, (b) MR valve
details.

upon application of a current. In addition to imposing constraints on valve
sizing, the current MR retrofit package creates unfavorable passive viscous
losses due to unavoidable fluid flow through small diameter passage ways and
90° turns, resulting from the existing Fluidlastic® lag damper design feature.
Future designs can avoid these unfavorable conditions by designing an MR
retrofit that can be fully embedded in the center wall of the FE damper. The
original FE damper is preloaded in compression and the viscous fluid inside the
chambers is sealed under pressure. However, during assembly of the prototype
parts, the MR fluid was injected under no pressure and the elastomer was
approximately 18% pre-compressed, which is less than the roughly 20% pre-
compression of the FE damper.

To evaluate the MRFE damper, the same procedures as for the FE damper
discussed above were applied. In addition, for the MRFE damper, a DC power
supply was employed to provide a controlled current input to the MR valve
during testing. The applied current varied from 0–1 A, in increments of 0.25 A.

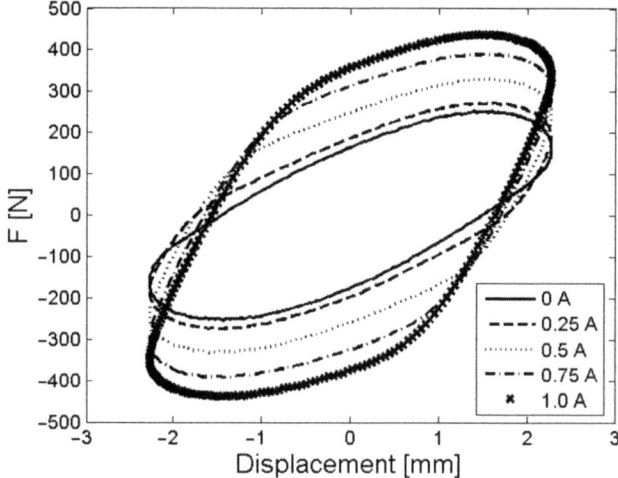

Figure 14.13   Typical force-displacement hysteresis results at fixed amplitude.

Figure 14.13 shows the force versus displacement plots at constant amplitude of the MRFE damper at current values varying from 0 A to 1 A. The input displacement is given by:

$$x(t) = X_s \sin(\omega t) + X_c \cos(\omega t) \qquad (14.25)$$

The measured force data was not filtered and was used as recorded. At an amplitude of 2.286 mm, Figure 14.13 shows the hysteretic behavior of the MRFE damper. In each force–displacement plot, the area enclosed by the hysteresis loop is proportional to the amount of energy dissipated per cycle. The damping available is in turn proportional to the energy dissipated per cycle. From Figure 14.13, the MRFE damper at zero applied current has an elliptical force-displacement plot. Under the application of a magnetic field, the MRFE damper behavior transforms substantially, exhibiting significant non-linearity. This property, demonstrated in Figure 14.13, shows the hysteretic plots of the MRFE damper under varying input currents. As control current is applied, the dissipated energy per cycle, thus the available damping in the MRFE device, increases significantly. This is due to the additional controllable damping contribution from the MR effect, on top of the viscous and bulge damping.

14.4.1   FE Damper Characterization

A typical approach for characterization of damper performance is the complex modulus method, which was discussed above. Based on this model, the damping coefficient and in-phase stiffness of the FE and MRFE dampers at lag and rotor frequencies are shown in Figures 14.14 and 14.15, respectively. Figure 14.14(a) shows the equivalent damping of the FE and MRFE dampers at the lag frequency as a function of lag motion for different constant applied

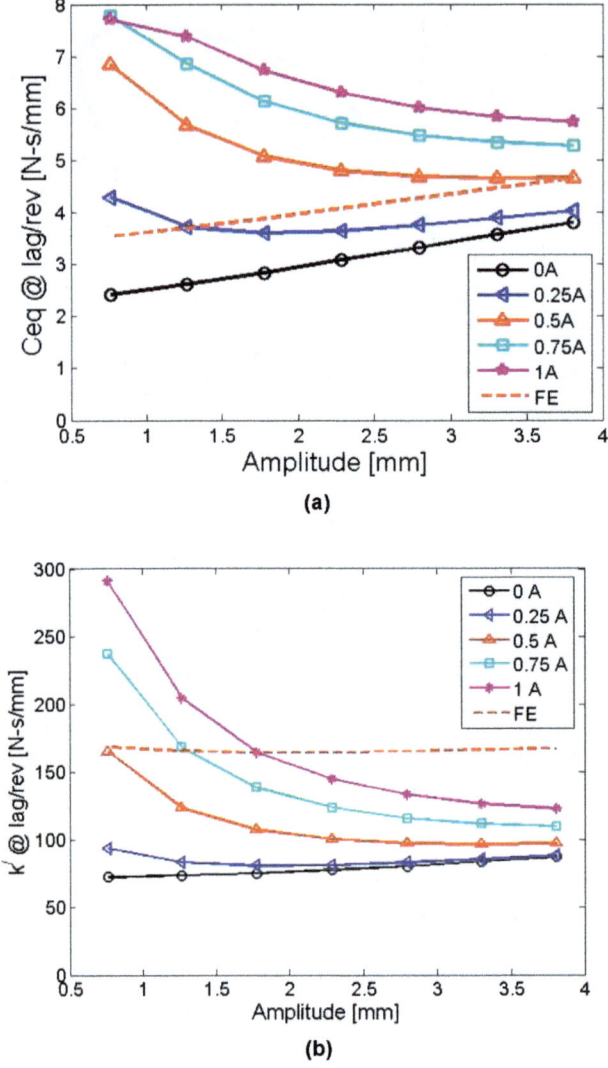

Figure 14.14 Complex modulus characterization of MRFE damper at lag/rev.
(a) Equivalent damping, (b) In-phase stiffness.

currents. The dotted line in Figure 14.14(a) is the equivalent damping of the FE
damper. Although the viscosity of the MR fluid is lower than the silicon fluid
used in the FE damper, the passive damping component of the MRFE damper
has additional viscous losses due to the retrofit components. However, the field-
off (0 A) equivalent damping of the MRFE damper is still lower than that of the
FE damper, which was a goal of this design. This is beneficial for reducing
helicopter hub loads since high damping is not required for most flight conditions. Comparatively, the maximum field-on equivalent damping (1 A) of the
MRFE damper is higher than the FE damper such that the required lag

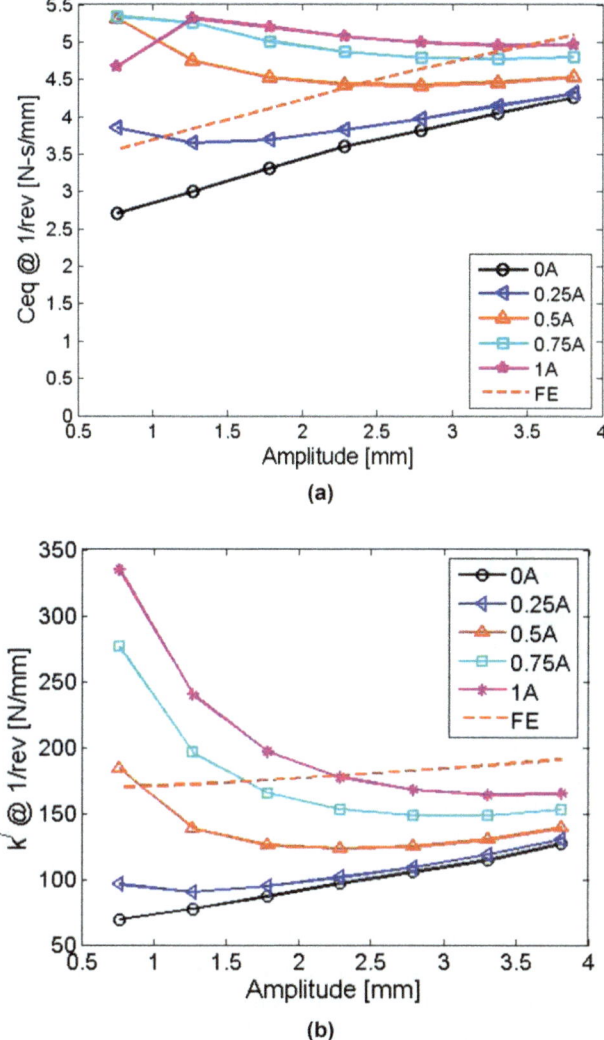

Figure 14.15 Complex modulus characterization of MRFE damper at 1/rev. (a) Equivalent damping, (b) in-phase stiffness.

damping at certain flight conditions can be achieved. Notably, the equivalent damping of the MRFE damper can be varied as a function of the applied current, and the damping increment can be more than three-fold that of the field-off damping at the same lag motion condition. This allows a large damping controllable range and, thus, the ability to optimize damping at different flight conditions. As shown in Figure 14.14(b), the in-phase stiffness of the field-on MRFE damper is higher at low displacements and high fields, while it is fairly level for FE and the field-off MRFE dampers. Note that the off-state in-phase stiffness of the MRFE dampers is much lower than the FE damper, which is attributed to the damaged elastomeric center wall, which occurred during retrofitting, as mentioned in the previous section.

Having determined that the MRFE can augment damping at the lag/rev frequency, the next set of results will compare the damping and in-phase stiffness of this damper at the primary rotor frequency of 5.8 Hz. Evaluation at 1/rev is important in the development of the MRFE damper because during high speed forward flight, the forced lag motion occurs at the 1/rev rotor frequency. Figure 14.15 displays these results. The equivalent damping at 1/rev resembles the response at lag/rev, in that the damping range of the MRFE device extends above and below the FE damper performance. However, the dynamic range of the MRFE damper is observed to decrease at this frequency. It is believed that this is due to choked flow in the MR valves. During field-on condition, as the excitation frequency rises, fluid velocity in the MR valve tends to further increase, enhancing viscous flow effects. However, it can still be seen that the MRFE damper has a much better controllable damping range than the FE lag damper as shown in Figure 14.15(a).

14.5   MRFE Damper Modeling

Under the application of a magnetic field, the MRFE damper provides controllable damping that can be adjusted as a function of flight condition. However, the MRFE damper exhibits a non-linear behavior (Figure 14.13), and the complex modulus method cannot accurately represent this behavior. In addition, the physics of the fluid flow in the MRFE lag dampers, as was the case for the FE damper, is not well represented. Thus, to describe the flow dynamics and predict the non-linear hysteretic damping force of the MRFE damper, we will modify the hydro-mechanical analysis developed for the FE damper by introducing a magnetorheological component. For clarity, the details of the MRFE hydro-mechanical model is discussed henceforth. The lumped parameter control volumes are shown in Figure 14.16, and the two MR valves are identified with subscripts a and b. There are two volume chambers (1 and 5) which are lumped into pressures P_1 and P_5 and compliances S_1 and S_5. Both S_1 and S_5 represent the compliances of the fluid and the chamber. However, since the elastomeric cylinder and center wall are much more compliant than the MR fluid, they dominate the compliances of the chambers. Control volumes $a2$, $a3$, $b2$ and $b3$ are lumped into fluid inertances I_{a2}, I_{a3}, I_{b2} and I_{b3} and passive flow resistances R_{a2}, R_{a3}, R_{b2} and R_{b3}. Control volumes $a4$ and $b4$ are also lumped into fluid inertances I_{a4} and I_{b4} and passive flow resistances R_{a4} and R_{b4}, respectively, while ΔP_{a4} and ΔP_{b4} are the pressure drops due to the yield stress of the MR fluid in the corresponding control volumes. A one-dimensional fluid flow is assumed such that the velocity and pressure of a control volume are span-wise constant at a particular time, but are time varying. In addition, incompressible and laminar flow is again assumed.

Similar to the FE damper case, the model is first formulated for a general case where the two valves and two chambers are assumed to be dissimilar. The Kelvin chain model is assumed for the elastomer shear deformation. The two

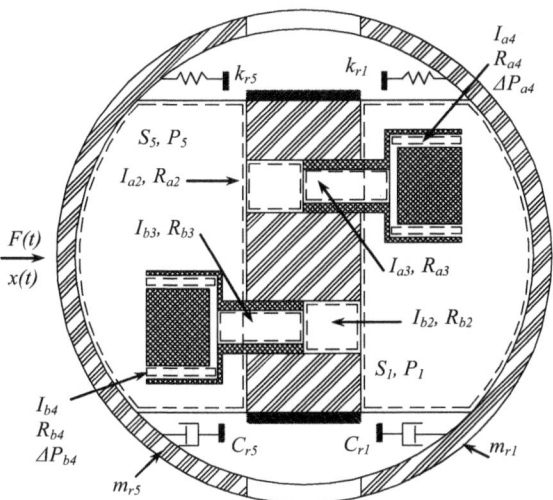

Figure 14.16 Proposed hydro-mechanical model of MRFE damper.

elastomeric chambers are broken into two semi-circular walls connected at the center, and they are assigned masses m_{r1} and m_{r5}. The shear mode in-phase stiffness and equivalent damping of the elastomeric body from the complex modulus method are utilized as the spring constants k_{r1} and k_{r5} and damping coefficients C_{r1} and C_{r5}. These values were measured by testing the empty damper, as was done for the FE damper.

Eqn (14.7), which is the governing equation of the elastomer in the FE damper modeling, remains the same for the MRFE damper. However, in order to account for the flow resistance due to the addition of the MR valves, eqn (14.8) and (14.9) change to:

$$A_p \dot{x}(t) - q_a(t) - q_b(t) = S_5(\dot{P}_5(t) - R_5(A_p \ddot{x}(t) - \dot{q}_a(t) - \dot{q}_b(t))) \quad (14.26)$$

$$-A_p \dot{x}(t) + q_a(t) + q_b(t) = S_1(\dot{P}_1(t) - R_1(-A_p \ddot{x}(t) + \dot{q}_a(t) + \dot{q}_b(t))) \quad (14.27)$$

The total pressure drop, from eqn (14.13) and (14.14) now accounts for the pressure drop due to the addition of the MR valves. The total pressure drop in the two port holes (control volumes 2 and 3) and MR valve annular gap (4) due to flow of the MR fluid is given by:

$$(P_5(t) - P_1(t)) = I_a \dot{q}_a(t) + R_a q_a(t) + \Delta P_{a4} \text{sgn}(q_a(t)) \quad (14.28)$$

$$(P_5(t) - P_1(t)) = I_b \dot{q}_b(t) + R_b q_b(t) + \Delta P_{b4} \text{sgn}(q_b(t)) \quad (14.29)$$

The terms I_a, R_a and ΔP_a are given by:

$$I_a = \rho \sum_{i=2}^{4} \frac{l_{ai}}{A_{ai}} \quad (14.30)$$

$$R_a = \frac{128\mu_o}{\pi} \sum_{i=2}^{3} \left(\frac{l_i}{D_i^4}\right)_a + 12\mu_{MR}\left(\frac{l_4}{A_d d_4^2}\right)_a + R_{\text{minor losses}} \tag{14.31}$$

$$\Delta P_{a4} = 2\frac{l_{a4}}{d_{a4}}\tau_{ya}(H) \tag{14.32}$$

Under the Bingham-plastic assumption, the fluid viscosity and the post-yield viscosity are assumed equal ($\mu_o = \mu_{MR}$). The yield stress of the MR fluid at the applied magnetic field is represented by $\tau_y(H)$. The term ΔP can be interpreted as the minimum required differential pressure below which flow is fully restricted. The equation for the resistance is derived from laminar flow analysis. The above equations also hold true for the second flow path by replacing the subscript a with b. Due to the laminar flow assumption justified by the low Reynold's number flow here, minor losses are neglected.

By eliminating the internal variables P_1 and P_5 from eqn (14.7), and (14.26)–(14.29), we obtain the following two equations of motion in terms of $x(t)$ and $x_A(t)$:

$$(m_{r1} + m_{r5})\ddot{x}(t) + (C_{r1} + C_{r2})\dot{x}(t) + (k_{r1} + k_{r2})x(t)$$
$$+ (C_{c1} + C_{c5})(\dot{x}(t) - (1+\psi)\dot{x}_A(t)) + (k_{c1} + k_{c5})(x(t) - (1+\psi)x_A(t)) = F(t) \tag{14.33}$$

$$m_A\ddot{x}_A(t) + C_A\dot{x}_A(t) - (C_{c1} + C_{c5})(\dot{x}(t) - (1+\psi)\dot{x}_A(t))$$
$$- (k_{c1} + k_{c5})(x(t) - (1+\psi)x_A(t)) + F_{yA}\text{sgn}(\dot{x}_A(t)) = 0 \tag{14.34}$$

In this case, the valve factor, ψ, is given by:

$$\psi(i\omega) = \frac{x_A}{x_B}(i\omega) = \frac{\left(m_B - \frac{F_{yA}-F_{yB}}{\omega^2 x_B(i\omega)}\right)i\omega + C_B}{m_A i\omega + C_A} \tag{14.35}$$

The new term F_{yA} for valve A is expressed as:

$$F_{yA} = \Delta P_{a4}A_p \tag{14.36}$$

Replacing subscripts A and a with B and b in the above equations will give the corresponding terms for the second flow valve. Here, F_{yA} and F_{yB} represent the controllable yield forces in each MR valve and are a function of the applied magnetic field. The valve factor, ψ, which relates the generalized inertance variables x_A and x_B, varies from zero (no flow in valve b) to infinity (no flow in valve a). The valve factor now includes additional head losses encountered in the MR valves as shown in eqn (14.35).

Figure 14.17 gives the mechanism of the proposed hydro-mechanical model, which is composed of parallel and series combinations of linear springs, dashpots, and Coulomb friction elements. The Coulomb friction element is the new element that is added to the FE damper model to account for the MR effect. The analogous mechanical system, shown in Figure 14.17, and the governing eqn (14.33)–(14.35) show that in the pre-yield region, the hysteresis

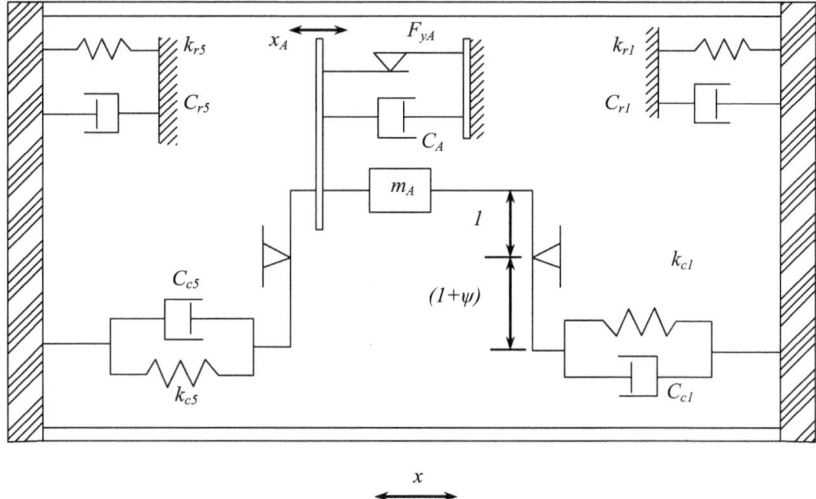

Figure 14.17 Analogous mechanical system of the hydro-mechanical MRFE damper analysis.

behavior is evident in the chamber compliance effects (C_{c1}, C_{c5}, k_{c1}, k_{c5}). The Coulomb elements (F_{yA}, F_{yB}), generalized inertance masses (m_A, m_B) and viscous damping (C_A, C_B) terms represent the post-yield behavior. Coupling of the mechanical and hydraulic systems of the MRFE damper is well illustrated by these governing equations.

Similar to the FE damper analysis, by assuming identical flow ports, the following assumptions will hold true: $F_{yA} = F_{yB} = F_y$, $k_{c1} = k_{c5} = k_c$, $C_{c1} = C_{c5} = C_c$, $m_{r1} = m_{r5} = m_r$, $k_{r1} = k_{r5} = k_r$, $C_{r1} = C_{r5} = C_r$ and $\psi = 1$. Thus, eqn (14.33)–(14.35) can be simplified to:

$$2m_r\ddot{x}(t) + 2C_r\dot{x}(t) + 2k_rx(t) + 2C_c(\dot{x}(t) - 2\dot{x}_A(t)) + 2k_c(x(t) - 2x_A(t)) = F(t) \tag{14.37}$$

$$m_A\ddot{x}_A(t) + C_A\dot{x}_A(t) - 2C_c(\dot{x}(t) - 2\dot{x}_A(t)) - 2k_c(x(t) - 2x_A(t))$$
$$+ F_{yA}\mathrm{sgn}(\dot{x}_A(t)) = 0 \tag{14.38}$$

Prior to initiating the parameter identification procedure, some parameters were measured or estimated *a priori*, as shown in Table 14.1, to reduce the number of parameters that must be identified using the error minimization routine. In the same fashion as for the FE damper, a constrained least-mean-squared (LMS) error minimization routine was employed to identify the equivalent piston area A_p, bulge stiffness k_c, bulge damping C_c and yield force F_y. The rubber mass was neglected in the parameter identification procedure.

The error function used to minimize the error of the hydro-mechanical analysis is expressed as:

$$E(A_p, k_c, C_c, F_y) = \sum_{j=1}^{n} \left(F(t_j) - F^*(t_j) \right)^2 \qquad (14.39)$$

Figure 14.18(a)–(b) show the equivalent piston areas obtained from the error minimization procedure at different amplitudes and frequencies. The shear stiffness and bulge stiffness of the MRFE damper at lag/rev and 1/rev are given in Figure 14.19(a)–(b) respectively. The damping contributions from the hydraulic and mechanical components are given in Figure 14.19(c)–(d). The plots show that the majority of the passive damping is provided through viscous damping, while elastomeric shear deformation contributes minimal damping. Figure 14.20 shows the variation of yield force, F_y, as obtained from the parameter identification procedure. The yield force increases with increasing applied current, and this trend is similar at both frequencies.

The damping force *vs.* velocity cycles at lag/rev and 1/rev frequencies and different applied fields are given in Figure 14.21. Figure 14.21(a) shows the force *vs.* velocity at the lag/rev frequency and Figure 14.21(b) shows the same at 1/rev. The displacement amplitude for each case is 2.286 mm (0.09 in). The figures show that the hydro-mechanical analysis is able to capture the nonlinear dynamic behavior and closely estimates the actual hysteretic damping force behavior of the MRFE damper from experimental measurements.

The hydro-mechanical model is formulated on the basis that it can describe or estimate the physical flow phenomena that is inherent in the MRFE damper system. Based on this assumption, the contribution of the MR effect to the damper system is extracted from the model. At both frequencies, the input amplitude was chosen to be 2.286 mm (0.09 in), while the current is varied from 0 A, 0.5 A and 1 A. The MR effect in each valve of the damper system, as

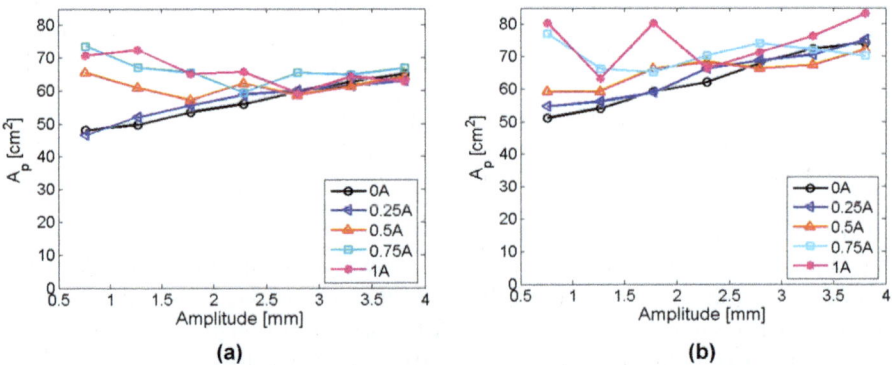

Figure 14.18 Optimized equivalent piston area A_p. (a) Equivalent area at lag/rev, A_p, (b) Equivalent area at 1/rev, A_p.

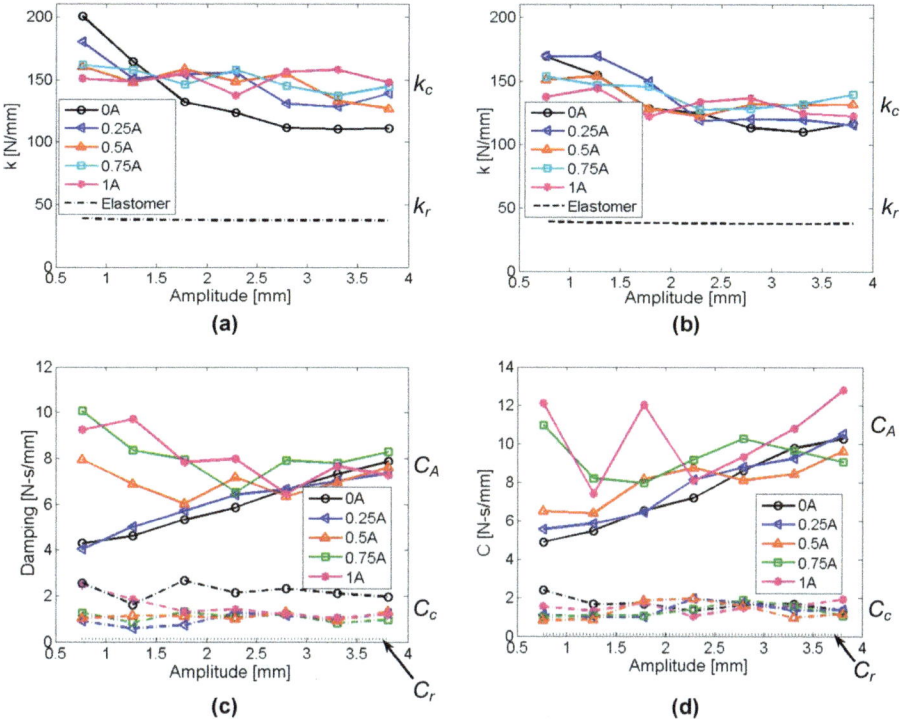

Figure 14.19 Stiffness and damping contribution of MRFE components. (a) Stiffness at lag/rev, (b) Stiffness at 1/rev (c) Damping at lag/rev (d) Damping at 1/rev.

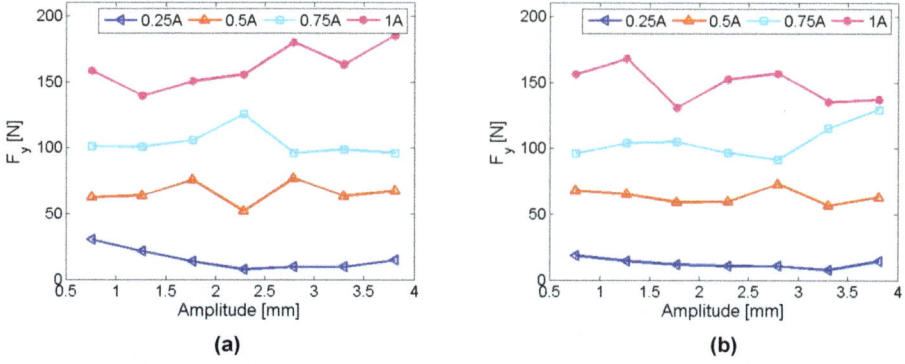

Figure 14.20 Optimized yield force, F_y. (a) Yield force at lag/rev, (b) yield force at 1/rev.

extracted from the model (eqn (14.38)), is assumed to be governed by a Bingham-plastic assumption:

$$F_{MR}(t) = C_A \dot{x}_A(t) + F_{ya} \text{sgn}(\dot{x}_A(t)) \qquad (14.40)$$

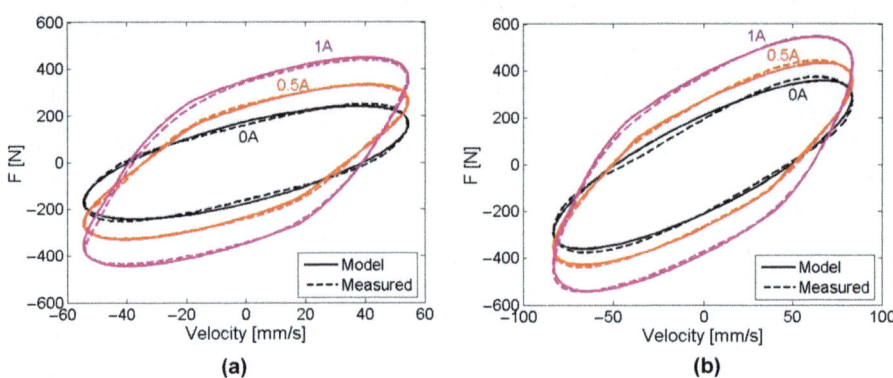

Figure 14.21   Measured and reconstructed damping force hysteresis. (a) Model performance at lag/rev, (b) model performance at 1/rev.

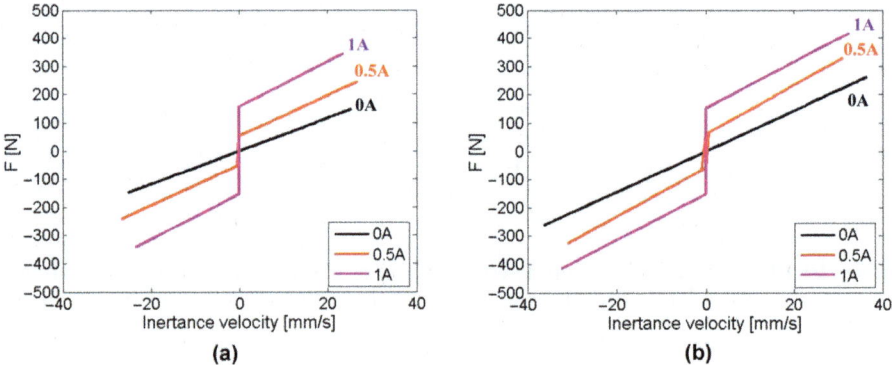

Figure 14.22   Model estimated MR valve damping force hysteresis at lag/rev. (a) Lag/rev F_{MR} *vs.* generalized inertance velocity, (b) 1/rev F_{MR} *vs.* generalized inertance velocity.

The first term in the above equation is related to the viscous effect and the second Coulomb term results from the magnetorheological (MR) effect. These results are shown in Figure 14.22. The analysis accounts for the Bingham-plastic flow typical of MR fluids, with the intercept of the force axis denoting the yield force of the MR damper, which is related to the yield stress of the MR fluid. The MR effect is activated *via* applied current. As the applied current is increased, the yield force increases, and thus damping increases.

Further, the total flow rate, $A_p \dot{x}(t)$, and flow rate through MR valves, $2A_p \dot{x}_A(t)$, at lag/rev frequency were plotted and compared at the field-on condition ($I = 1$ A), at two different input amplitudes. As shown in Figure 14.23(a), the MR valve flow rate is less than the total. This is manifested in the "No-flow interval" region in the figure. At this condition, the MR fluid has not yet yielded and volume compensation is attained through bulging of the elastomeric damper body. In addition, as the input amplitude increases, more of the total fluid flow goes through the valves. This is because the MR fluid

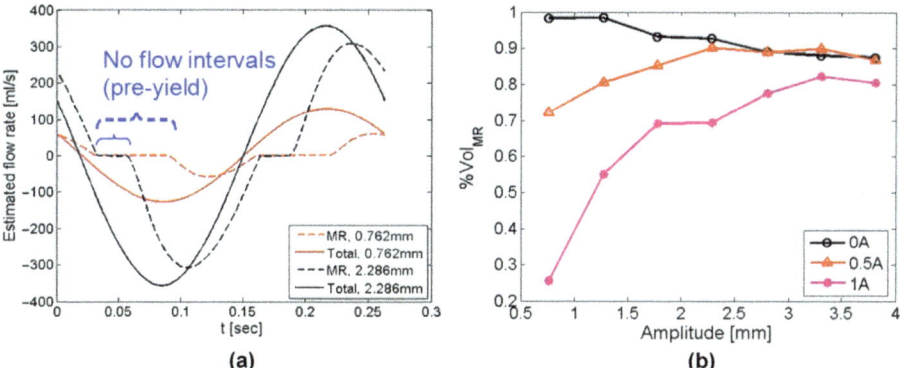

Figure 14.23   Total and MR valve flow rate comparisons at lag/rev. (a) Total and valve flow rates at $I = 1$ A, $X_{lag} = 2.286$ mm and $X_{lag} = 0.762$ mm, (b) Half cycle volume flow through MR valve: $\% Vol_{MR}$.

yields faster at higher velocity, giving the fluid more time to flow through the valves. The portion of the total displaced volume per half cycle that flows through the MR valve ($\% Vol_{MR}$) is estimated from the formulated model per the following equation:

$$\% Vol_{MR} = \frac{\int 2A_p \dot{x}_A(t)\mathrm{d}t}{\int A_p \dot{x}(t)\mathrm{d}t}\bigg|_{\text{halfcycle}} \times 100\% \tag{14.41}$$

These results are shown in Figure 14.23(b), which shows that, in the field-on condition, $\% Vol_{MR}$ increases with increasing amplitude. In addition, as the applied current is increased, $\% Vol_{MR}$ decreases due to the increased MR effects.

14.6   Conclusion

Most modern helicopters are equipped with lag dampers to mitigate instabilities such as air and ground resonance. Currently, some rotor systems utilize elastomeric lag dampers because they are lighter in weight, easier to maintain, and more reliable compared to a conventional hydraulic damper. However, elastomeric lag dampers exhibit strong amplitude and frequency dependent damping and stiffness losses, and are adversely affected by operating temperature. A fluid-elastomeric (FE) lead-lag damper, which combines elastomeric materials with viscous fluid damping, was developed to address the excitation dependent damping and stiffness behavior of elastomeric lag dampers. Unlike elastomeric lag dampers, FE dampers provide damping and stiffness that does not diminish with excitation amplitude. However, similar to elastomeric dampers, FE lag dampers tend to present damping loss as temperature increases, and provide a fixed level of damping; that is, these dampers are passive. Magnetorheological fluid-elastomeric (MRFE) lag dampers employing magnetorheological (MR) fluids were developed by the Smart Structures Laboratory at the University of Maryland[10,46] and the Advanced

Technology Division of Techno-Sciences, Inc. (now InnoVital Systems, Inc.) to provide adaptive or adjustable lead-lag damping augmentation for a helicopter rotor. Unlike passive dampers, an MRFE lag damper has the potential to compensate for performance losses due to input excitation, flight conditions, inherent material properties or extreme environmental conditions by the application of a controllable magnetic field.

In this chapter, FE and MRFE lag dampers, derived from a snubber damper used for the Bell 430 helicopter, were investigated and their performance presented. The MRFE damper was a retrofit design to an existing FE lag damper incorporating MR valves. The MRFE retrofit design does not introduce any moving parts to the system in an effort to preserve the proven reliability and maintenance free conditions of the original FE damper. Unlike the FE damper that produces a fixed level of damping for a particular operating condition, the MRFE damper provides a controllable damping force, so that damping levels can be adjusted as a function of operating condition. Moreover, a substantial increase in damping and a wide controllability range was gained by using the MRFE damper. Up to a 100% change in damping was obtained over the amplitude range tested. This damping and controllability range of the MRFE damper can be tuned to match actual damping requirements of specific flight and operational conditions where damping augmentation is required, to reduce damping, and associated hub loads, when high levels of damping are not required. Designing the device with reduced field-off damping helps decrease rotor hub loading because higher damping is only required at certain flight conditions. In addition, the MRFE damper can provide fail-safe capability in case of reduced or loss of power, albeit at a reduced level of effectiveness.

To characterize the behavior of the FE and MRFE lag dampers, a lumped parameter based, hydro-mechanical analysis was formulated and validated. The model describes the interaction of the mechanical and hydraulic flow phenomena within the FE and MRFE damper systems. The hydro-mechanical analysis is semi-empirical, so that most parameters can be measured *a priori*. Physical parameters such as fluid inertance and resistance, chamber compliance, and yield force (for MRFE damper) were considered in the development of the analysis. Model parameters at a given test condition are dependent on damper geometry, material properties, and excitation amplitude and frequency. The model was observed to have good correlation with the force *vs.* displacement and force *vs.* velocity behavior of both the FE and MRFE dampers for the range of amplitudes and frequencies tested. The hydro-mechanical analysis accurately predicted FE and MRFE lag damper loads, and will be used in the future for purposes of refined and improved designs of snubber dampers, and to assess rotor stability on existing and future rotor designs equipped such snubber dampers.

Nomenclature

| A | $=$ area |
| A_d | $=$ active gap area |

A_p	=	effective pressure area
C_c	=	bulge damping
C_eq	=	equivalent viscous damping
C_r	=	shear damping
D	=	diameter
D_o	=	flux return outer diameter
d	=	active gap thickness
E	=	error function
F	=	force
F_MR	=	MR valve force
F^*	=	model estimated force
F_c	=	first cosine force Fourier coefficient
F_s	=	first sine force Fourier coefficient
F_y	=	yield force
H	=	magnetic field
I	=	inertance
k_c	=	bulge stiffness
k_r	=	shear stiffness
k^*	=	complex modulus
k'	=	in-phase stiffness
k''	=	loss stiffness
L_c	=	active valve length
L_p	=	passive valve length
l	=	length
m	=	mass
m_r	=	elastomer mass
P	=	pressure
q	=	flow rate
R	=	resistance
r	=	bobbin radius
S	=	compliance
t	=	time
X_c	=	first cosine displacement Fourier coefficient
X_s	=	first sine displacement Fourier coefficient
X_lag	=	input amplitude at lag/rev frequency
X_pri	=	input amplitude at 1/rev frequency
x	=	input displacement
x_a, x_b	=	inertance displacements
x_A, x_B	=	generalized inertance displacements
A, B	=	generalized valve index
a, b	=	valve index
i	=	control volume index
j	=	sampling index
ΔP	=	pressure drop over MR valve
η	=	loss factor
μ_o	=	MR fluid viscosity

μ_{MR} = post yield MR fluid viscosity
ρ = MR fluid density
τ_y = yield shear stress
ψ = valve factor
ω = frequency
$\%Vol_{MR}$ = fraction of total displaced fluid volume flowing through MR valve

References

1. I. Chopra, Perspective in Aeromechanical Stability of Helicopter Rotors, *Vertica*, 1990, **14**(4), 457–508.
2. P. Jones, D. Russell and P. McGuire, Latest Development in Fluidlastic® Lead-Lag Dampers for Vibration Control in Helicopters, *Proceedings of the American Helicopter Society 59th Annual Forum*, AHS International, Alexandria, VA, 2003.
3. R. Snyder, R. Krishnan, N. Wereley and T. Seig, Mechanisms Based Analysis of Elastomeric Lag Damper Behavior Under Single and Dual Frequency Excitation Including Temperature Effects, *Proceedings of the American Helicopter Society 57th Annual Forum*, AHS International, Alexandria, VA, 2001.
4. B. Panda, E. Mychalowycz and F. Tarzanin, Application of Passive Dampers to Modern Helicopters, *Smart Mater. Struct.*, 1996, **5**(5), 509–516.
5. W. Hu and N. M. Wereley, "Distributed Rate-Dependent Elastoslide Model for Elastomeric Lag Dampers", *AIAA Journal of Aircraft*, 2007, **Vol. 44**, No. 6, pp. 1972–1984.
6. F. Felker, B. Lau, S. McLaughlin and W. Johnson, Nonlinear Behavior of an Elastomeric Lag Damper Undergoing Dual-Frequency Motion and its Effect on Rotor Dynamics, *J. Am. Helicopter Soc.*, 1987, **32**(4), 45–53.
7. W. Hu and N. M. Wereley, Hybrid Magnetorheological Fluid-Elastomeric Lag Dampers for Helicopter Stability Augmentation, *Smart Mater. Struct.*, 2008, **17**(4), 045021.
8. D. P. McGuire, Fluidlastic® Dampers and Isolators for Vibration Control in Helicopters, Proceedings of the *American Helicopter Society 50th Annual Forum*, AHS International, Alexandria, VA, 1994.
9. B. Panda and E. Mychalowycz, Aeroelastic Stability Wind Tunnel Testing with Analytical Correlation of the Comanche Bearingless Main Rotor, *J. Am. Helicopter Soc.*, 1997, **42**(3), 207–217.
10. G. M. Kamath, N. M. Wereley and M. R. Jolly, Characterization of Magnetorheological Helicopter Lag Dampers, *J. Am. Helicopter Soc.*, 1999, **44**(3), 234–248.
11. G. Hausmann and P. Gergely, Approximate Methods for Thermo-viscoelastic Characterization and Analysis of Elastomeric Lead-lag Dampers, *Proceedings of the 18th European Rotorcraft Forum*, Avignon, France, 1992, pp. 88–1.

12. R. Brackbill, G. Lesieutre, E. Smith and K. Govindswamy, Thermo-mechanical Modeling of Elastomeric Materials, *Smart Mater. Struct.*, 1996, **5**(5), 529–539.

13. M. Jolly, J. Bender and J. Carlson, Properties and Applications of Commercial Magnetorheological Fluids, *J. Intell. Mater. Syst. Struct.*, 1999, **10**(1), 5–13.

14. G. Ngatu and N. M. Wereley, High Versus Low Field Viscometric Characterization of Bidisperse MR Fluids, *Int. J. Mod. Phys. B*, 2007, **21**(28–29), 4922–4928.

15. S. Genc and P. Phule, Rheological Properties of Magnetorheological Fluids, *Smart Mater. Struct.*, 2002, **11**(1), 140–146.

16. J. Ginder and L. Davis, Shear Stress in Magnetorheological Fluids: Role of Magnetic Saturation, *Appl. Phys. Lett.*, 1994, **65**(26), 3410–3412.

17. Y. T. Choi, N. M. Wereley and Y. S. Jeon, Semi-active Vibration Isolation Using Magnetorheological Isolators, *AIAA J. Aircraft*, 2005, **42**(5), 1244–1251.

18. J. Carlson, Critical Factors for MR Fluids in Vehicle Systems, *Int. J. Vehicle Des.*, 2003, **33**(1–3), 207–217.

19. H. Sahin, Y. Liu, X. Wang, F. Gordaninejad, C. Evrensel and A. Fuchs, Full-scale Magnetorheological Fluid Dampers for Heavy Vehicle Rollover, *J. Intell. Mater. Syst. Struct.*, 2007, **18**(12), 1161–1167.

20. Y. T. Choi and N. M. Wereley, Vibration Control of a Landing Gear System Featuring Electrorheological/Magnetorheological Fluids, *AIAA J. Aircraft*, 2003, **40**(3), 432–439.

21. D. C. Batterbee, N. D. Sims, R. Stanway and Z. Wolejsza, Magnetorheological Landing Gear: 1. A Design Methdology, *Smart Mater. Struct.*, 2007, **16**(6), 2429–2440.

22. D. C. Batterbee, N. D. Sims, R. Stanway and M. Rennison, Magnetorheological Landing Gear: 2. Validation Using Experimental Data, *Smart Mater. Struct.*, 2007, **16**(6), 2441–2452.

23. S. J. McManus, K. A. St. Clair, P. E. Boileau, J. Boutin and S. Rakheja, Evaluation of Vibration and Shock Attenuation Performance of a Suspension Seat with a Semi-active Magnetorheological Fluid Damper, *J. of Sound Vib.*, 2002, **253**(1), 313–327.

24. S. B. Choi, M. H. Nam and B. K. Lee, Vibration Control of a MR Seat Damper for Commercial Vehicles, *J. Intell. Mater. Syst. Struct.*, 2000, **11**(12), 936–944.

25. G. J. Hiemenz, W. Hu and N. M. Wereley, Semi-active Magnetorheological Helicopter Crew Seat Suspension for Vibration Isolation, *AIAA J. Aircraft*, 2008, **45**(3), 945–953.

26. Y. T. Choi and N. M. Wereley, Mitigation of Biodynamic Response to Vibratory and Blast-induced Shock Loads Using Magnetorheological Seat Suspensions, *J. Automobile Eng.*, 2005, **219**(D6), 741–754.

27. G. J. Hiemenz, Y. T. Choi and N. M. Wereley, Semi-active Control of a Vertical Stroking Helicopter Crew Seat for Enhanced Crashworthiness, *AIAA J. Aircraft*, 2007, **44**(3), 1031–1034.

28. S. Marathe, F. Gandhi and K. W. Wang, Helicopter Blade Response and Aeromechanical Stability with a Magnetorheological Fluid Based Lag Damper, *J. Intell. Mater. Syst. Struct.*, 1998, **9**(4), 272–282.
29. Y. S. Zhao, Y. T. Choi and N. M. Wereley, Semi-active Damping of Ground Resonance in Helicopters Using Magnetorheological Dampers, *J. Am. Helicopter Soc.*, 2004, **49**(4), 468–482.
30. W. Hu, N. M. Wereley, L. Chemouni and P. C. Chen, Semi-active Linear Stroke Magnetorheological Fluid-Elastic (MRFE) Damper for Helicopter Main Rotor Blades, *AIAA J. Guidance, Control and Dynamics*, 2007, **30**(2), 565–575.
31. G. Kamath, N. Wereley and V. Madhavan, Hysteresis Modeling of Semi-Active Magnetorheological Helicopter Dampers, *J. Intell. Mater. Syst. Struct.*, 1999, **10**(8), 624–633.
32. B. Panda and E. Mychalowycz, Aeroelastic Stability Wind Tunnel Testing with Analytical Correlation of the Comanche Bearingless Main Rotor, *J. Am. Helicopter Soc.*, 1997, **42**(3), 207–217.
33. D. Kunz, Influence of Elastomeric Damper Modeling on the Dynamic Response of Helicopter Rotors, *AIAA J.*, 1994, **35**(2), 349–354.
34. F. Gandhi and I. Chopra, Analysis of Bearingless Main Rotor Dynamics with the Inclusion of an Improved Time-domain Non-linear Elastomeric Damper Model, *J. Am. Helicopter Soc.*, 1996, **41**(3), 267–277.
35. C. Brackbill, E. Smith, L. Ruhl and G. Lesieutre, Characterization and Modeling of the Low Strain Amplitude and Frequency Dependent Behavior of Elastomeric Damper Materials, *J. Am. Helicopter Soc.*, 2000, **45**(1), 34–42.
36. G. Kamath, M. Hurt and N. Wereley, Analysis and Testing of Bingham Plastic Behavior in Semi-Active Electrorheological Fluid Dampers, *Smart Mater. Struct.*, 1996, **5**(5), 576–590.
37. N. Wereley, Non-Dimensional Herschel-Bulkley Analysis of Magnetorheological and Electrorheological Dampers, *J. Intell. Mater. Syst. Struct.*, 2008, **19**(3), 257–268.
38. L. Pang, G. Kamath and N. M. Wereley, Idealized Hysteresis Modeling of Electrorheological and Magnetorheological Dampers, *J. Intell. Mater. Syst. Struct.*, 1998, **9**(8), 642–649.
39. B. Spencer, B. Dyke, M. Sain and J. Carlson, Phenomenological Model of a Magnetorheological Damper, *J. Eng. Mech.*, 19997, **123**(3), 230–238.
40. W. Hu and N. Wereley, Rate-Dependent Elasto-Slide Model for Single and Dual Frequency MR Lag Damper Behavior, *Int. J. Modern Phys. B*, 2005, **19**(7 9), 1527–1533.
41. R. Singh, G. Kim and P. V. Ravindra, Linear Analysis of Automotive Hydro-Mechanical Mount with Emphasis on De-coupler Characteristics, *J. Sound Vib.*, 1992, **158**(2), 219–243.
42. A. Geisberger, A. Khajepour and F. Golnaraghi, Non-linear Modelling of Hydraulic Mounts: Theory and Experiment, *J. Sound Vib.*, 2002, **249**(2), 371–397.

43. S. R. Hong, S. B. Choi, Y. T. Choi and N. M. Wereley, A Hydro-mechanical Model for Hysteretic Damping Force Prediction of ER Damper: Experimental Verification, *J. Sound Vib.*, 2005, **285**(4–5), 1180–1188.

44. S. R. Hong, G. Wang, W. Hu and N. M. Wereley, Hydro-mechanical Analysis of a Magnetorheological Bypass Damper, *Proceedings of the 10th International Conference on Electrorheological Fluids and Magnetorheological Suspensions*, Lake Tahoe, USA, June 18–22, 2006, pp. 438–444.

45. J. Colgate, T. Chang, C. Chiou, K. Liu and M. Kerr, Modeling of a Hydraulic Engine Mount Focusing on Response to Sinusoidal and Composite Excitations, *J. Sound Vib.*, 1995, **184**(2), 503–528.

46. N. M. Wereley, W. Hu, G. T. Ngatu, C. S. Kothera and P. C. Chen, Magnetorheological Fluid Elastic Lag Damper for Helicopter Rotors, U.S. Patent 8 413 772, filed: Feb. 12, 2009, issued: April 9, 2013.

CHAPTER 15

Magnetorheological Devices with Multiple Functions

WEI-HSIN LIAO,* CHAO CHEN AND HONGTAO GUO

Department of Mechanical and Automation Engineering, The Chinese University of Hong Kong, Shatin, Hong Kong, China
*Email: whliao@cuhk.edu.hk

15.1 Introduction

15.1.1 Functional Integration of Linear Magnetorheological Dampers

Linear magnetorheological (MR) dampers are one of the most popular MR devices. The schematic of a typical MR damper based semi-active control system is illustrated in Figure 15.1. As shown in the figure, to implement an MR damper system, a power supply is needed for driving the current through electromagnetic coils inside the MR damper in order to provide the magnetic field for the MR fluid. To utilize the controllable damping characteristics of an MR damper, sensors are also needed to measure the dynamic responses, which may include the relative displacement or velocity across the MR damper. The extra power supply and sensors for the conventional MR damper system would have concerns as follows: (i) difficulties of system installation and maintenance, (ii) waste of dissipated energy, (iii) space of installation, (iv) weight penalty, (v) addition of cost, and (vi) degradation of reliability.

The multifunctional integration of MR dampers provides a promising solution to the above mentioned problems of conventional MR damper systems. Recently, functional integration of MR dampers has been considered.

RSC Smart Materials No. 6
Magnetorheology: Advances and Applications
Edited by Norman Wereley
© The Royal Society of Chemistry 2014
Published by the Royal Society of Chemistry, www.rsc.org

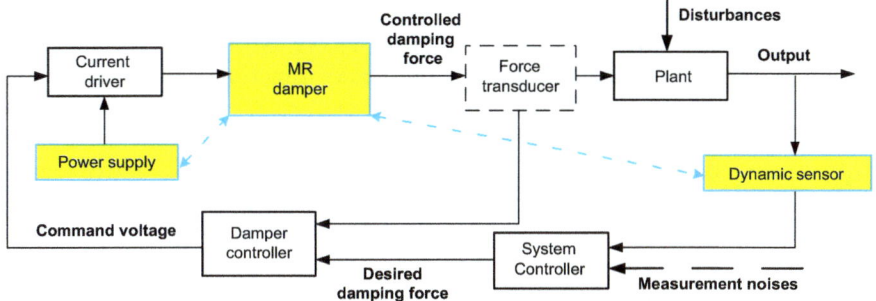

Figure 15.1 Schematic of MR damper based semi-active control system.[1]

Especially, efforts to eliminate the need for sensors and power supply have been made. If the mechanical energy resulting from vibration and shock is converted into the electrical energy for the power source of an MR damper, and the dynamic responses can be obtained without an extra sensor, the separate power supply and sensor would not be needed anymore.

Integration of the displacement/velocity sensing function with MR dampers was realized recently by several methods, to obtain the dynamic information between the two ends of MR dampers. These methods are: magnetostrictive displacement sensing by Gloden *et al.*,[2] an electromagnetic induction method by Wang *et al.*[3] and Chen and Liao,[4] and an indirect velocity extraction method by Chen and Liao.[5] The integration of the force sensing function with the MR damper was investigated by Or *et al.*[6] A piezoelectric element was used to measure the damping force generated by the MR damper. Integration of power generation function with MR damper has been focused on the electromagnetic power generation in recent years. Different designs of electromagnetic power generators were explored by Choi and Wereley,[7] Liao and Chen,[8] and Sapiński.[9]

The most multifunctional MR damper was developed by Liao and Chen.[8] It integrates dynamic sensing, energy harvesting, and MR damping functions within the same device, to obtain a self-powered, self-sensing MR damper. The developed multifunctional MR dampers are expected to be applied to vehicles and buildings.[10,11]

15.1.2 Functional Integration of Rotary Magnetorheological Actuators

The most commonly used rotary actuators are electric motors, which are widely used in many applications such as household appliances and machine tools. By converting electrical power to mechanical power, electric motors can provide active torque. On the other hand, there are different types of rotary actuators that can provide controllable torque with semi-active means. These actuators can utilize smart materials such as MR fluid in devices, for examples, brakes, clutches, and valves.[12–14] These devices can provide controllable resistance while consuming little power. For controllable energy dissipative devices, these smart fluid based actuators would be a good choice. Moreover, for active actuators, in particular electric motors, the brake function usually consumes much power and might have safety concern. Therefore, it would be beneficial to use smart fluid based actuators. However, in some situations that need active torque, these smart fluid based actuators might not be helpful in active means.

Assistive knee braces use actuation devices to provide adequate supporting force/torque. To assist the wearer in various postures and prevent the knee brace from exceeding the restricted motion, it would be desired to have the actuator that can also function as a brake/clutch with safety interlock. Additionally, power consumption in the actuation devices should be considered in extending the working time of batteries after fully charged. For the braking function, MR brakes would be more energy efficient than electric motors. However, in some situations such as climbing upstairs, the conventional MR actuators will not be able to provide active torque to assist the wearer. Therefore, it is desired to design actuators with better energy efficiency and safety for assistive knee braces. To utilize the advantageous features of smart fluids and active actuators, a hybrid assistive knee brace using an MR actuator together with an electric motor was developed.[15] The MR actuator can function as a brake when controllable dissipative torque is preferred; or work as a clutch to transmit torque from the motor to the brace. With control, the actuation system can provide desired torque with better safety and energy efficiency. However, the actuator seemed a bit bulky to be used on human body. A more compact, multifunctional actuator would be beneficial for assistive knee braces. Besides, in some applications with limited space, an integrated actuation device with multiple functions would also be desired.

With this motivation, we aim to develop a multifunctional actuator that could be used for assistive knee braces.[16] Such a multifunctional rotary actuator could also be used in other machinery, such as metal processing, as the key for tension control while saving space that additional braking parts will otherwise occupy. This multifunctional actuator also has potential to be applied in household appliances such as washing machines.

15.2 Self-Sensing Magnetorheological Dampers with Power Generation

15.2.1 Concepts and Key Issues of Multifunctional Integration

The self-sensing MR dampers with power generation will be discussed in this section in detail. It combines the advantageous features of energy harvesting – to scavenge wasted energy, MR damping – to provide controllable damping force, and sensing – to provide dynamic information for system control.

Taking the car suspension system as an example, Figure 15.2 shows the configurations of suspension systems based on a conventional MR damper and self-sensing MR damper with power generation. Ideally, utilizing the multifunctional MR damper, a separate power supply and sensor in the conventional MR

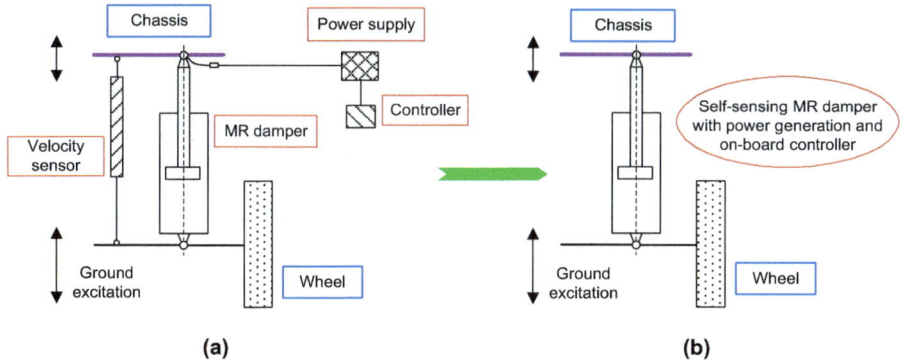

Figure 15.2 Configurations of MR suspension systems: (a) based on a conventional MR damper; (b) based on a self-sensing MR damper with power generation.

damper system would not be needed anymore. The suspension system can be greatly simplified as shown in Figure 15.2(b). This multifunctional integration would bring out great benefits such as energy saving, size and weight reduction, lower cost, high reliability, and less maintenance for MR damper systems.

To integrate MR damper with power generation and sensing abilities, there are four key issues:

The first issue is the size limitation. The multifunctional damper is expected to be an integrated device to reduce the installation space. Thus, the size of each function part, especially the power generation part, should not be too large. It could be possible that some components and space of the three different functions can be shared.

The second issue is about the power generation part. Considering the working conditions of MR dampers, the input excitations for the power generation part are generally small. For most of the MR damper applications, the frequencies of input excitations for MR dampers are low (several hertz) compared with those of most energy harvesting devices (hundreds or kilos hertz). With the size limitation of the whole device, the power generation part is preferred to have high energy conversion efficiency.

The third one is the interaction issue; that is, the performances of three functions should not adversely affect one another, in particular, when different functions share some common space.

The fourth issue is how to extract the dynamic information, such as velocity and displacement from the generated voltage of power generator. For ordinary electromagnetic power generator (open loop without guidance of the magnetic flux), based on the Faraday's law, the induced voltage is proportional to the velocity. However, for the power generator that has high energy conversion efficiency with guided magnetic flux, the relation between velocity and voltage is not clear. The velocity-extraction algorithms should be developed to obtain the dynamic information.

Table 15.1   Configurations of functional components for self-sensing MR damper with power generation.

MR damping	Single-ended		
	Double-ended		
Power generation	Electromagnetic	Linear	Multi-pole slotless
			Multi-pole slotted
			Single-pole
		Rotary	
	Non-electromagnetic: thermoelectric, piezoelectric *etc.*		
Velocity sensing	Direct	Independent sensing mechanism	
		Velocity extraction from power generation	
	Indirect from displacement, acceleration		

There are different methods in order to provide three functions: MR damping, power generation and dynamic sensing. The self-sensing MR damper with power generation is composed of an MR damping part, power generator, and sensing part.

For the MR damping part, there are two configurations: single-ended and double-ended structures. The double-ended structure of MR damper does not need an accumulator to compensate the rod volume when the MR damper is in operation.

For the power generator, there are three mechanisms by which mechanical energy due to vibrations can be converted into electrical energy: thermoelectric, piezoelectric, and electromagnetic. The electromagnetic mechanism employs electromagnetic induction arising from the relative motion of the magnetic flux and a conductor. There are two types of electromagnetic power generators: linear and rotary. To power a linear MR damper, utilizing a linear generator could simplify the mechanical design and eliminate the transmission of linear-to-rotary motion. For the linear power generator, the multi-pole design has higher energy conversion efficiency than that of single-pole design.

For the velocity-sensing part, there are two kinds of methods to obtain the velocity information: direct and indirect. The indirect measurement is to use the displacement or acceleration information. In either case, perfect differentiation or integration is difficult to realize. The direct velocity acquisition is usually based on electromagnetic induction. The acquisition could either have an independent sensing mechanism, or share the hardware with the power generator. Configurations of functional components are given in Table 15.1.

15.2.2   Design of Multifunctional Magnetorheological Dampers

The structure of a novel self-sensing MR damper with power generation is shown in Figure 15.3. It is composed of the MR damping part, power generator (also acts as the hardware of the sensing function), interaction components, mounting and motion guidance. The power generator is concentric and radially outside the MR damping part. The mounting components are for connection

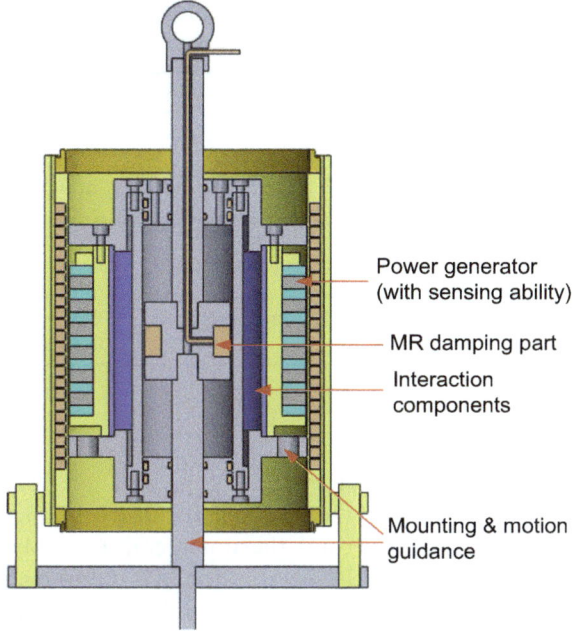

Figure 15.3 Sectional view of a self-sensing MR damper with power generation.

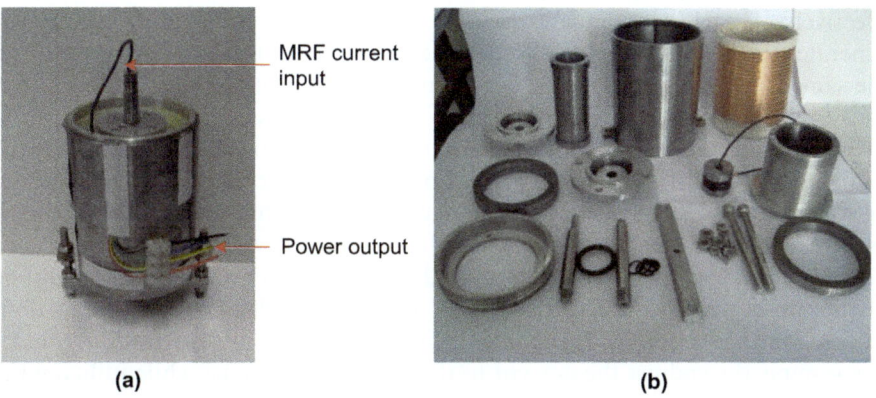

Figure 15.4 Photos of MR damper prototype: (a) assembled prototype; (b) components of the prototype.

with external parts. Motion guidance components are used to guide the linear motion of the relative movement between the inner and outer assemblies of the power generator. The interaction components are used to prevent the magnetic-field interference between the MR damping part and power generator. To avoid the use of an accumulator, the double-ended structure of MR damping part is adopted. Photos of the prototype of self-sensing MR damper with power generation are shown in Figure 15.4. The assembled prototype is shown in

Table 15.2 Specifications of multifunctional MR damper.

Parameter	Value
Length	168 mm
Outer diameter	110 mm
Stroke	±30 mm
Weight	5 kg
Thickness of magnetic flux guided layer	2 mm
Magnet quantity of power generator	6
Electrical wire gauge of power generator coil	Φ 0.16 mm
Electrical wire gauge of MR damping coil	Φ 0.5 mm
MR fluid flow gap	0.8 mm

Figure 15.4(a), and the components of the prototype are shown in Figure 15.4(b). Specifications of the developed prototype are given in Table 15.2.

The inner magnet assembly of the power generator is connected to the cylinder of the MR damping part, and the outer coil assembly of the power generator is connected with the piston rod of the MR damping part. When the piston rod moves under external excitation, two relative movements occur. The outer assembly of the power generator moves with the piston rod, so there is a relative movement between the inner and outer assemblies. As the magnet groups and coil groups are installed on the inner and outer assemblies, respectively, the electromagnetic induction caused by this relative movement generates the electrical voltage in the outer assembly of the power generator. At the same time, the piston of the MR damping part also moves with the piston rod, so there is another relative movement between piston and cylinder of the MR damping part; thus, the MR damping part generates the damping force while having the MR fluid flow in and out the annular fluid gap (valve mode) between the piston and cylinder. Some of the coils wound on the power generator simultaneously act as the sensing coils, and the voltages of sensing coils are outputted together with the electrical power. The sensing voltage would be further processed by a sensing estimator under a velocity-extraction algorithm to provide the velocity information. With the sensing information, a control law is applied to adjust the current to the MR damping part while utilizing the harvested electrical energy. As the yield strength of the MR fluid is changed by the current to the MR damping part, the damping force can be controlled.

The linear multi-pole slotless power generator is shown in Figure 15.5(a). The term "multi-pole" means that there are several pole pairs. Each pole pair includes one permanent magnet, coil and pole piece.

The induced voltage E_i (open circuit) in the ith coil is:

$$E_i = -N\phi_g \frac{\pi}{\tau} \sin\left(\frac{\pi}{\tau}z + \varphi_i\right) \frac{\mathrm{d}z}{\mathrm{d}t} \qquad (15.1)$$

$$E_1 = -N\phi_g \frac{\pi}{\tau} \sin\left(\frac{\pi}{\tau}z\right) \frac{\mathrm{d}z}{\mathrm{d}t} \qquad (15.2)$$

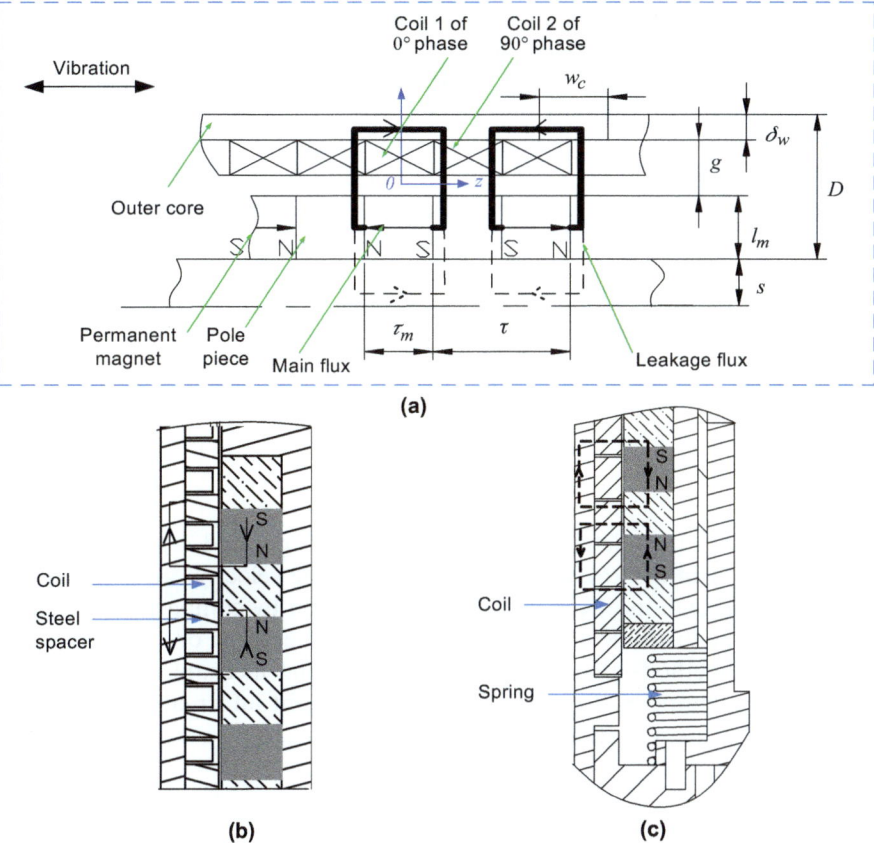

Figure 15.5 Schematic of the linear multi-pole power generators: (a) slotless; (b) slotted; (c) spring-based slotless.

$$E_2 = -N\phi_{\mathrm{g}}\frac{\pi}{\tau}\cos\left(\frac{\pi}{\tau}z\right)\frac{\mathrm{d}z}{\mathrm{d}t} \qquad (15.3)$$

where E_1, E_2 are the voltages for the coils of $0°$, $90°$ phase respectively, N is the number of turns of the coil, ϕ_{g} is the air-gap magnetic flux considering the leakage, τ is the pole pitch, z is the displacement, $\mathrm{d}z/\mathrm{d}t$ is the velocity, and ϕ_i is the phase angle of the ith coil. Z_{coil} is the impedance of the power generation coil, and Z_{load} is the impedance of the external load. When $Z_{\mathrm{coil}} = Z_{\mathrm{load}}$, the power of load P is:

$$P = \frac{N^2\phi_{\mathrm{g}}^2\pi^2 n}{4Z_{\mathrm{coil}}\tau^2}\left(\frac{\mathrm{d}z}{\mathrm{d}t}\right)^2 \qquad (15.4)$$

where n is the number of coil groups. Slotted and spring-based slotless multi-pole power generators are shown in Figures 15.5(b) and (c), respectively. The analysis for the slotless generator is also applicable to the slotted one. For the slotted power generator, there is a steel spacer between two adjacent coils. Thus, the equivalent air gap g of the slotted structure is smaller than that of the slotless structure. For the spring-based slotless power generator, there is a spring connected with the moving magnet. That is, the base excitation for the power generator could be amplified under the natural frequency.

For the multi-pole power generator, the relation between velocity and voltage is not clear. From the generated voltages of two adjacent coils, such as coil 1 and coil 2, the velocity could be extracted to be self-sensing. From eqn (15.2) and (15.3),

$$|\dot{z}| = \sqrt{\frac{E_1^2 + E_2^2}{K^2}} \qquad (15.5)$$

$$K = N\phi_g \frac{\pi}{\tau} \qquad (15.6)$$

where K is a constant determined by structure design, and $\dot{z}$ is the velocity. From eqn (15.5) and (15.6), the absolute value of the velocity can be obtained. To decide the direction of the velocity, an algorithm by comparing the difference between the theoretically estimated value and measured one, can be used.[8] Other sensing methods that use independent sensing mechanisms could also make the MR damper self-sensing.[8] The displacement information could also be obtained by integrating the velocity.

15.2.3 Magnetic Field Interactions

For the design of multifunctional MR dampers, one important issue is that the performances of three functions should not adversely affect one another. The power generator and the MR damping part have their own magnetic fields while sharing some common space. An ordinary magnetic-field isolation method is to use a non-magnetic component with low magnetic permeability. To minimize the magnetic-field interference, a combined magnetic-field isolation approach is utilized. The combined components include the magnetic flux shield layer and flux guided layer. The magnetic flux shield layer can be made of non-magnetic materials, such as stainless steel, aluminium, or air. The magnetic flux guided layer is made of material with high magnetic permeability, such as electrical pure iron. The shield layer is used to prevent the magnetic flux, and the guided layer is used to guide the magnetic flux.

Two different cases of the magnetic-field distributions are compared, one is with the ordinary magnetic flux shield layer, and the other is with the combined flux shield and guided layers. Finite element analysis (FEA) results are shown in Figure 15.6. As shown in Figure 15.6(a), with the ordinary magnetic flux shield layer, the magnetic-field interferences between the power generator and MR damping part are not completely prevented. The magnetic flux of the power generator goes through the cylinder, and affects the magnetic field of the

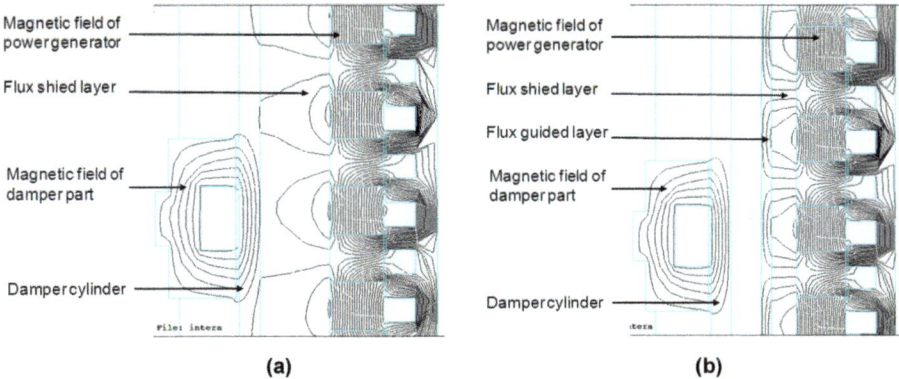

Figure 15.6 Magnetic flux distributions of the power generator and MR damping part: (a) with the ordinary magnetic flux shield layer; (b) with the combined flux shield and guided layers.

MR damping part. It shows that only using the magnetic flux shield layer would not effectively prevent the interference. As shown in Figure 15.6(b), with the combined flux shield and guided layers, the magnetic-field interferences between the power generator and MR damping part are effectively prevented.

15.2.4 Testing of Prototype of Multifunctional Linear MR Damper

When external excitation is applied, the multifunctional MR damper will simultaneously generate the controllable damping force, electrical power, velocity and displacement signals. The prototype of the self-sensing MR damper with power generation is tested to evaluate those functions. The corresponding specifications are provided in Table 15.2. The experimental results of multiple functions under 1 Hz, 5 mm harmonic excitation are shown in Figure 15.7.

For the power generation function, as shown in Figure 15.7(a), the open circuit voltages of theoretical analysis and experimental results match very well. For the dynamic sensing function, a laser vibrometer (OFV303, Polytec) that has a high velocity-measurement resolution up to 1 μm s^{-1} is used for comparison. As shown in Figure 15.7(b), the self-sensing velocities match well with those measured by the laser vibrometer. As shown in Figure 15.7(c), the self-sensing displacements also match well with those measured by the laser vibrometer. As shown in Figure 15.7(d), the controllable MR damping forces are in the force-displacement loops. As the current applied to the MR damper increases, the area of the force-displacement loop significantly increases. That is, the dissipated energy increases as the applied current increases. The damping force increases from 175 to 660 N when the current is increased from 0 to 1.2 A.

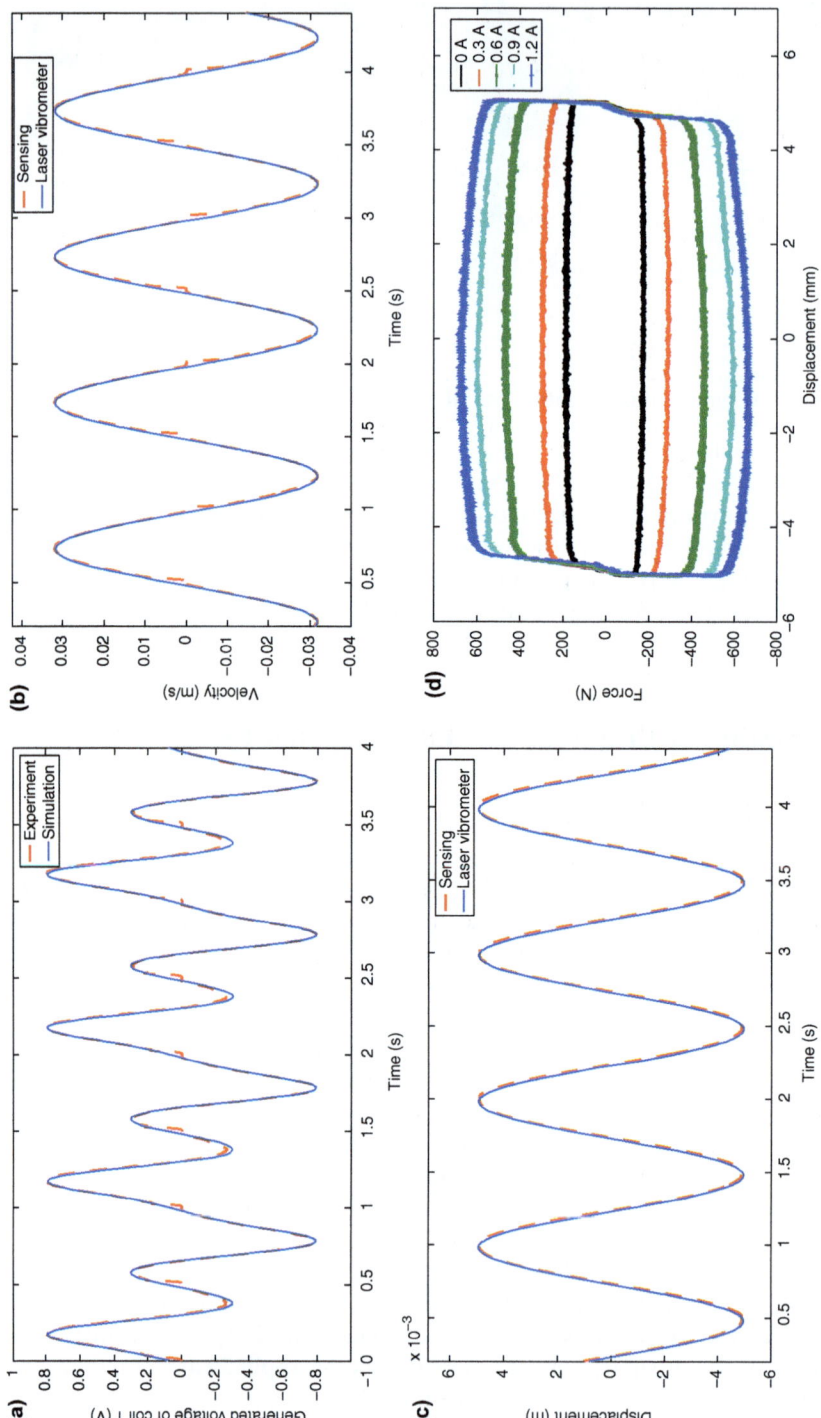

Figure 15.7 Responses of the multifunctional MR damper prototype under 1 Hz, 5 mm excitation: (a) simulation and experimental results of open circuit voltages of coil 1; (b) self-sensing velocity and velocity measured by the laser vibrometer; (c) self-sensing displacement and displacement measured by the laser vibrometer; (d) force *versus* displacement with different applied current.

15.3 Multifunctional Rotary Magnetorheological Actuators

15.3.1 Design Considerations of Multifunctional Rotary Actuators

Multifunctional rotary MR actuators are also presented in the second part of this chapter. The new actuator is comprised of two main parts into a single device: motor part and clutch/brake part. Each part is associated with the corresponding coils, the outer and inner coils. MR fluid is filled inside the motor part along with the inner clutch/brake part. The motor part converts electric power into mechanical power to provide active torque. Changing the properties of the MR fluid, the clutch/brake part can transmit the torque generated from motor part as a clutch or provide controllable torque as a brake with less power consumption than conventional electric motors. These multiple functions can be realized by controlling currents on the corresponding coils. Figure 15.8 shows the schematic of the multifunctional actuator.

In this multifunctional rotary actuator, the motor part is based on the design of the Brushless Permanent Magnet (BLPM) DC motor, which is essentially configured as a stator around which windings are wound and a rotor with permanent magnets. The windings provide an electromagnetic field to drive the permanent magnets fixed on the rotor.

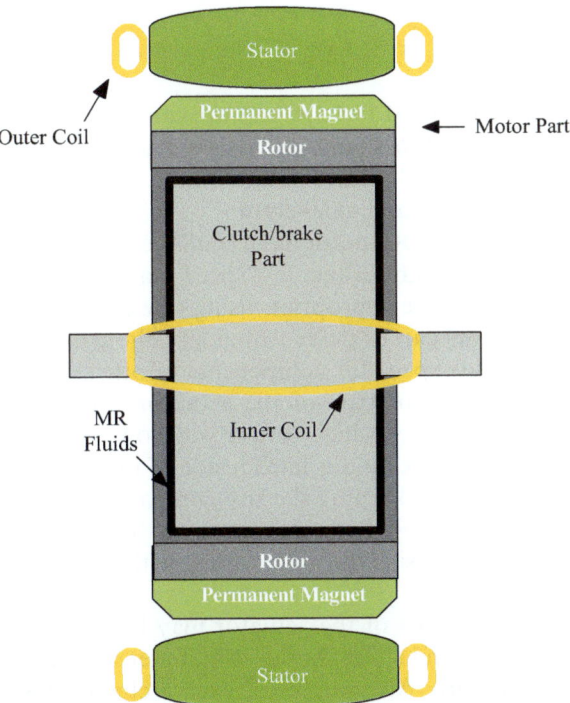

Figure 15.8 Schematic of the multifunctional rotary MR actuator.

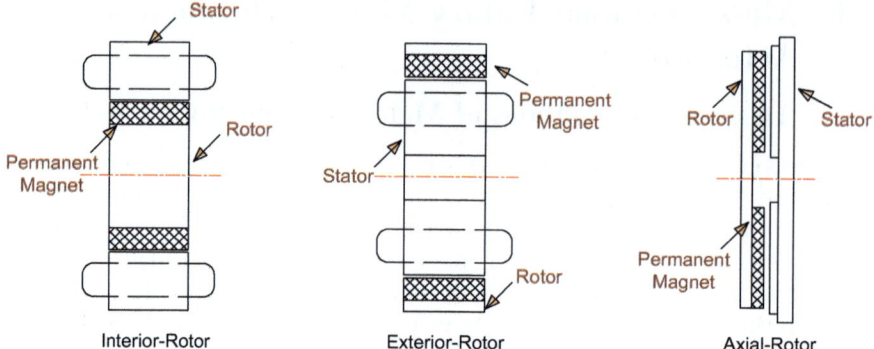

Figure 15.9 Three basic BLPM DC motor configurations.

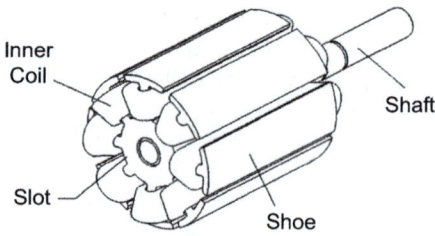

Figure 15.10 Clutch/brake part in form of inner armature.

Various configurations of **BLPM DC** motors can be modified as the motor part in designing the multifunctional actuator. Generally, **BLPM DC** motors are made in three different physical configurations, as shown in Figure 15.9, interior-rotor, exterior-rotor and axial-rotor.

The interior-rotor **BLPM DC** motor has a similar configuration to that of the AC induction motor. If a high torque, low speed machine is required, then an interior-rotor design would be appropriate using rare earth magnets and a high pole count.[17] Such motors can be made with a large hole through the center of the rotor so as to provide space for components of other functions. Thus, for assistive knee braces, a motor part of the actuator with interior-rotor configuration is a suitable choice, as high torque with low speed is needed.

In this rotary actuator, MR fluid is filled inside of the motor part to provide clutch or brake torque while keeping the size compact. The motor part is operating in rotation, thus the MR fluid is working in shear mode. Here, the clutch/brake part is placed inside the motor along with the MR fluid to provide torque.

The inner clutch/brake part of the actuator may be realized in the form of inner armature with slots and shoes. In this case, inner coils are wound on each shoe of the inner armature. An example of such a clutch/brake element is illustrated in Figure 15.10. As shown, the clutch/brake part comprises an inner coil, an inner armature and a shaft. A plurality of turns of wire is wound on the

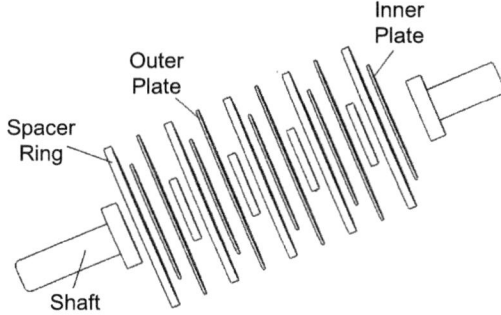

Figure 15.11 Clutch/brake part in form of input/output plates.

inner armature, which is made of high magnetic permeable material in form of lamination with shoes and slots. The shaft is fixed into the inner armature to transmit torque. An additional guide layer made of high magnetic permeable material may be utilized to guide the magnetic flux produced by the inner coil through the MR fluid.

Alternatively, the clutch/brake part can also be realized in form of a plurality of input/output plates spaced apart by nonmagnetic spacer rings forming gaps between them to carry MR fluid. The output plates and spacer rings may be clamped by shafts with flanges. An example of such a clutch/brake part is illustrated in Figure 15.11. As shown, the clutch/brake part includes a set of input plates and a set of output plates, which are separated by input spacer rings and output spacer rings. Two shafts are used to fix these plates and transmit torque. In this rotary MR actuator, the inner clutch/brake part in form of input/output plates is utilized into the design. The sectional view of the multifunctional actuator is shown in Figure 15.12.

15.3.2 Finite Element Analysis of Rotary MR Actuator

In this section, the motor part and clutch/brake part in different forms are analyzed. The interaction of the magnetic force between the stator and permanent magnets is investigated for the motor part. The influence of permanent magnets on the MR fluid, magnetic flux distribution and dynamic sealing are analyzed, and the shear stress provided by the inner clutch/brake part is simulated as well.

Figure 15.13 shows the contour plot of the magnetic flux density in the motor part when no current is applied on the outer coil. The torque between the stator and permanent magnets can be calculated based on the flux density distribution. The simulation result is about 0.733 Nm. This magnetic interaction torque between stator and permanent magnets can be used to hold the rotor statically and then play a role in the brake operation.

For the clutch/brake part with input/output plates, a 2D model is built to analyze the electromagnetic flux from the inner coil in the MR fluid and the yield stress produced in the clutch/brake part. It was illustrated that the

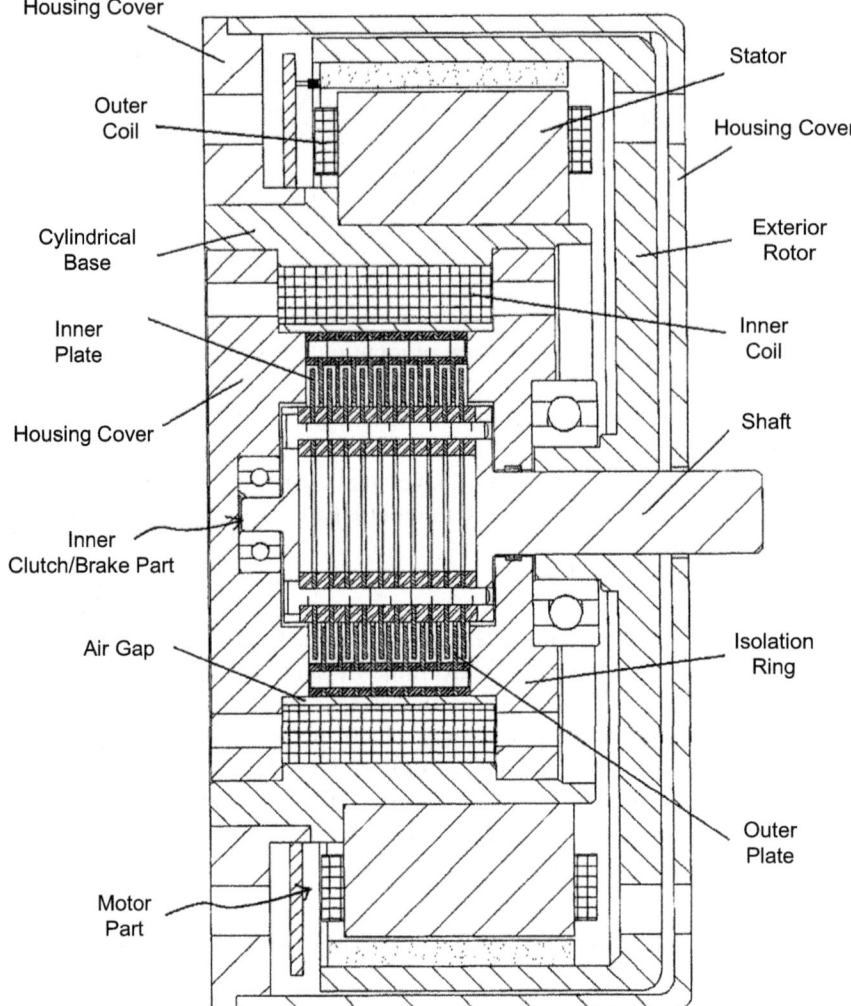

Figure 15.12   Sectional view of the multifunctional MR actuator.

maximum stress occurred when the direction of the magnetic field is perpendicular to the shear motion of the MR fluid.[18] As shown in Figure 15.14, it can be seen that the direction of the magnetic flux in the MR fluid are along the normal direction of the input/output plates with the result that the yield shear stress reaches the maximum value.

15.3.3   Selection of Brake/Clutch Part

Under the geometric constraints, the main factor affecting the output torque in the clutch/brake part is the magnetic circuit. In the design of an MR fluid based clutch and brake part with a plurality of input/output plates, inner coils have

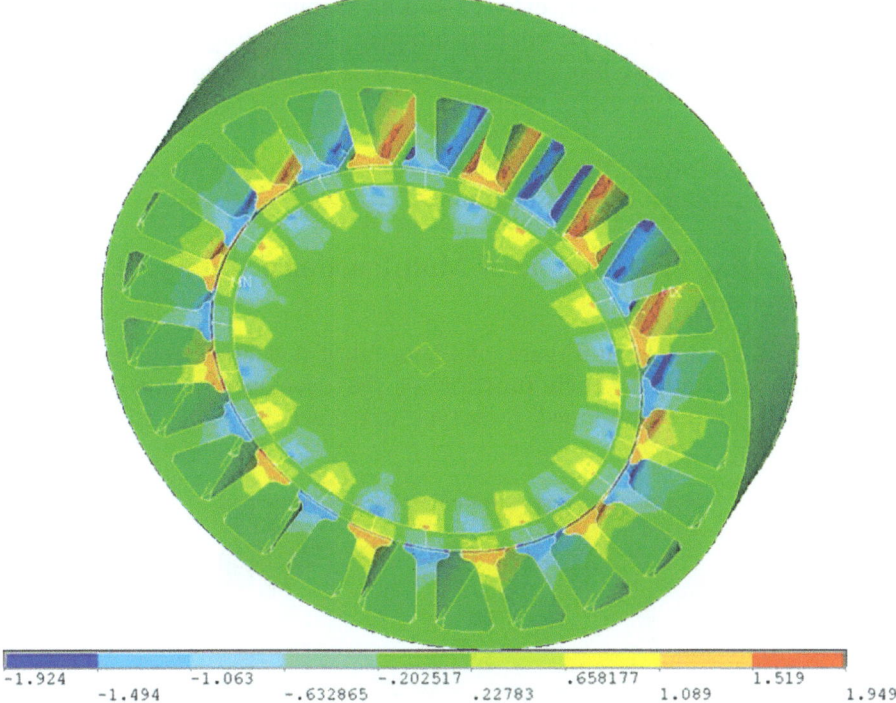

Figure 15.13 Contour plot of the magnetic flux density in the motor part.

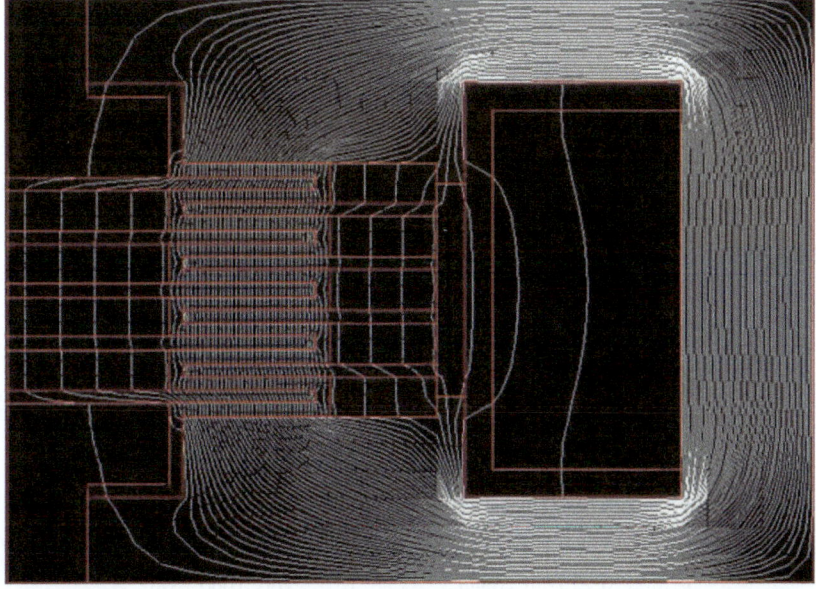

Figure 15.14 Magnetic flux in the clutch/brake part.

three different forms: interior, exterior and axial inner coils. Although these three configurations can all be possibly adopted in designing the clutch/brake part, the electromagnetic flux produced by these magnetic circuits should be considered. For the purpose of comparison, clutch/brake models with various inner coils were built.[19]

Through optimization, the maximum torque produced by the clutch/brake part with interior inner coil can be found.[19]

15.3.4   Testing of Prototype of Multifunctional Rotary MR Actuator

A prototype, as shown in Figure 15.15, is fabricated and experiments are conducted to investigate each function of the developed rotary actuator. Specifications of the developed prototype are given in Table 15.3. In the

Figure 15.15   Prototype of the multifunctional rotary MR actuator.

Table 15.3   Specifications of multifunctional rotary MR actuator.

Parameter	Value
Width	30–40 mm
Diameter	80–100 mm
Weight	1–2 kg
Rated voltage	24 V
Range of output torque	0.25–1.0 Nm
Range of output speed	800–1000 rpm
Maximal input power	100–200 W

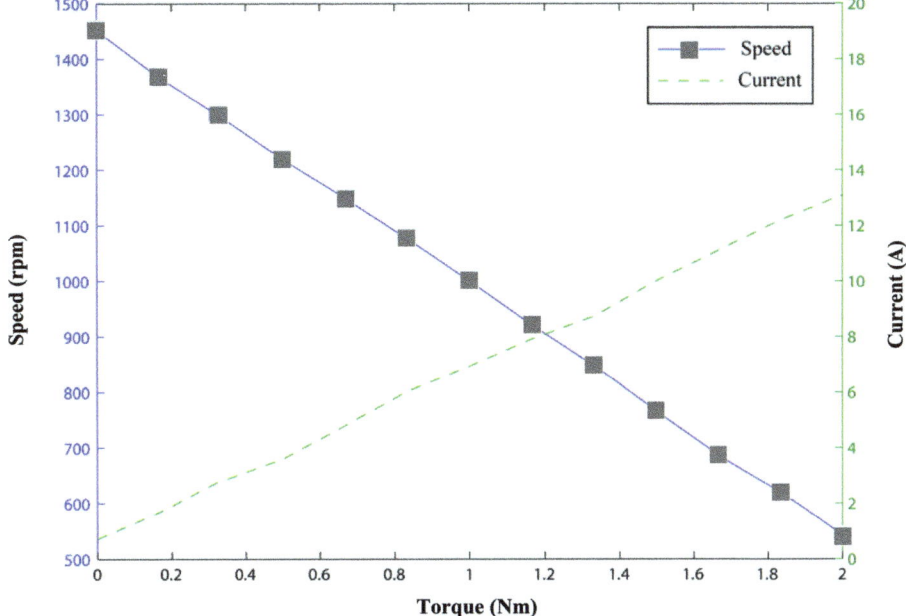

Figure 15.16 Motor function testing results for the prototype.

experimental setup, a dynamic torque sensor (Model RST-C4A-30-1-A, RSTSensor Inc.) is used to measure the output torque produced by the prototype. By changing the payload, for the motor function, the output torque *versus* applied stator current, and the output torque *versus* output speed can then be investigated. If the output torque of the payload is kept constant and the rotor is driven by the motor part at a constant speed, by changing the current on the inner coil, then the output torque of the brake function is measured. In this case, if a current with a pulse signal is applied to the inner coil, the response of the clutch function is then tested. In the experiments, the signals are processed by the dSPACE system (DS 1104, dSPACE Inc.), which also commands suitable control voltage signal to the actuator.

For the motor function, Figure 15.16 shows the measured torque at different currents and speeds. The torque/current as well as torque/speed lines are nearly straight.

For the brake function, the torque generated from the clutch/brake part is shown in Figure 15.17. The rotor is rotated at a speed of 600 rpm and a step current is applied on the inner coil gradually. With the current increasing from 0 A to 2.5 A, the measured torque increases until reaching a maximum value 0.48 Nm.

For the clutch function, the response under a pulse current is shown in Figure 15.18. Considering the reaction time of a normal human (0.15 to 0.4 second), the response time of the clutch/brake part is about 0.1 second, so it is capable of providing safety for stopping the torque transfer from the motor to the device, such as in a knee brace in case of emergency.

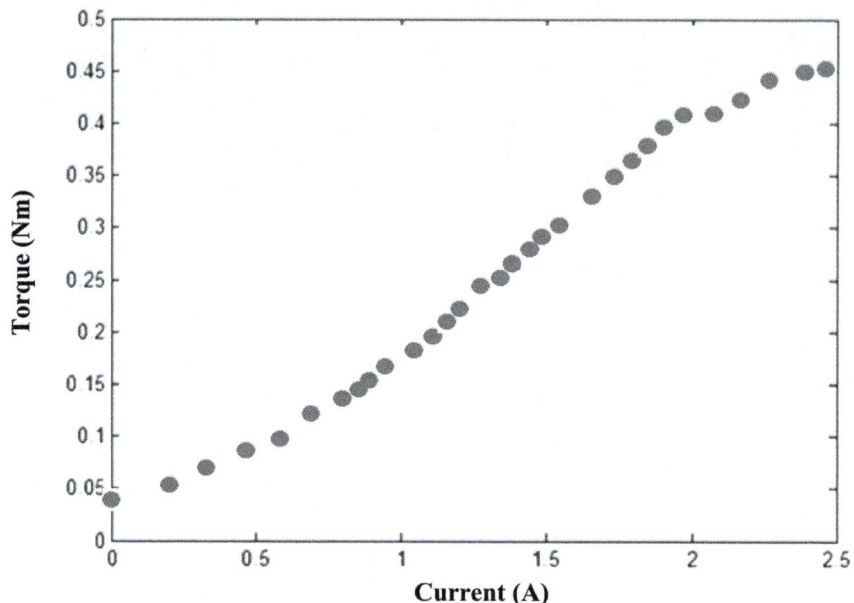

Figure 15.17   Measured torque *versus* applied current in the brake function.

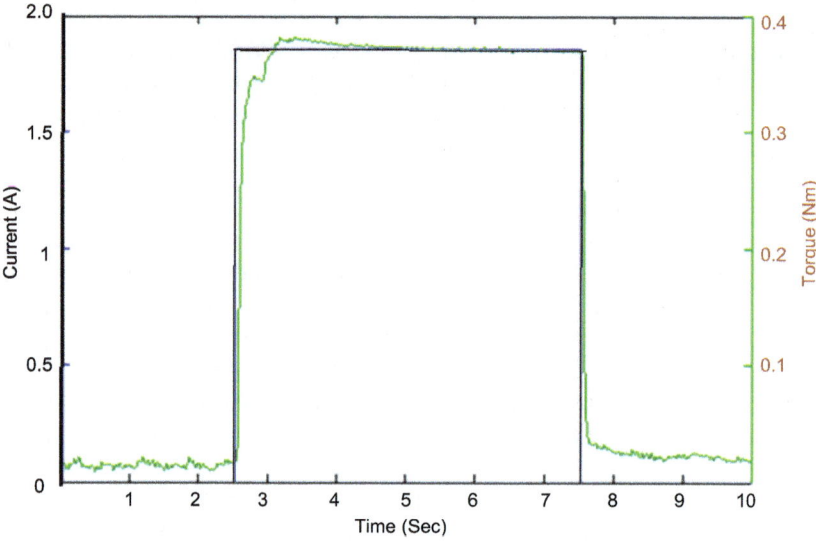

Figure 15.18   Clutch function testing results for the prototype.

15.4   Conclusions

There are two kinds of magnetorheological devices: linear and rotary. MR devices with multiple functions, including linear MR dampers and rotary MR actuators would be beneficial to be implemented as compared to the conventional MR systems. In the first part of this chapter, self-sensing MR dampers with power generation were proposed. Special issues should be taken into consideration when integrating the multiple functions, such as size limitation, power generation efficiency and functional interference. The multi-pole electromagnetic power generator could provide high power generation efficiency. Self-sensing could utilize algorithms to extract the dynamic information, including velocity and displacement, from the signal of power generation. Testing results of a fabricated prototype showed that the self-sensing MR damper with power generation could simultaneously perform several functions including MR damper, velocity/displacement sensor, and power supply. It is promising to realize self-sensing self-powered MR applications in the future.

In the second part, a novel MR fluid based multifunctional rotary actuator was designed. To decrease the dimension of the actuation device while enhancing its performance, a motor and MR fluid were integrated into a single device. By applying current on different coils in each part, the actuator possesses multiple functions as motor, clutch, and brake. Configurations and designs of the motor and clutch/brake parts were discussed. In the design, the motor part was based on interior-rotor BLPM DC motors, and the clutch/brake part was in the form of input/output plates with an interior inner coil. Since the actuator was comprised of MR fluid and permanent magnets, the influence of permanent magnets on MR fluid, magnetic interaction force, and magnetic flux distribution were analyzed using the finite element method. A corresponding prototype of the multifunctional rotary actuator was fabricated and tested, and characteristics of each function were experimentally investigated. The power generation capability with the motor part of this rotary actuator could also be explored. The new multifunctional rotary MR actuator would be beneficial to be used in situations that prefer compact actuators with multiple functions, such as assistive knee braces, tension control systems, and household appliances.

Acknowledgements

The work described in this chapter was supported by grants from the Research Grants Council (Project Nos. CUHK 415208 and 414810), and the Start-up Support Scheme, Shenzhen Research Institute, The Chinese University of Hong Kong.

References

1. D. H. Wang and W. H. Liao, *J. Intell. Mater. Syst. Struct.*, 2005, **16**, 983.
2. M. L. Gloden, W. D. Peterson and L. J. Russell, U. S. Patent 5313160, 1994.
3. D. H. Wang and T. Wang, *Smart Mater. Struct.*, 2009, **18**, 095025.
4. C. Chen and W. H. Liao, *Proc. IEEE ICMA*, 2010, 1364.

5. C. Chen and W. H. Liao, *Proc. SPIE Smart Structures/NDE*, 2011, **7977**, 797716.
6. S. W. Or, Y. F. Duan, Y. Q. Ni, Z. H. Chen and K. H. Lam, *J. Intell. Mater. Syst. Struct.*, 2008, **19**, 1327.
7. Y. T. Choi and N. M. Wereley, *J. Vib. Acoust.*, 2009, **131**, 44.
8. W. H. Liao and C. Chen, U. S. Patent Application 12/896 760, 2010.
9. B. Sapiński, *Smart Mater. Struct.*, 2010, **19**, 105012.
10. C. Chen and W. H. Liao, *Proc. SPIE Smart Structures/NDE*, 2012, **8341**, 83410Q.
11. I. H. Kim, H. J. Jung and J. H. Koo, *Smart Mater. Struct.*, 2010, **19**, 115027.
12. Q. H. Nguyen and S. B. Choi, *Smart Mater. Struct.*, 2010, **19**, 115024.
13. B. M. Kavlicoglu, F. Gordaninejad, C. Evrensel, A. Fuchs and G. Korol, *J. Vib. Acoust.*, 2006, **128**, 604.
14. D. H. Wang, H. X. Ai and W. H. Liao, *Smart Mater. Struct.*, 2009, **18**, 115001.
15. J. Z. Chen and W. H. Liao, *Smart Mater. Struct.*, 2010, **19**, 035029.
16. W. H. Liao and H. T. Guo, U. S. Patent 8 193 670, 2012.
17. J. Hendershot and T. Miller, Design of Brushless Permanent-Magnet Motors, *Magna Phys. Pub.*, 1994.
18. V. Kordonsky, Z. Shulman, S. Gorodkin, S. Prokhorov, E. Zaltsgendler and B. Khusid, *J. Mag. Mag. Mat.*, 1990, **85**, 1–3.
19. H. T. Guo and W. H. Liao, *Proc. IEEE/ASME International Conference on Advanced Intelligent Mechatronics*, 2011, 67–72.

CHAPTER 16

A Novel Medical Haptic Device Using Magneto-rheological Fluid

SEUNG-BOK CHOI,*[a] PHUONG-BAC NGUYEN[b] AND JONG-SEOK OH[a]

[a] Smart Structures and Systems Laboratory, Department of Mechanical Engineering, Inha University, Korea; [b] Department of Mechanical Engineering, Industrial University of Ho Chi Minh City, Vietnam
*Email: seungbok@inha.ac.kr

16.1 Introduction

Minimally invasive surgery (MIS) is a procedure that is carried out through tiny anatomical incisions instead of one large opening. This procedure involves use of laparoscopic maneuvers and remote-control manipulation of instruments with indirect observation with the aid of an endoscope attached at the head of the laparoscopic device. Because the body is invaded minimally, this procedure is expected to cause less injury and less pain for patients than the conventional surgery. As a result, it is becoming more and more popular in hospitals. Commonly, a robot system for MIS consists of a master system with which the surgeons manipulate and a slave system which works under the guidance signal from the master one. Images from the endoscope are projected onto monitors so surgeons can get a clear view of the surgical area.[1] In some cases, this visual information is not enough for decision of surgeons. In particular, they cannot sense what is happening on site beneath the skin. Consequently, an inaccurate decision of the surgeon may occur. Therefore, it is vital to develop the haptic master device whose output not only supplies the

RSC Smart Materials No. 6
Magnetorheology: Advances and Applications
Edited by Norman Wereley
© The Royal Society of Chemistry 2014
Published by the Royal Society of Chemistry, www.rsc.org

visual information but also reflects the physical phenomena to the surgeon such as stiffness and temperature of organs.

In the literature, there are an enormous number of haptic devices proposed in a variety of fields. However, most of them lack of versatility and robustness. Moreover, they are bulky and complicated because they are normally manufactured with conventional components such as motors, links or wires.[2,3] Recently, smart materials have also been applied in haptic applications thanks to their outstanding properties, such as fast response, wide bandwidth and miniaturization *etc.* Among them, electro-rheological (ER) and magneto-rheological (MR) fluids have shown great potential in applications that acquire high robustness, safety and versatility, such as rehabilitation and haptic devices.[4-7]

In this article, a novel 4-DOF haptic master device for MIS featuring MR fluid is presented. This device consists of two bi-directional magneto-rheological (BMR) brakes incorporated with a gimbal mechanism plus a conventional magneto-rheological (MR) brake for rotational motion and torque generation. In addition, a linear BMR actuator is used for translational motion and force generation. The benefits of the BMR brake and actuator compared to conventional ones are the possibility of compensating of undesired friction and consistency of reflecting the torque and force in different environments from soft (free) to hard (collision) stiffness ones. Therefore, these BMR devices are expected to be effective in high accurate force feedback systems, such as medical haptic devices. In our work, the description, modeling of BMR brake and linear actuator as well as its driving mechanism is proposed first. After that, a 4-DOF haptic master device using these BMR devices is developed. Finally, a prototype of the device is manufactured and an experiment in virtual environments is undertaken to investigate the effectiveness of the haptic master device.

16.2   Bi-directional MR Brake and Driving System

16.2.1   Configuration and Mechanical Model of a Bi-directional MR Brake

The configuration of the proposed BMR brake is demonstrated in Figure 16.1. It consists of two coils, two rotors, one outer casing and MR fluid filling the gap between the rotors and casing. In order to avoid the two magnetic fields interfering with each other, a non-magnetic partition is inserted at the middle location of the casing. The two power supplies to supply to the coils are distinct such that the current magnitudes of these coils can be controlled independently. Unlike other (conventional) MR brakes which only have one rotor and one stator (casing), the casing in this brake is not stationary. Indeed, it is fixed to a driving shaft, which might be connected to a 1-dimensional handle in haptic applications. Moreover, two rotors are fixed to their respective shafts, which are transmitted from a driving bi-output source so that they rotate counter each other. This assures that there exist two relative shear motions between the surfaces of two rotors and the outer casing even when the casing is at a stop.

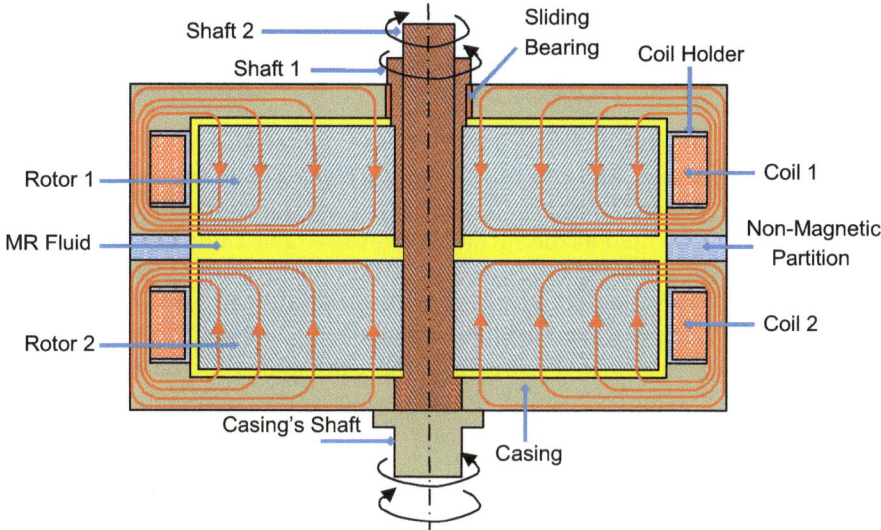

Figure 16.1 Configuration of the proposed BMR brake.

As the current sources are applied to the coils, magnetic fields are generated in two separate zones as shown in Figure 16.1. Consequently, the solidification of the MR fluid at the gap between the rotors and outer casing occurs promptly. The shear friction between the casing and the rotors provides the resultant torque. The function of this torque can be either resistive or repulsive due to the schemes of applying the current sources to the coils as well as the rotation of the casing. Assume a case that the outer casing is motionless while two rotors rotate counter each other. When only coil 1 (or 2) is excited, due to the solidification of MR fluid between the surfaces of rotor 1 (or 2) and casing, the outer casing tends to be pulled to rotate along with rotor 1 (or 2). The braking torque is the necessary torque to keep the casing still stationary and possesses the opposite direction with rotor 1 (or 2). In summary, the direction of the torque can be changed according to the excitation scheme of coil 1 or 2. Moreover, in the case where the casing rotates with same direction as rotor 1, if only coil 1 is excited, the torque resulted from the yield stress in MR fluid between the casing and rotor 1 is repulsive. Or in other words, the BMR brake works as a clutch. Otherwise, if only coil 2 is excited while the casing and rotor 1 rotate with same direction, the generated torque is resistive and the brake works as a pure brake.

The total torque of the BMR brake is generated from three sources: the friction between the end-faces of the rotors and the casing; the friction between the annular faces of the rotors and the casing; the dry friction resulted from the sealing scheme. The torque induced from these sources can be expressed in detail as follows:

$$T = \left| \overrightarrow{T_1} - \overrightarrow{T_2} \right| \tag{16.1}$$

where T_1, T_2 are the induced torques contributed from the rotors 1 and 2 whose expressions are given by

$$T_i = T_{ai} + T_{ei} + T_{fi}, \quad i = 1, 2 \tag{16.2}$$

where T_{fi} are the torques due to dry friction between the surfaces of the rotor's shafts and casing's shaft and from the sealing scheme which can be determined *via* experiment; T_{ai} and T_{ei} are the induced torques transmitted from the friction between the MR fluid at the surfaces of rotors 1 and 2 and the faces of the casing, respectively, whose magnitude depends on the property of the MR fluid and applied field. In principle, by applying appropriate current sources to the coils, the total induced torque can be eliminated completely. With the geometric parameters shown in Figure 16.2, the expressions for field-dependent torques T_{ai} and T_{ei} can be given in the following forms:

$$T_{ai} = 2\pi \left(\frac{D_R}{2}\right)^2 \int_0^{b_R} \tau_{ai} dz, \quad i = 1, 2 \tag{16.3}$$

$$T_{ei} = 2\pi \int_{\frac{D_{S2}}{2}}^{\frac{D_R}{2}} r^2 \tau_{ei} dr, \quad i = 1, 2 \tag{16.4}$$

where τ_{ai} and τ_{ei} are the shear stresses acting on the MR fluid at the surfaces of rotors i ($i = 1, 2$) and the faces of the casing respectively whose values can be mathematically expressed by Bingham's model as follows:

$$\tau_{ai} = \tau_{yai} + K\dot{\gamma}_{ai} \tag{16.5}$$

$$\tau_{ei} = \tau_{yei} + K\dot{\gamma}_{ei} \tag{16.6}$$

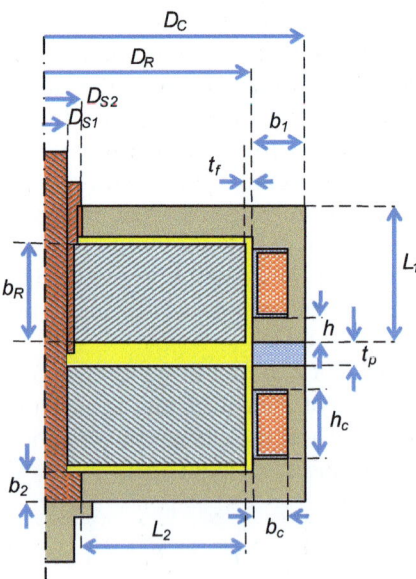

Figure 16.2   Significant geometric dimensions of the BMR brake.

where, K is called the consistency; τ_{yai} and τ_{yei} are the yield stresses of MR fluid at the surfaces of rotors i ($i = 1, 2$) and casing respectively. The variation of these yield stresses depends on the property of the MR fluid, which is available from the manufacturer's datasheet, and the magnitude of the applied currents to the coils. In eqn (16.5) and (16.6), $\dot{\gamma}_{ai}$ and $\dot{\gamma}_{ei}$ are the shear rate of MR fluid at the gap between the annular faces and end faces whose values can be determined as follows:

$$\dot{\gamma}_{ai} = \frac{D_R\left|\overrightarrow{\Omega_i} - \overrightarrow{\Omega_c}\right|}{2t_f}, \quad i = 1, 2 \tag{16.7}$$

$$\dot{\gamma}_{ei} = \frac{r\left|\overrightarrow{\Omega_i} - \overrightarrow{\Omega_c}\right|}{t_f}, \quad i = 1, 2 \tag{16.8}$$

where, Ω_1, Ω_2 and Ω_c are the angular velocities of the rotors 1, 2 and casing. By substituting eqn (16.5)–(16.8) to eqn (16.3) and (16.4), the yield stresses of MR fluid at the surfaces of the rotors and casing can be expressed as follows:

$$T_{ai} = 2\pi\left(\frac{D_R}{2}\right)^2 \int_0^{b_R} \tau_{yai}dz + \frac{\pi K D_R^3 b_R\left|\overrightarrow{\Omega_i} - \overrightarrow{\Omega_c}\right|}{4t_f}, \quad i = 1, 2 \tag{16.9}$$

$$T_{ei} = 2\pi \int_{\frac{D_{S2}}{2}}^{\frac{D_R}{2}} r^2\tau_{yei}dr + \frac{\pi K\left[\left(\frac{D_R}{2}\right)^4 - \left(\frac{D_{S2}}{2}\right)^4\right]\left|\overrightarrow{\Omega_i} - \overrightarrow{\Omega_c}\right|}{2t_f}, \quad i = 1, 2 \tag{16.10}$$

In eqn (16.5) and (16.6), τ_{yai} and τ_{yei} are the field-dependent yield stresses of MR fluid at the surfaces of rotors i and the casing, respectively. The variation of these yield stresses depends on the property of the MR fluid, which is available from the manufacturer's datasheet, and the magnitude of the applied currents to the coils. In the off-state, where there is no magnetic field applied to the coils, these yield stresses are insignificant. Consequently, there are only exists the viscosity-dependent components in eqn (16.9) and (16.10).

$$T_{ai0} = \frac{\pi K D_R^3 b_R}{4t_f}\left|\overrightarrow{\Omega_c} - \overrightarrow{\Omega_i}\right|, \quad i = 1, 2 \tag{16.11}$$

$$T_{ei0} = \frac{\pi K}{32t_f}\left[D_R^4 - D_{S2}^4\right]\left|\overrightarrow{\Omega_c} - \overrightarrow{\Omega_i}\right|, \quad i = 1, 2 \tag{16.12}$$

where, T_{ai0} and T_{ei0} are off-state induced torques contributed by rotors i. By substituting eqn (16.2), (16.11) and (16.12) into eqn (16.1), the off-state resultant torque can be obtained in detail as follows:

$$T_0 = \left|\frac{\pi K}{4t_f}\left(D_R^3 b_R + \frac{D_R^4 - D_{S2}^4}{8}\right)\left(\left|\overrightarrow{\Omega_c} - \overrightarrow{\Omega_1}\right| - \left|\overrightarrow{\Omega_c} - \overrightarrow{\Omega_2}\right|\right) + T_{f1} - T_{f2}\right| \tag{16.13}$$

where, T_0 is the off-state resultant torque. In our work, two rotors are preset so that they rotate counter to each other with the same magnitude of the angular velocity. In addition, the angular velocity of the casing is less than that of these rotors. Hence, eqn (16.13) can be simplified as follows:

$$T_0 = \left| \pm \frac{\pi K}{2t_f} \left(D_R^3 b_R + \frac{D_R^4 - D_{S2}^4}{8} \right) \Omega_c + T_{f1} - T_{f2} \right| \tag{16.14}$$

where the symbol "plus" is corresponding to the case that the casing and rotor 1 rotate counter to each other; the symbol "minus" is used in the case that the casing and rotor 2 rotate counter to each other. In eqn (16.14), the effect of the component T_{f1} is more dominant than the others due to large dry friction induced by sealing at the contact surface between the casing and rotor 1. Therefore, eqn (16.4) can be simplified as follows:

$$T_0 = \pm \frac{\pi K}{2t_f} \left(D_R^3 b_R + \frac{D_R^4 - D_{S2}^4}{8} \right) \Omega_c + T_{f1} - T_{f2} \tag{16.15}$$

It is noteworthy that the off-state resultant torque is different from zero, in general. Moreover, it possesses the same direction as rotor 1. Therefore, in order to eliminate this torque completely, the torque induced by the field-dependent field stress contributed by rotor 2 can be used to compensate for it.

16.2.2 Gear System for Motion Transmission of BMR Brake

As mentioned above, the BMR brake possesses two rotors rotating counter to each other with the same magnitude of rotation. Hence, in order to drive these rotors, a driving gear system is proposed as shown in Figure 16.3. In this gear system, a driving motion from a motor is split into two rotary motions of shafts

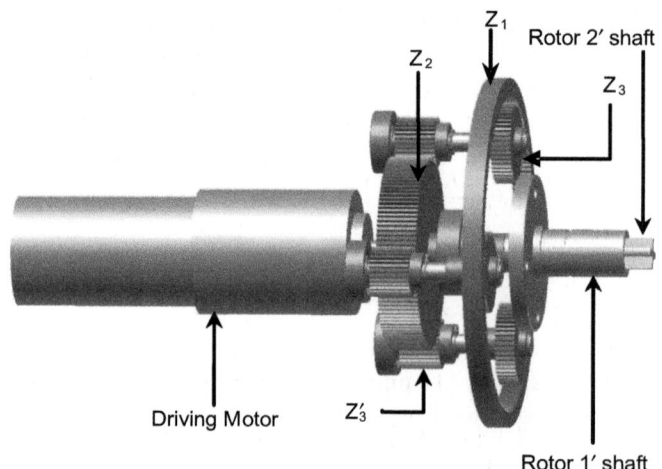

Figure 16.3 Gear system for motion transmission to the BMR brake.

1 and 2. In fact, the gear system just involves the transmission between the motor and shaft 1. Its function is to make them to rotate counter each other. On the other hand, shaft 2 is driven directly from the motor as shown in Figure 16.3. It is noteworthy that multiple intermediate gears are used in the system instead of a single one, and are located in symmetric positions to enhance the strength of the system. The speed ratio of shaft 1 to shaft 2 is computed as follows:

$$i_{12} = \frac{n_1}{n_2} = \left(\frac{n_1}{n_3}\right)\left(\frac{n_3}{n_2}\right) = \left(\frac{Z_3}{Z_1}\right)\left(-\frac{Z_2}{Z'_3}\right) \tag{16.16}$$

where, n_1, n_2, n_3 are the revolutions of shaft 1, 2, 3, respectively; Z_1, Z_2, Z_3 and Z'_3 are the numbers of teeth of the gears 1, 2, 3 and 3′, respectively as shown in Figure 16.4. In order for this gear system to work, the geometric constraint of shaft distances should be followed by:

$$\frac{d_2 + d'_3}{2} = \frac{d_1 - d_3}{2} \tag{16.17}$$

where, d_1, d_2, d_3 and d'_3 are the diameters of the gears 1, 2, 3 and 3′, respectively. These diameters relate to the numbers of teeth *via* a factor called the module. In this work, all gears used have the same module. Therefore, eqn (16.17) can be rewritten as follows:

$$m(Z_2 + Z'_3) = m(Z_1 - Z_3) \Leftrightarrow Z_2 + Z'_3 = Z_1 - Z_3 \tag{16.18}$$

where, m is the module of the gears. In addition, in order for the gear Z_3 to be located inside the internal gear Z_1, the diameter of the gear Z_3 should be less

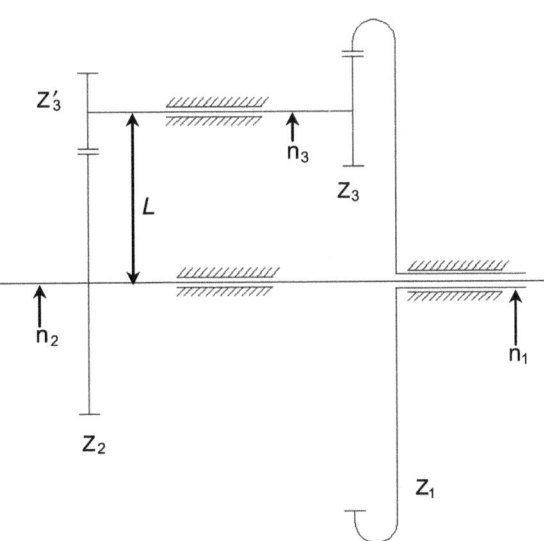

Figure 16.4 Schematic of the gear system.

than the radius of the gear Z_1. Or in other words, this constraint can be expressed by:

$$Z_3 < \frac{Z_1}{2} \tag{16.19}$$

Because two rotors are supposed to rotate counter to each other with the same magnitude of rotation, the speed ratio in must be -1. Consequently, from eqn (16.16) the ratio constraint can be expressed as follows:

$$\frac{Z_3}{Z_1} = \frac{Z'_3}{Z_2} \tag{16.20}$$

With three constraints (18)–(20), there are indefinite solutions for gear selection that satisfy them. However, if the sizes of the gears are chosen to be too small, the strength of the gears and their shafts may not be assured. Hence, in order to improve the performance of the gear system, in our work, an optimization procedure was carried out with the aid of commercial software, Autodesk's Inventor.

16.3   Bi-directional MR Linear Actuator

Figure 16.5 shows the configuration of the proposed BMR linear actuator. The actuator consists of two MR clutches with respective motors for torque

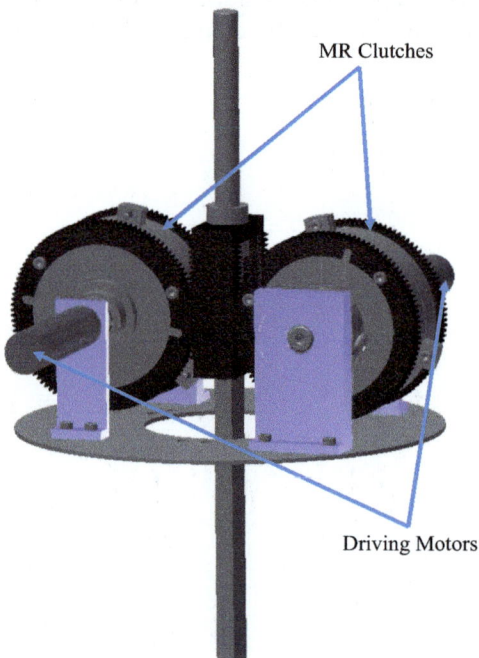

MR Clutches

Driving Motors

Figure 16.5   Configuration of the proposed BMR linear actuator.

generation in two counter directions. The translational motion and corresponding force are created *via* a toothed rack-gear mechanism. The two power supplies to supply to the coils of the clutches are distinct so that the generated torques of the clutches can be controlled independently. The resultant force is indeed proportional to the difference of these torques. Consequently, either resistive or repulsive force can be created with this actuator based on which torque is dominant. The magnitude of the resultant force of the BMR linear actuator is given as follows:

$$F = \frac{2\left|\overrightarrow{T_1} - \overrightarrow{T_2}\right|}{D_C} \qquad (16.21)$$

where T_1, T_2 are the induced torques contributed from the clutches 1 and 2 whose expressions are given by

$$T_i = T_{ei} + T_{fi}, \quad i = 1, 2 \qquad (16.22)$$

where T_{fi} are the torques due to dry friction due to sealing which can be determined *via* experiment; T_{ei} are the induced torques transmitted from the friction between MR fluid at the surfaces of rotors of the clutches 1 and 2 and the faces of the casing, respectively, whose magnitude depends on the property of the MR fluid and applied field. With the proposed configuration of the clutches as shown in Figure 16.6 and its geometric parameters shown in

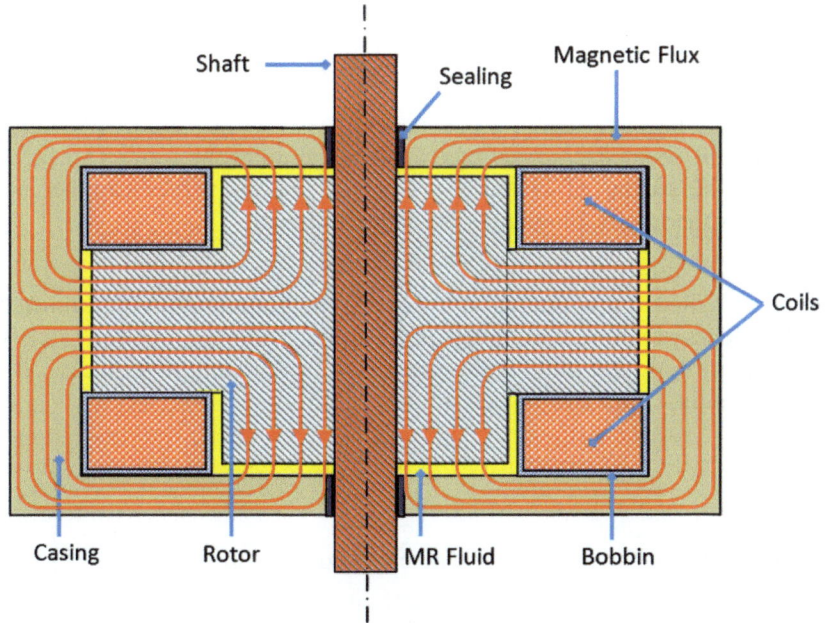

Figure 16.6 Configuration of the proposed MR clutch for the BMR linear actuator.

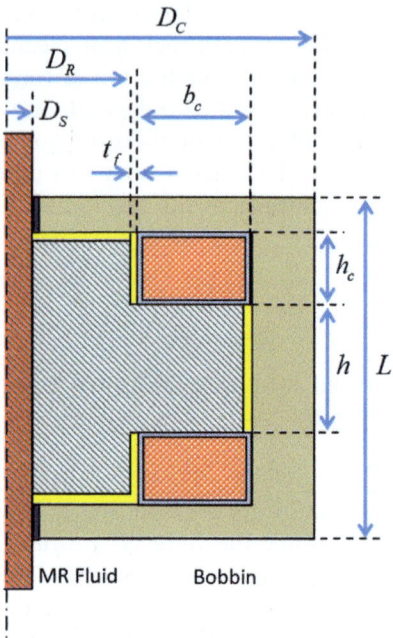

Figure 16.7   Significant geometric dimensions of the proposed MR clutch.

Figure 16.7, the expressions for field-dependent torques and T_{ei} can be given in the following forms:

$$T_{ai} = 2\pi \left[\left(\frac{D_R}{2} + t_f + b_c \right)^2 \int_0^h \tau_i dz + \int_{\frac{D_S}{2}}^{\frac{D_R}{2}} r^2 \tau_i dr \right], \quad i = 1, 2 \qquad (16.23)$$

where τ_i are the shear stresses acting on the MR fluid at the contact surfaces of rotors and the casing of clutches i $(i = 1, 2)$, whose values can be mathematically expressed by Bingham's model as follows:

$$\tau_i = \tau_{yi} + K\dot{\gamma}_i, \quad i = 1, 2 \qquad (16.24)$$

where, K is called the consistency; τ_{yi} is the yield stress of MR fluid at the contact surfaces of rotors and casing of the clutch i $(i = 1, 2)$. In eqn (16.24), $\dot{\gamma}_i$ is the shear rate of MR fluid at the gap between the contact surfaces. At the annular faces its values can be determined as follows:

$$\dot{\gamma}_i = \frac{\left(\frac{D_R}{2} + t_f + b_c \right) \left| \overrightarrow{\Omega_i} - \overrightarrow{\Omega_c} \right|}{t_f}, \quad i = 1, 2 \qquad (16.25)$$

At the end faces its values can be determined as follows:

$$\dot{\gamma}_i = \frac{r \left| \overrightarrow{\Omega_i} - \overrightarrow{\Omega_c} \right|}{t_f}, \quad i = 1, 2 \qquad (16.26)$$

where, Ω_1, Ω_2 are the angular velocities of the rotors of the clutches 1 and 2 respectively, and Ω_c is the resultant angular velocities of the casings of the clutches driven by a translational motion of the handle. By substituting eqn (16.24)–(16.26) into eqn (16.23), the field-dependent torque of the clutch can be expressed as follows:

$$
T_{ai} = 2\pi \left(\frac{D_R}{2} + t_f + b_c\right)^2 \left[\int_0^h \tau_{yi} dz + \frac{Kh\left(\frac{D_R}{2} + t_f + b_c\right)\left|\overrightarrow{\Omega_i} - \overrightarrow{\Omega_c}\right|}{t_f}\right] +
$$

$$
+ 2\pi \int_{\frac{D_S}{2}}^{\frac{D_R}{2}} r^2 \tau_{yi} dr + \frac{\pi K\left[\left(\frac{D_R}{2}\right)^4 - \left(\frac{D_S}{2}\right)^4\right]\left|\overrightarrow{\Omega_i} - \overrightarrow{\Omega_c}\right|}{2t_f}, \quad i = 1,2
$$

(16.27)

In above equation, τ_{yi} is the field-dependent yield stress of MR fluid at the contact surfaces of the rotors and casing of the clutches. In the off-state, where there is no magnetic field applied to the coils, this yield stress is insignificant. Consequently, there only exists the viscosity-dependent components in the following equation.

$$
T_{0i} = \frac{\pi K}{2t_f}\left|\overrightarrow{\Omega_i} - \overrightarrow{\Omega_c}\right|\left[4h\left(\frac{D_R}{2} + t_f + b_c\right)^3 + \left(\frac{D_R}{2}\right)^4 - \left(\frac{D_S}{2}\right)^4\right], \quad (16.28)
$$
$$
i = 1,2
$$

where, T_{0i} is the off-state induced torque contributed by clutch i. By substituting eqn (16.22) and (16.28) into eqn (21), the off-state resultant force can be obtained in detail as follows:

$$
F_0 = \frac{\pi K}{D_C t_f}\left(\left|\overrightarrow{\Omega_1} - \overrightarrow{\Omega_c}\right| - \left|\overrightarrow{\Omega_2} - \overrightarrow{\Omega_c}\right|\right)\left[4h\left(\frac{D_R}{2} + t_f + b_c\right)^3 + \left(\frac{D_R}{2}\right)^4 - \left(\frac{D_S}{2}\right)^4\right]
$$
$$
+ \frac{2(T_{f1} - T_{f2})}{D_C}
$$

(16.29)

In our work, the angular velocity of the casing is less than that of these rotors. Moreover, two rotors rotate counter to each other with the same angular velocity. Hence, eqn (16.29) can be simplified as follows:

$$
F_0 = \frac{2}{D_C}\left|\pm\frac{\pi K}{t_f}\Omega_c\left[4h\left(\frac{D_R}{2} + t_f + b_c\right)^3 + \left(\frac{D_R}{2}\right)^4 - \left(\frac{D_S}{2}\right)^4\right] + T_{f1} - T_{f2}\right|
$$

(16.30)

where, the symbol "plus" is corresponding to the case that the casing and rotor of clutch 1 rotate counter to each other; the symbol "minus" is used in the case that the casing and rotor of clutch 2 rotate counter to each other.

16.4 Modeling of the Haptic Master Device

16.4.1 Kinematics of the Haptic Master Device

Our 4-DOF haptic master device is configured as shown in Figure 16.8. In order to model the system, it is necessary to introduce a coordinate system $\{O, x_4, y_4, z_4\}$ that is fixed with the body in addition to the fixed coordinate system $\{O, X, Y, Z\}$. The body-fixed $x_4 - y_4 - z_4$ system is related to the fixed $X - Y - Z$ system upon three rotations (ξ, η, ζ) about the body-fixed axes and one translation z_h, as shown in Figure 16.9. At the initial position, the fixed system and the body-fixed one coincide. When frame I is rotated an angle ξ about the body-fixed axis X, the body-fixed system goes to the new position $\{O, x_1, y_1, z_1\}$ with x_1 axis retained ($x_1 \equiv X$). When frame II is then rotated an angle η about the body-fixed axis y_1, the body-fixed system goes from position $\{O, x_1, y_1, z_1\}$ to $\{O, x_2, y_2, z_2\}$ with y_2 axis retained ($y_2 \equiv y_1$). In the same manner, y_1, the body-fixed system goes from position $\{O, x_2, y_2, z_2\}$ to $\{O, x_3, y_3, z_3\}$ with z_3 axis retained ($z_3 \equiv z_2$) when the handle is rotated an angle ζ about the body-fixed axis z_2. Finally, when the handle is moved a linear translation z_4 along the z_3 axis,

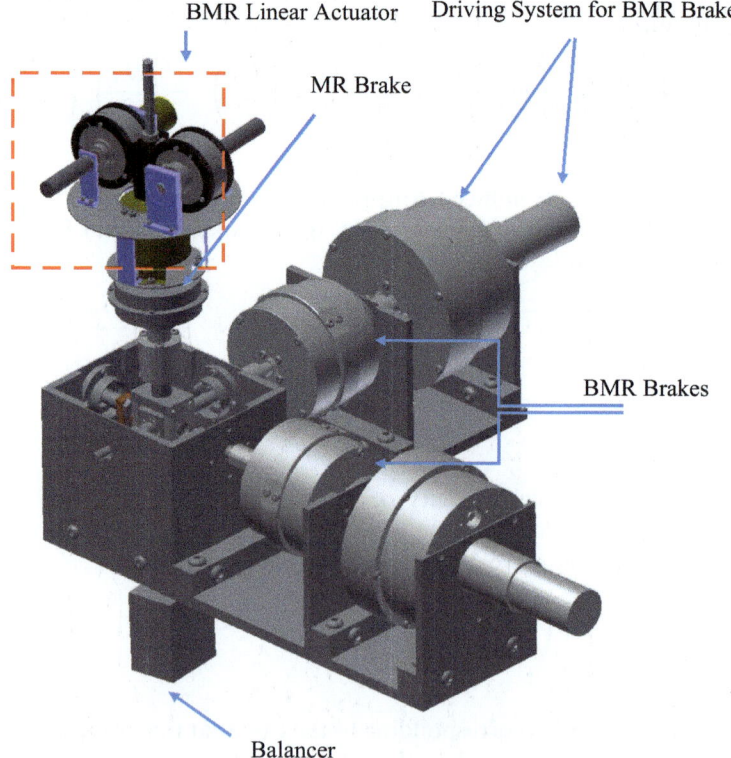

Figure 16.8 Proposed 4-DOF Haptic Master Device.

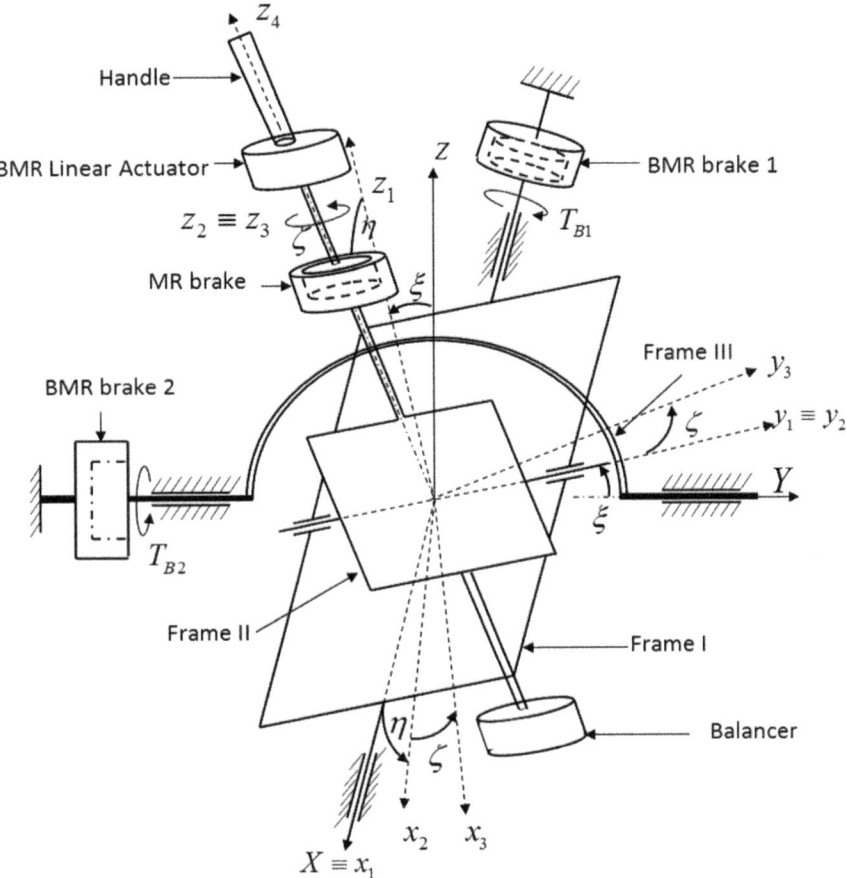

Figure 16.9 Schematic of the 4-DOF haptic master device.

the body-fixed system goes to position $\{O, x_4, y_4, z_4\}$. The position of an arbitrary point in the handle can be obtained using the following transformation:

$$\begin{bmatrix} \mathbf{L} \\ 1 \end{bmatrix} = \begin{bmatrix} \mathbf{A}_1 & \mathbf{0} \\ \mathbf{0} & 1 \end{bmatrix} \begin{bmatrix} \mathbf{A}_2 & \mathbf{0} \\ \mathbf{0} & 1 \end{bmatrix} \begin{bmatrix} \mathbf{A}_3 & \mathbf{0} \\ \mathbf{0} & 1 \end{bmatrix} \mathbf{A}_T \begin{bmatrix} \overline{\mathbf{L}} \\ 1 \end{bmatrix} \tag{16.31}$$

where, $\mathbf{L}$ and $\overline{\mathbf{L}}$ are the vectors of position with respect to the directions of the fixed coordinate system and of the fixed-body coordinate system, respectively; $\mathbf{A}_T$ is the translational transformation matrix (4×4); $\mathbf{A}_1, \mathbf{A}_2, \mathbf{A}_3$ are the rotational transformation matrices about the body-fixed axes. All these matrices are defined as follows:

$$\mathbf{A}_T = \begin{bmatrix} 0 & 0 & 0 & 0 \\ 0 & 0 & 0 & 0 \\ 0 & 0 & 0 & -z_h \\ 0 & 0 & 0 & 1 \end{bmatrix} \tag{16.32}$$

$$A_1 = \begin{bmatrix} 1 & 0 & 0 \\ 0 & C\xi & -S\xi \\ 0 & S\xi & C\xi \end{bmatrix}; \quad A_2 = \begin{bmatrix} C\eta & 0 & S\eta \\ 0 & 1 & 0 \\ -S\eta & 0 & C\eta \end{bmatrix}; \quad A_3 = \begin{bmatrix} C\zeta & -S\zeta & 0 \\ S\zeta & C\zeta & 0 \\ 0 & 0 & 1 \end{bmatrix} \quad (16.33)$$

The components of the vector of angular velocity of the handle base with respect to the fixed-coordinate system can be obtained as follows:

$$\omega = \begin{bmatrix} \omega_X \\ \omega_Y \\ \omega_Z \end{bmatrix} = \begin{bmatrix} \dot{\xi} + (S\eta)\dot{\zeta} \\ (C\xi)\dot{\eta} - (C\eta)(S\xi)\dot{\zeta} \\ (S\xi)\dot{\eta} + (C\eta)(C\xi)\dot{\zeta} \end{bmatrix} \quad (16.34)$$

The components of the vector of angular velocity of the handle base with respect to the fixed-body system can be obtained as follows:

$$\overline{\omega} = \begin{bmatrix} \omega_\xi \\ \omega_\eta \\ \omega_\zeta \end{bmatrix} = (A_1 A_2 A_3)^T \omega = \begin{bmatrix} (C\eta)(C\zeta)\dot{\xi} + (S\zeta)\dot{\eta} \\ -(C\eta)(S\zeta)\dot{\xi} + (C\zeta)\dot{\eta} \\ (S\eta)\dot{\xi} + \dot{\zeta} \end{bmatrix}$$

16.4.2 Dynamics of the Haptic Master Device

For simplicity, the overall system is divided into two sub-systems: the rotational and linear ones. The rotational system consists of three frames and the base of the linear system (including the rotor of the MR brake and linear motor). The kinetic energy of the rotational system is the sum of the kinetic energies of the components of the system:

$$W_{sys} = W_1 + W_2 + W_3 + W_{le} \quad (16.35)$$

where, W_1, W_2, W_3 and W_{le} are the kinetic energies of the frames and linear element, respectively. They are determined in the same manner as follows:

$$W_i = \frac{1}{2}\overline{\omega}_i \overline{J}_i \overline{\omega}_i \quad i = 1,2,3 \text{ and linear element} \quad (16.36)$$

where, $\overline{J}_i$ are the matrices of principal moments of inertia tensors in the frame i-fixed systems and in the linear element-fixed system. $\overline{\omega}_i$ can be obtained due to the transformation between the frame-fixed systems and the fixed coordinate system as follows:

$$\overline{\omega}_1 = A_1^T \omega_1 = \begin{bmatrix} \dot{\xi} & 0 & 0 \end{bmatrix}^T \quad (16.37)$$

$$\overline{\omega}_2 = (A_1 A_2)^T \omega_2 = (A_1 A_2)^T [\omega_1 + A_1 \omega_{21}]$$

$$= \begin{bmatrix} (C\eta)\dot{\xi} & \dot{\eta} & (S\eta)\dot{\xi} \end{bmatrix}^T \quad (16.38)$$

Because the frames II and III are constrained such that there only exists a relatively rotary motion about the Y-axis between them, the vector of angular velocity of frame III is determined as follows:

$$\Phi_3 = \begin{bmatrix} 0 & (C\xi)\eta & 0 \end{bmatrix}^T \Rightarrow \omega_3 = \dot{\Phi}_3 = \begin{bmatrix} 0 & (C\xi)\dot{\eta} - \eta(S\xi)\dot{\xi} & 0 \end{bmatrix}^T \quad (16.39)$$

Therefore, the vector of angular velocity of frame III with respect to the frame II-fixed system is obtained using the rotary transformation matrix.

$$\overline{\omega}_3 = A_2^T \omega_3 = \begin{bmatrix} 0 & (C\xi)\dot{\eta} - \eta(S\xi)\dot{\xi} & 0 \end{bmatrix}^T \tag{16.40}$$

From the velocity vectors obtained above, the kinetic energies of the components constituting the system can be determined as follows:

$$W_1 = \frac{1}{2}\overline{\omega}_1 \overline{J}_1 \overline{\omega}_1 = \frac{1}{2}J_{1I}\dot{\xi}^2 \tag{16.41}$$

$$W_2 = \frac{1}{2}\overline{\omega}_2 \overline{J}_2 \overline{\omega}_2 = \frac{1}{2}J_{2I}\left[(C\eta)\dot{\xi}\right]^2 + \frac{1}{2}J_{2II}\dot{\eta}^2 + \frac{1}{2}J_{2III}\left[(S\eta)\dot{\xi}\right]^2 \tag{16.42}$$

$$W_3 = \frac{1}{2}\overline{\omega}_3 \overline{J}_3 \overline{\omega}_3 = \frac{1}{2}J_{3II}\left[(C\xi)\dot{\eta} - \eta(S\xi)\dot{\xi}\right]^2 \tag{16.43}$$

$$W_{le} = \frac{1}{2}\overline{\omega}_{le} \overline{J}_{le} \overline{\omega}_{le} = \frac{1}{2}J_{leIII}\dot{\zeta}^2 \tag{16.44}$$

where $\overline{J}_1, \overline{J}_2, \overline{J}_3, \overline{J}_{le}$ are matrices of principal moments of inertia with respect to the body-fixed coordinates of the components, which are expressed as follows:

$$\overline{J}_1 = \begin{bmatrix} J_{1I} & 0 & 0 \\ 0 & J_{1II} & 0 \\ 0 & 0 & J_{1III} \end{bmatrix}; \quad \overline{J}_2 = \begin{bmatrix} J_{2I} & 0 & 0 \\ 0 & J_{2II} & 0 \\ 0 & 0 & J_{2III} \end{bmatrix}$$

$$\overline{J}_3 = \begin{bmatrix} J_{3I} & 0 & 0 \\ 0 & J_{3II} & 0 \\ 0 & 0 & J_{3III} \end{bmatrix}; \quad \overline{J}_{le} = \begin{bmatrix} J_{leI} & 0 & 0 \\ 0 & J_{leII} & 0 \\ 0 & 0 & J_{leIII} \end{bmatrix}$$

By applying Lagrange's equation of the second type, the governing equation for this system can be expressed as follows:

$$\begin{cases} \kappa_{11}\ddot{q}_1 + \kappa_{12}\dot{q}_1\dot{q}_2 + \kappa_{13}\dot{q}_1^2 + \kappa_{14}\ddot{q}_2 + \kappa_{15}\dot{q}_2\dot{q}_3 + \kappa_{16}\ddot{q}_3 = T_{B1} \\ \kappa_{21}\ddot{q}_1 + \kappa_{22}\dot{q}_1\dot{q}_2 + \kappa_{23}\dot{q}_1^2 + \kappa_{24}\dot{q}_1\dot{q}_3 + \kappa_{25}\ddot{q}_2 = T_{B2}\cos q_1 \\ \kappa_{31}\ddot{q}_1 + \kappa_{32}\dot{q}_1\dot{q}_2 + \kappa_{33}\ddot{q}_3 = T_{B3} \end{cases} \tag{16.45}$$

where q_1, q_2 and q_3 stand for angles ξ, η and ζ, respectively. κ_{ii} are coefficients constituting the governing eqn (45). The dynamics of the linear system is obtained as follows:

$$m_L\ddot{z}_L = F \tag{16.46}$$

where, $\ddot{z}_L$ is the translational acceleration of the handle; m_L is the equivalent mass of the handle plus the casings of the MR brakes in the linear system.

16.5 Experimental Results of Virtual Collision Simulation

In this section, several experiments are conducted to investigate the performance of the device. The transducer used to collect the force information is a

6-axis force sensor type (ATI, Nano17) assembled underneath the handle. From the force information, the necessary torque and force can be obtained.

As the first stage of the work, the individual components of the system are investigated independently. Open-loop control experiments for three BMR devices are undertaken to assess the performance in sensing a process of touching a soft tissue. The tip of the handle is assumed to be moved from the initial position at 0 to touch a surface of a soft tissue at certain position (angle $\theta = 20°$ with the BMR brake and 10 mm with BMR linear actuator) as shown in Figure 16.10. After that, it is continued, penetrating into the tissue (with stiffness K) to a destination position (angle $\theta = 50°$ with the BMR brake and 25 mm with BMR linear actuator) and then retreated slightly back to the initial position. It can be observed in Figure 16.11 that at the stage of free motion of

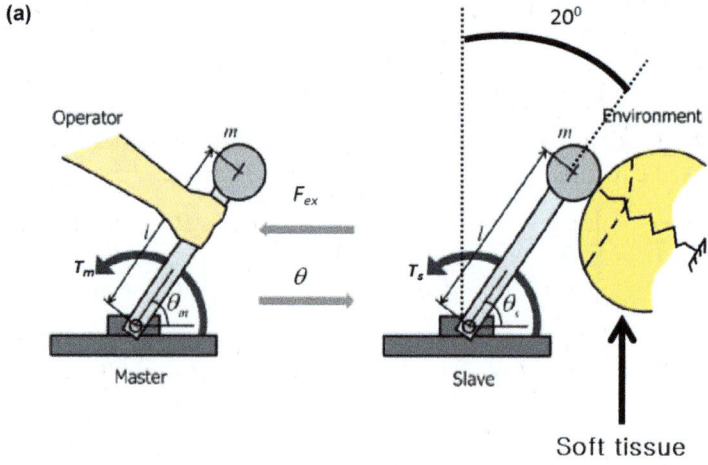

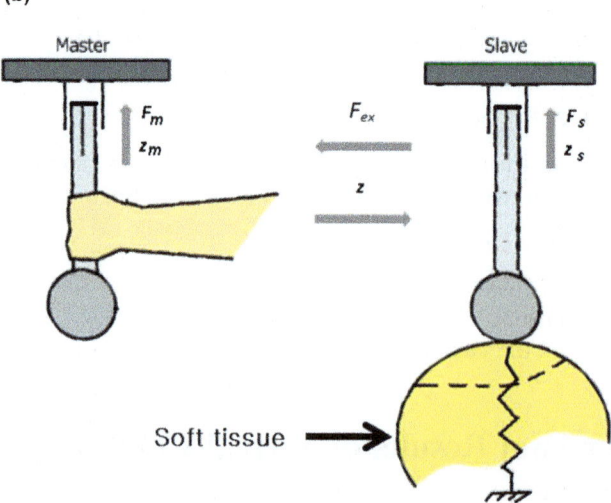

Figure 16.10   Soft tissue simulation (a) haptic torque (b) haptic force.

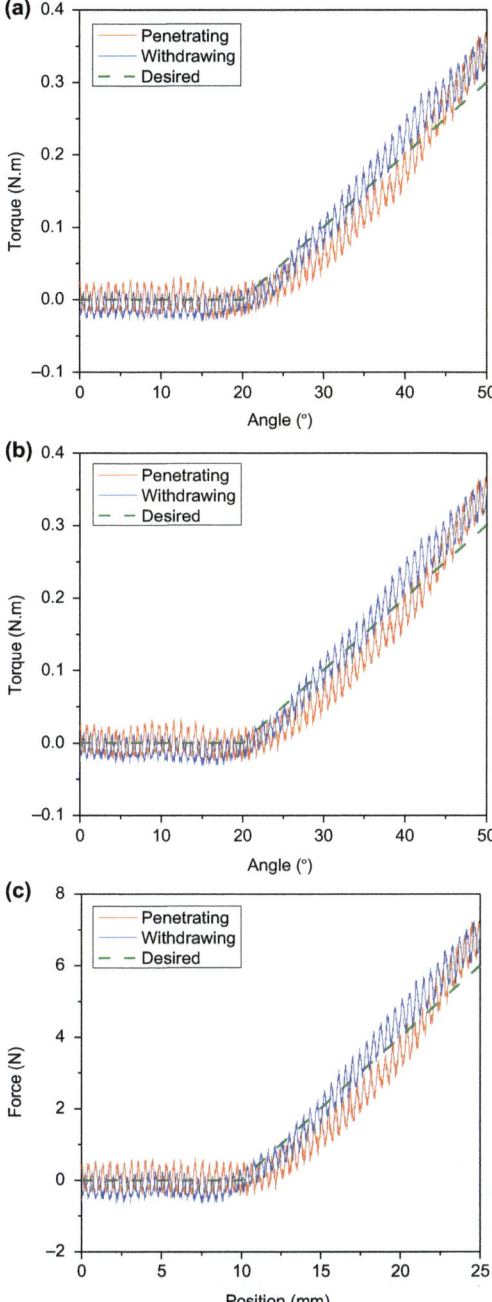

Figure 16.11 Experimental result of soft tissue simulation without friction compensation.

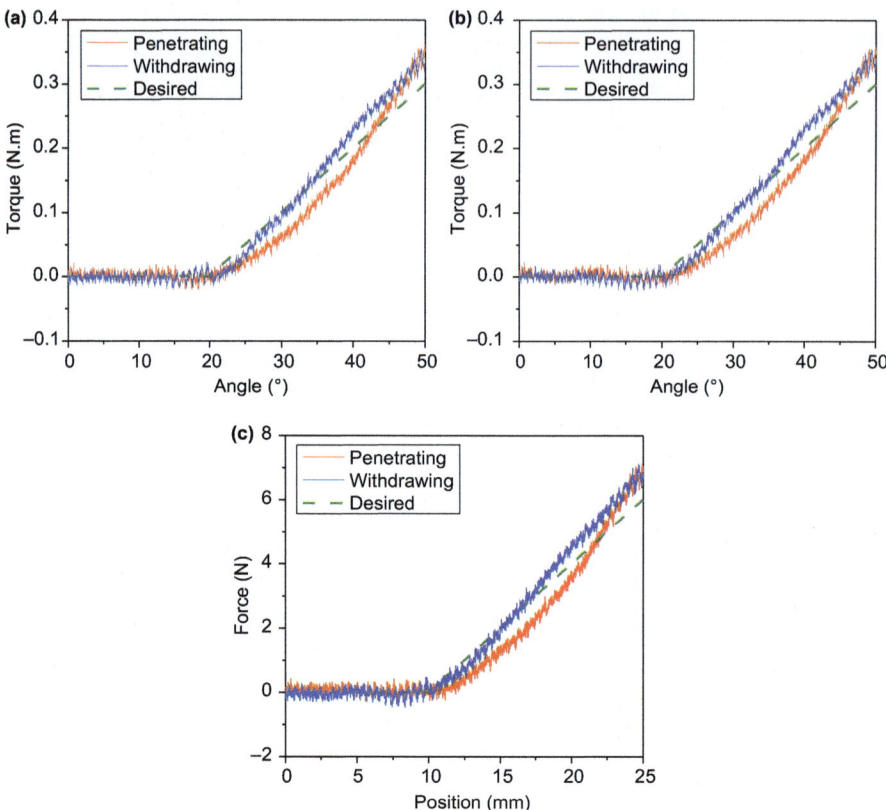

Figure 16.12   Experiment of soft tissue simulation with friction compensation.

the penetrating and withdrawing process, the haptic torque and force are approximately zero. When the handle touches the soft body, the sensing torque and force increase corresponding to the position of the handle. It is realized from the figure that the measured torque is fluctuating periodically resulting from the residue friction due to the imperfections of the manufacturing of the BMR brakes and linear actuator. The amplitude of this friction is significant. Therefore, it should be compensated. In order to compensate for the residue friction torque of the BMR brake, one of the two coils of the brake is used as a compensator. Similarly, in order to compensate for the residue friction force of the BMR linear actuator, one MR brake of the actuator is used as a compensator. The sensing force result after residue friction compensation is given in Figure 16.12. It is observed that with the device the sensing force agrees with the desired one fairly well. The difference of sensing forces between the penetrating and withdrawing processes, as well as the reference, is due to the effect of the nonlinear behavior and hysteresis of MR fluid. Taking account to these nonlinearities, it is expected that the proposed device performs with significant accuracy and can be a good candidate for application in MIS.

16.6 Conclusion

In this article, a novel 4-DOF medical haptic master device for MIS was introduced. In the device, two bi-directional MR (BMR) brakes and one BMR actuator, whose braking torque varies from negative to positive values, were used. Or in other words, they can generate the force actively. Therefore, the device is expected to be able sense a wide variety of environments from soft tissues to bones. Consequently, it may be a potential candidate in high accurate force feedback systems, such as medical haptic devices. In order to assess the ability of the device, several experiments in simulation of a virtual elastic body collision of the components of the device were carried out. The experimental results showed that the sensing force agrees with the desired one fairly well.

Acknowledgements

This work was financially supported by the National Research Foundation of Korea (NRF) grant funded by the Korea government (MEST) (No. 2010-0015090).

References

1. C. M. Cavusoglu, F. Tendick, M. Cohn and S. S. Sastry, *IEEE Trans. Robotics Autom.*, 1999, **15**, 728–739.
2. B. D. Adelstein, P. HoH. Kazerooni, *Proceedings of the ASME Dynamic Systems and Control Division*, 1996.
3. J. Arata, H. Kondo, N. Ikedo and H. Fujimoto, *IEEE Trans. Robotics*, 2011, **27**, 201–214.
4. Y. M. Han and S. B. Choi, *Smart Mater. Struct.*, 2008, **17**, 065012.
5. Y. M. Han, P. S. Kang, K. G. Sung and S. B. Choi, *J. Intell. Mater. Syst. Struct.*, 2007, **18**, 1149–1154.
6. W. H. Li, B. Liu, P. B. Kosasih and X. Z. Zhang, *Sens. Actuators, A*, 2007, **137**, 308–320.
7. D. Senkal and H. Gurocak, *Mechatronics*, 2011, **21**, 951–960.

Subject Index